T0245293

Lifestyle Medicine

Lifestyle Medicine

Lifestyle, the Environment and Preventive Medicine in Health and Disease

Third Edition

Edited by

Garry Egger
Andrew Binns
Stephan Rössner
Michael Sagner

Academic Press is an imprint of Elsevier
125 London Wall, London EC2Y 5AS, United Kingdom
525 B Street, Suite 1800, San Diego, CA 92101-4495, United States
50 Hampshire Street, 5th Floor, Cambridge, MA 02139, United States
The Boulevard, Langford Lane, Kidlington, Oxford OX5 1GB, United Kingdom

Library of Congress Cataloging-in-Publication Data
A catalog record for this book is available from the Library of Congress

British Library Cataloguing-in-Publication Data
A catalogue record for this book is available from the British Library

ISBN: 978-0-12-810401-9

For information on all Academic Press publications visit our website at https://www.elsevier.com/books-and-journals

Publisher: Mica Haley
Acquisition Editor: Tari Broderick
Editorial Project Manager: Joslyn Chaiprasert-Paguio
Production Project Manager: Edward Taylor
Designer: Mark Rogers

Typeset by TNQ Books and Journals

Contents

Section IV
The Future of Health

About the Editors

Professor Garry Egger, AM, MPH, PhD, is the director of the Center for Health Promotion and Research in Sydney and Adjunct Professor in Lifestyle Medicine at Southern Cross University in Northern NSW, Australia. He has worked in clinical and public health since 1972 and has been a consultant to the World Health Organization, industry, and several government organizations, as well as being involved in medical and allied health education. He is the author of 30 books, over 200 peer-reviewed scientific articles, more than 1000 popular press articles, and has made numerous media appearances.

Dr. Andrew Binns, AM, BSc, MBBS, DRCOG, DA, FACRRM, is a general practitioner in Lismore in rural NSW, Australia. He has a special interest in lifestyle medicine and its relevance to primary care. He is adjunct professor with the Division of Health and Applied Sciences, Lismore Campus, Southern Cross University. He is also medical editor of *GP Speak*, a bimonthly magazine for the Northern Rivers General Practice Network.

Professor Stephan Rössner, MD, PhD, was Professor of Health Behavior Research at the Karolinska Institute in Stockholm and director of the Obesity Unit at the Karolinska University Hospital. He has worked with international obesity-related matters for more than 20 years and served as the secretary, vice president, and president of the International Association for the Study of Obesity. He has published more than 500 scientific articles on nutrition and lifestyle-related matters, written more than 20 books for the lay press, appeared repeatedly in media, and promoted health as a stand-up comedian in "science theater".

Michael Sagner, MD, specializes in sports medicine and preventive medicine. He is also a certified fitness trainer and certified sports nutritionist. He is founding head of several international projects focused on clinical chronic disease and risk factor treatment and prevention. Michael Sagner is actively involved in scientific research in the field of preventive medicine and has developed the "cell to community" approach, which aims at connecting basic research, clinical medicine, and public health to address chronic diseases. His research interests are the underlying mechanisms and causes of chronic diseases and interdisciplinary approaches in clinical treatment. His research considers the complex interactions within the human body in light of a patient's lifestyle, genomics, and environment.

About the Authors

Dr. Julia Anwar-McHenry, BA, BAHons, PhD, is the Evaluation Officer for Mentally Healthy WA's Act-Belong-Commit campaign at Curtin University. Her professional and research interests include mental health and wellbeing, social determinants of health, the arts and health, and regional development.

Ross Arena, PhD, PT, FAHA, FESC, FESPM, FACSM, is professor and head of the Physical Therapy Department at the University of Illinois at Chicago. He is also chairman of the European Society of Preventive Medicine. Dr. Arena's scholarly interests include healthy lifestyle interventions across the prevention spectrum, as well as exercise testing and training in patients diagnosed with cardiopulmonary disease/dysfunction. He is also a fellow of the American Heart Association, European Society of Cardiology, and American College of Sports Medicine, as well as a founding fellow of the European Society of Preventive Medicine.

Malcolm Battersby, PhD, FRANZCP, FAChAM, MBBS, is professor of psychiatry and director of the Flinders Human Behavior and Health Research Unit and course leader of the Mental Health Science programs at Flinders University, Adelaide. He is also director of the State-wide Gambling Therapy Service. He trained with Professor Isaac Marks at the Institute of Psychiatry, London, in behavioral treatment of anxiety disorders and severe neurotic conditions. He was awarded a Harkness Research Fellowship in the study of chronic conditions self-management in the United States during 2003–04 and has led the development of the Flinders program of chronic condition management, now provided across Australia and internationally.

Dr. Mike Climstein, PhD, FASMF, FACSM, FAAESS, DE, is a clinical exercise physiologist, director of Chronic Disease Rehabilitation at Freshwater Rehabilitation (one of Australia's largest community rehabilitation programs), and director of Diabetes Rehabilitation at the Vale Medical Clinic. Mike is adjunct clinical associate professor with the Faculty of Health Sciences, Australian Catholic University (NSW). He is also on the editorial board for the Australian Journal of Science and Medicine in Sport and a columnist for "Research Reviews" in the Australian Fitness Network magazine.

Dr. Maximilian de Courten, MD, MPH (University Basel, Switzerland), is Professor of Global Public Health with Victoria University in Melbourne. His medical education started in Switzerland where he undertook clinical research in the area of hypertension and insulin resistance. In the United States he worked at the National Institutes of Health on diabetes. His international experience was enhanced through positions at the World Health Organization as a scientist at HQ and as Medical Officer, Chronic Disease Control, for the South Pacific.

Barbora de Courten is associate professor at Monash University, National Heart Foundation Future Leader Fellow with the Monash Center for Health Research and Implementation, and a general physician (specialist in internal medicine) at Monash Health. She has extensive experience in clinical research, clinical trials, epidemiology, and public health in the area of noncommunicable diseases, in particular, obesity, type 2 diabetes, and cardiovascular diseases.

Rosanne Coutts, BHM, ExSci (Hons), AEp Sp MAAESS, is an accredited exercise physiologist and sports scientist. She is a member of the Australian Psychological Society and lectures at Southern Cross University in sports and exercise psychology. Her interests include the support of a lifestyle approach to the enhancement of health and well-being. She also coordinates the Exercise Physiology Clinic within the School of Health and Human Sciences at Southern Cross University.

Associate Professor John Dixon, MBBS, FRACGP, PhD, is an NHMRC Senior Research Fellow with a combined position at both Baker IDI Heart and Diabetes Institute and Monash University in Melbourne, Australia. With over 120 papers published in peer-reviewed journals, his research bridges gaps in communication between advances in surgery for weight loss, metabolic research, clinical practice, and evidence-based medicine. He is the immediate past president of the Australian and New Zealand Obesity Society, an executive member of the Obesity Surgeons Society of Australia and New Zealand, and serves on both the research and program committees of the American Society for Metabolic and Bariatric Surgery as a physician member.

Professor Rob Donovan, PhD (Psychology), is professor of behavioral research at the Center for Behavioral Research in Cancer Control in the Division of Health Sciences at Curtin University and professor of social marketing and director of the Social Marketing Research Unit in the Curtin Business School. Current interests include regulation of the marketing of "sin" products, mental health promotion, racism, violence against women, Indigenous issues, and doping in sport.

Dr. Bernadette Drummond, BDS, MS, PhD, FRACDS, is a specialist in and associate professor of pediatric dentistry in the Faculty of Dentistry at the University of Otago in Dunedin. She is an active clinician, teacher, and researcher with a particular interest in understanding and preventing dental problems in young people.

Dr. Michael Gillman, MBBS, FRACGP, is the director of the Brisbane Health Institute for Men. He has worked in the area of men's health for over 15 years and sits on several national advisory boards concerned with male sexual health. He has given presentations extensively throughout Australia and internationally to medical and community groups, has a regular statewide radio talkback segment, and is actively involved in clinical work and clinical research.

Dr. Ross Hansen, BSc, MPH, PhD, is a clinical physiologist in the Gastrointestinal Investigation Unit of the Department of Gastroenterology, Royal North Shore Hospital, Sydney. He is also a clinical associate professor in the Sydney Medical School, University of Sydney. His clinical and research interests include physiological monitoring of gut function, interactions between lifestyle and function, and body composition and metabolism in health and disease.

Dr. Chris Hayes, FFPMANZCA, FANZCA, MMed, is a pain medicine specialist working at Hunter Integrated Pain Service and based at John Hunter Hospital in Newcastle, Australia. With a background in anesthetics, he is now working exclusively in pain management. He is interested in holistic models of health care delivery and the emerging paradigm in pain medicine.

Dr. Julian Henwood, BSc, PhD, is a medical writer and managing director of PharmaView, a medical education agency based in Sydney. He trained as a pharmacologist in the United Kingdom and has spent 20 years researching and writing about drugs. He has written for independent international publications aimed at specialists (e.g., the journal *Drugs*) or at general practitioners (e.g., the monthly magazine *Medical Progress*), as well as on numerous publications supported by the pharmaceutical industry.

Dr. David L. Katz, MD, MPH; FACPM, FACP, FACLM, is the Founding Director of Yale University's Prevention Research Center. He is Past-President of the American College of Lifestyle Medicine and Founder of the True Health Initiative.

Dr. Neil King, BSc, PhD, is a senior lecturer at Queensland University of Technology in Brisbane. He is an expert on the relationships between physical activity, appetite control, and energy balance. He has published extensively on physical activity, diet, and obesity.

Rob Lawson, BSc, MBChB, MRCGP, FRCGP, is a UK GP and a long-time proponent of Lifestyle Medicine. In 2016, he cofounded the British Society of Lifestyle Medicine, a registered charity and is its current Chairman.

Dr. John Litt, MBBS Dip RACOG MSc (Epid) FRACGP, FAFPHM, PhD, is an academic general practitioner and public health physician at Flinders University in Adelaide. He is also the Deputy Chair of the Royal Australian College of General Practitioners National Quality Committee. He has been involved in prevention research and teaching for 15 years and was a key contributor to the RACGP monograph Guidelines for prevention in general practice. He has been involved in running lifestyle intervention workshops for many years.

Dr. Joanna McMillan Price, BSc (Hons), PhD, is a nutrition scientist and fitness leader, based in Sydney. She is a popular media spokesperson, appearing regularly on the Today show, a health writer for *Life etc* magazine, and an author of several nutrition books. Her PhD looked at the role of the glycemic index and protein intake in weight loss, body composition, and cardiovascular disease risk.

Amy McNeil, BA, has been an instructor of writing in the Department of Nutrition at the University of Illinois at Chicago for five years. She is a student in the PhD program in the same department and studies health literacy.

Dr. Hamish Meldrum, MBChB, FRACGP, DRANZCOG, is the President of the Australasian Society of Lifestyle Medicine and cofounder of the Ochre Health Group.

Dr. Hugh F. Molloy*, MRCS (Eng), LRCP (Lond), DObst RCOG (Lond), DDM (Syd), FACD, is a retired Sydney dermatologist. He began his medical career as a GP in the United Kingdom, before becoming a surgeon at sea. He then moved to Australia where he trained and practiced in dermatology. In the 1980s, he spent time in Oxford investigating the effects of overheating, especially at night in bed using doonas (duvets). He has a particular interest in the effects of the local environment on skin and general body functions. (*Dr. Molloy passed away in 2012.)

Dr. Bob Morgan, EdD, is an Aboriginal educator with over 40 years' experience in Australia and internationally. He is a Conjunct Professor with the Wollotuka Institute, University of Newcastle, and is Chair of the Board of Aboriginal and Torres Strait Islander Education and Research. He is a Visiting Professor with Minzu University, China, Co-Chair of the National Indigenous Elders Alliance, and is also Chair of the World Council of Indigenous People's Education.

Dr. Suzanne Pearson, BBiolSc (Hons), MNutrDiet, is a practicing dietitian and consultant working in Sydney. She completed her undergraduate training in Canada and her PhD in nutrition at the University of Sydney.

Dr. Robert Reznik, MBBS, MPH, MSc, MD, FAFPHM, FRANZCP, is a consultant psychiatrist in private practice in Sydney. He also works as a consultant psychiatrist for a rural area of NSW, is part of the chronic pain management team at the Prince of Wales Private Hospital, and is a senior forensic psychiatrist for justice health within NSW Correctional Services. He was previously the director of community medicine at Royal Prince Alfred Hospital and part of the WHO Collaborative Center for Early Intervention for Alcohol Abuse at Royal Prince Alfred Hospital and had a major interest in the effects of cardiovascular disease in the community.

John Stevens, RN, PhD, FACN, is a health scientist and Associate Professor with Southern Cross University's School of Health and Human Sciences in Australia. He is also the director of a number of companies engaged in health education and research. In the past John has been Head of the School of Nursing and Health Practices (SCU), Director of Postgraduate Studies (which included convening the first ever Master's Award in Lifestyle Medicine), and Director of Professional Development and Enterprise. He has over 60 peer-reviewed publications including a book on dementia. In 2008 John cofounded the Australasian Society of Lifestyle Medicine and remains a member of the board.

Dr. Caroline West, MBBS, specializes in healthy lifestyle medicine. Her areas of interest include preventative health, weight management, smoking cessation, mental well-being, and sleep medicine. She combines her clinical work in a busy Sydney city medical practice with her role in the media. She is a well-known print and television health journalist, and her credits include presenting for the television shows *Beyond Tomorrow*, *Beyond 2000*, and *Good Medicine*.

Dr. Kevin Wolfenden, PHD, MHA, has over 30 years of practice and policy experience in mental health and population health. He has worked as a senior clinical psychologist and been a mental health consultant to the Australian Government. As a population health practitioner, his specialist area has been injury prevention. He has held senior population health positions in Australia and been a long-term international consultant.

Preface

With more than 120 years of combined experience, ranging from clinical practice and research to epidemiology and public health intervention, we have become increasingly frustrated with the inability of established primary care to deal adequately with modern disease conditions. Far from creating the health utopia described glowingly in the popular press, advances in health care have been limited and confusing. Decreases in deaths from heart disease, for example, seem set to be reversed by the march of obesity and its associated metabolic disorders, for which the most pessimistic of us can see no solution, save a major economic slowdown, energy crisis, or world event.

In *The Rise and Fall of Modern Medicine*, author James Le Fanu states that medical advances have been vastly exaggerated (by the public), as well as confused by the changing nature of modern disease. It is true that we have managed to delay mortality, but it may be argued that this has come with increases in many diseases that diminish life's exuberance. Diabetes, heart disease, cancers, fatty liver, sleep apnea, sexual dysfunction, and many other conditions are problems related not to microorganisms but to the lifestyles we now lead in unhealthy environments—although perhaps only through the blinkered eyes of economic rationalists rather than those of the concerned health practitioner. Ironically, the material advances achieved over the past century, which suggest to the populace that simplistic, effort-free technological fixes will solve all our health woes, in fact advance the need for greater patient involvement in their own health, suggesting a revolution in primary care.

Preventing chronic diseases—the diseases of the 21st century—has to involve improving lifestyle for the individual (clinical medicine) and populations (public health).

This completely revised third edition has been written for translation into several languages to assist in the spread of the knowledge about chronic disease prevention and treatment throughout the developed world. The book draws much of its knowledge from the American, European, and Australian experience. This is seen as one of its advantages, as these countries are the richest and most advanced technological countries of the modern world, without the immediate influences of other societies. It thus provides a windscreen through which lifestyle and environmental problems arising from modern development can be viewed.

We hope that through this book we have at least made a start in doing so. It is certain that there will be much more to come.

Garry Egger
Andrew Binns
Stephan Rössner
Michael Sagner

Section I

BACKGROUND AND BASIS FOR THE ROLE OF LIFESTYLE FACTORS IN CHRONIC DISEASE PREVENTION AND TREATMENT

Chapter 1

Introduction to the Role of Lifestyle Factors in Medicine

Garry Egger, Andrew Binns, Stephan Rössner, Michael Sagner

Life is a sexually transmitted disease with 100% mortality and inevitable intermittent morbidity.

R. D. Laing

INTRODUCTION: WHAT IS LIFESTYLE MEDICINE?

The rise in obesity worldwide has focused attention on lifestyle as a prominent cause of disease in modern times. However, obesity is just one manifestation, albeit an obvious one, of a range of health problems that have arisen from the environment and behaviors associated with our modern way of living. Inactivity, poor and overnutrition, smoking, drug and alcohol abuse, inappropriate medication, stress, sexual behavior, inadequate sleep, risk taking, and environmental exposure (sun, chemical, the built environment) are significant modern determinants of disease that call for a modified approach to health management.[1] "Lifestyle medicine" is a relatively new and different approach to doing this.

In the first edition of this book, we defined lifestyle medicine as "the application of environmental, behavioral, medical and motivational principles to the management of lifestyle-related health problems in a clinical setting." To this we have now added "…including self-care and self-management" to take account of the increased recent emphasis on self-management now seen as necessary for the proper treatment of chronic disease.

This book summarizes aspects of lifestyle medicine, examining the determinants ("causes" is a difficult concept in chronic disease; see Chapter 3), measurement, and management of a range of modern health problems with predominantly lifestyle-based etiologies. In this third edition we have included extra chapters, which is indicative of the rapid changes being made in the management of chronic disease and the appreciation of lifestyle factors determining these. The need for an

1. For a more structured list of determinants, see Chapter 4.

Lifestyle Medicine. http://dx.doi.org/10.1016/B978-0-12-810401-9.00001-2

approach like this stems from rapid economic advancement, which has changed the environments in which humans live (both "macro" and "micro") and, in doing so, the lifestyles and behaviors associated with those environments. Up to 70% of all visits to a doctor are now thought to have a predominantly lifestyle-based etiology (AIHW, 2006), hence the need for a new approach to dealing with the problem. As we will see, an appropriate response leads to a convergence of a range of diverse issues confronting humankind in the third millennium, from personal and population health and self-management to pollution, climate change, social equity, population stabilization, and globalization. We concentrate largely on the contribution that can be made directly by the clinician at the personal level and conclude in Chapter 30 with a discussion of personal carbon trading schemes and economic reform as a potential encompassing "distal" approach to health and the environment.

THE SCOPE OF LIFESTYLE IN MEDICINE

The practice of lifestyle interventions in medicine extends from that of primary prevention (preventing a disease from developing by modifying the behavioral or environmental cause) to secondary prevention (modifying risk factors to avert the disease) to tertiary prevention (rehabilitation from a disease state and prevention of recurrence). For example, helping prevent a person become overweight by implementing lifestyle changes is primary prevention; helping an overweight person with prediabetes avoid diabetic complications is secondary prevention; and advising a morbidly obese patient with poor diabetic control to undergo bariatric surgery to avoid the need for insulin is tertiary prevention.

Although it is a clinical discipline, lifestyle medicine forms a bridge with public health and health promotion, where the latter is defined as "the combination of educational and environmental supports for actions and conditions of living conducive to health" (Greene and Kreuter, 1991).

In contrast to population and environmental interventions, lifestyle medicine focuses on individuals (and, in some cases, small groups), where interventions are typically administered in a primary care setting. Just as in any specialized area, there is a body of knowledge, skills, tools, and procedures that need to be mastered to become proficient in lifestyle medicine (see Chapter 4). However, the involvement of different disciplines, as is necessary in a field like this, ensures a greater availability of these skills within a practicing team.

HISTORICAL BACKGROUND

Around 500 BC, the Greek philosopher and founder of modern medicine, Hippocrates, first hinted at the notion of lifestyle medicine by suggesting that in order to keep well, one should simply "avoid too much food, too little toil." For the ensuing two and a half millennia, humans had little difficulty conforming to Hippocrates' lifestyle recommendations. Indeed, the problem was getting *enough* food and having to toil *too much* in order to survive the trials and tribulations of human evolution. Their problem was more the acute ravages of infection.

A change came with the industrial revolution of the late 19th century. Machines began to replace people, thus making physical effort in the gathering of food less necessary, as well as increasing the availability (and density) of energy in such food. Leisure time was more at a premium, population and work stresses increased, access to ingestible and mind-altering substances increased, and social and community structures changed.

As a result, the Hippocratic prescription now needs to be expanded. To the previous quote, we could add: "and don't smoke, don't eat too much (or eat or drink too much in general), don't drink too much alcohol (while having a couple of alcohol-free days a week), try not to get anxious or depressed, get just the right amount of stress, don't do too many drugs (of all kinds), don't have unsafe sex, eat breakfast, keep regularly active, sleep well and for long enough, do some stretching and strength work every other day, wear sunscreen, use a moisturizer, avoid air-conditioning where possible, keep the skin well hydrated, floss regularly, and, remember, moderation in all things—including moderation!"

Lifestyle interventions include these preventive tactics. They can combine changes in behavior or the environment with standard medical management and, at another level, do this in the absence of conventional medical intervention.

LIFESTYLE IN THE CONTEXT OF CHRONIC AND ACUTE DISEASE

Acute diseases are those that usually have a short prodrome and cause debilitating symptoms but in an otherwise healthy person resolve completely and relatively quickly, without sequelae. These are usually caused by infectious agents and include diseases such as rubella, measles, and influenza. Chronic diseases are usually those with a long prodrome and that require ongoing management over an extended period—sometimes a lifetime.

In earlier times, the acute, infectious diseases often resulted in death because there was no treatment for the secondary complications of illnesses, such as pneumonia or rheumatic fever. People who developed more serious illnesses, such as insulin-dependent diabetes, tuberculosis, or typhoid fever, usually died from the disease itself.

With the development of vaccinations, improved living standards, better public health and hygiene, and antibiotic medications, people with illnesses that once would have caused death were able to live with their diseases into later life. Crohn's disease, ulcerative colitis, multiple sclerosis, ischemic heart disease, and diabetes are all examples of diseases that could previously have been fatal but that are now manageable. HIV and hepatitis B and C are more recent examples. At the other extreme, some chronic diseases (e.g., cystic fibrosis,

A Lifestyle Parable: How Health Has Changed

Bill Blogg's parents, born in the 1920s, were poor but hard working. Having grown up in a rural region they spent their days physically active in growing and cultivating food for themselves and for sale at the local markets. Such food was "natural" and healthy as humans had evolved eating this type of unprocessed product for thousands of years. As a result, Bill Sr. and his wife, Mary, were lean and healthy (although the effects of smoking taken up since WWII were starting to take its effects on Bill and it was a constant battle to avoid serious infections such as polio and diphtheria). They visited the doctor irregularly, and then usually for set procedures such as childbirth, or the treatment of an infection or injury. They lived into their late sixties, with a short period of illness before death. Bill Sr. and Mary's offspring, Bill Jr. and Barbara, achieved much greater wealth in their lifetime than Bill and Mary could ever have imagined. As a result they had energy-rich, processed food on the table every day without much (physical) effort. And technology that meant they hardly had to lift a finger to enjoy the good life. But around 1980 they both started to get fat. As a result, Bill Jr. developed sleep apnea and angina and Barbara developed type 2 diabetes and arthritis. They're still alive and in their seventies, but Bill Jr. has now had a stroke and is confined to a wheelchair. Barbara is also feeling the side effects of the good life and both are not enjoying the benefits of the higher standard of living that has come with development.

hemochromatosis, polycystic kidneys) have a significant genetic component and cannot be classified as lifestyle based. Consequently, the boundary between chronic and acute diseases and the scope of lifestyle medicine is much less discrete than it may have been in the past (Fig. 1.1).

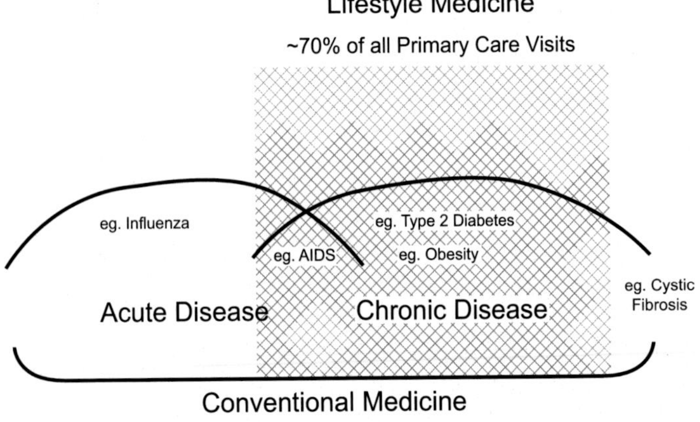

FIGURE 1.1 The place of lifestyle medicine.

DIFFERENCES BETWEEN TRADITIONAL/CONVENTIONAL AND THE MODEL OF LIFESTYLE MEDICINE

A lifestyle medicine approach differs from traditional/conventional medicine (hereafter called "conventional") in that it is aimed at modifying the behavioral and lifestyle bases of disease, rather than simply treating the disease. As such, it requires the "patient" to be an active partner in the process rather than a passive recipient of medical care. In fact, the term "patient" is really no longer appropriate in this setting; unfortunately, no fully acceptable alternative term has presented itself, hence the need for a comprehensive approach to understanding how to involve the patient in his or her self-management, as discussed in Chapter 6.

Lifestyle medicine differs from nonmedical clinical practice in that it may include medication (e.g., for smoking cessation or hunger control) and even surgery where this is appropriate (e.g., for weight control), and it differs from purely behavioral approaches in that it examines environmental etiologies as well as individual behaviors. Some of the other main differences between the two approaches are shown in Table 1.1.

THE LINKS THAT HIGHLIGHT LIFESTYLE FACTORS IN MEDICINE

Until recently, most aspects of lifestyle-related health were seen as being on a "linear" scale, for example,

$$\text{Specific disease reduction} = \text{Specific lifestyle change.}$$

This is most typically illustrated by body weight, which has for years been thought of as a "physics- type" notion that occurs outside the human body, for example,

$$\begin{aligned}\text{Weight} = {} & \text{Energy in (food/drink)} \\ & - \text{Energy out (exercise/metabolism/thermogenesis)}.\end{aligned}$$

This ignores the fact that the mammalian body is a reactive biological (not a physics) organism with complicated feedback mechanisms that more appropriately require a "systems" than a "linear" approach (see Chapter 3). Hence, aspects of lifestyle covered in this book are closely linked. Poor sleep, for example, can lead to fatigue; fatigue to inactivity; inactivity to poor nutrition or overeating; and all these can exacerbate obesity and depression, leading to metabolic syndrome, type 2 diabetes, sex and mood problems, and possibly heart disease. In reverse, poor nutrition, inactivity, and smoking or alcohol abuse may result in injury, poor sleeping habits, or disease proneness that can make the process cyclical. Medication can help manage this but may also, if misused, have side effects such as weight gain and sexual dysfunction. Antidepressant medication can cause weight gain, which may lead to greater psychological disturbances than the original depression. All of this—the predisposing factors, determinants, the disease, and the treatment—are part of lifestyle medicine.

TABLE 1.1 Differences Between Traditional/Conventional and Lifestyle Medicine Approaches in Primary Care

Traditional/Conventional Medicine	Lifestyle Medicine
Treats individual risk factors	Treats lifestyle/environmental causes
Patient is a more passive recipient of care	Patient is an active partner in care
Patient is not required to make big changes	Patient is often required to make big changes
Treatment is often short term	Treatment is usually long term
Responsibility is on the clinician	Responsibility is also on the patient
Medication is often the "end" treatment	Medication may be necessary, but the emphasis is on lifestyle/environmental change
Emphasis on diagnosis and prescription	Emphasis on motivation and compliance
Goal is disease management	Goal is primary/secondary/tertiary prevention
Less consideration of environment	More consideration of environment
Side effects are balanced by the benefit	Side effects that impact on lifestyle require greater attention
Involves other medical specialties	Involves allied health professionals as well
Doctor generally acts independently on a one-to-one basis	Doctor is part of a team of health professionals

Exercise and nutrition are the "penicillin" of lifestyle interventions; hence, the early chapters are devoted to these, albeit modestly, as both topics are covered in much greater detail in other individual texts. Psychology is the "surgery" by which homeostasis is restored in lifestyle-based chronic diseases; hence, psychological processes litter the text. Together with other more specific treatment approaches, these make up the bulk of lifestyle medicine.

WHO IS BEST QUALIFIED TO PRACTICE LIFESTYLE INTERVENTIONS?

In our view, there is currently no single discipline equipped to practice preventive or therapeutic lifestyle changes alone. Medical practitioners have the training to diagnose and treat disease but little time to supervise and motivate patients through complex lifestyle change. Nurses are equipped for ongoing management and care of patients, but they often lack specific knowledge about nutrition, exercise, and behavior changes (although the emerging breed of practice nurses may be able to fill this gap).

Dietitians can advise on the basis of existing nutritional knowledge but the ground is constantly changing under their feet and they often have a minimal background in exercise or behavior change. Exercise physiologists are knowledgeable about the physical activity requirements for specific problems but have less complete knowledge of nutrition and psychology. Psychologists and complementary medicine practitioners bring their own advantages but lack the prescribing ability and often the physiological background of the doctor and other mainstream disciplines.

University courses are now being tailored to try to fill this gap. Courses are being offered in the combination of nutrition, exercise, and psychology (i.e., metabolic health). Many medical degrees now also include a related nonmedical discipline as a prerequisite and are increasingly emphasizing the role of prevention, as well as treatment, and an understanding of psychology and behavior. However, it is unlikely in the near future that any one professional will be able to assume the mantle of the fully fledged lifestyle medicine "expert." As a result, the practice is likely to be carried out by a team of health professionals with the doctor, very often a general practitioner, as the coordinator of the team.

THE EVIDENCE BASE

Where possible, we use evidence-based information to make recommendations for lifestyle-based prescriptions. However, as lifestyle interventions in medicine are a growth area, there is still much to be learned and much that is difficult to objectify. This becomes particularly apparent when discussing exercise prescription for various ailments (Chapter 12) or learning positive psychology to combat mood disorders (Chapter 16). Although specific prescriptions for different lifestyle-based ailments are now emerging, there has only recently been an appreciation of the general importance of lifestyle in what have often otherwise been thought of as medical issues. Hence, evidence-based prescription is not yet available in several areas. Where this is the case, we have attempted to compensate by providing recommendations based on the available data and the careful reasoning of our experienced contributors.

Other publications cover the more theoretical rationales for lifestyle medicine, its evidence base, and specific areas of lifestyle medicine such as sports medicine, women's health, and pulmonary and pediatric medicine (e.g., Knight, 2004; Rippe, 2013), so these are not dealt with in detail here. We have also not covered each area of lifestyle medicine exhaustively, given that most have extensive literature of their own elsewhere. After defining a structure for lifestyle medicine in Chapter 4, the idea is to synthesize these topics into one unified concept.

FRAMEWORKS FOR LIFESTYLE CHANGES IN MEDICINE

Because structured lifestyle interventions in medicine are in the early stages, this is still a work in progress, influenced by its developing status, the norms and aspirations of the population, and ongoing feedback. This book represents an attempt to capture the best in an embryonic field, accepting that this is only a beginning in a developing process. We have adopted and adapted existing initiatives to suit changing conditions. The five A's of lifestyle prescription (Ask, Assess, Advise, Assist, Arrange), for example, have been reduced to three A's (Assess, Assist, Arrange) for easier recall, and are provided at the end of each chapter, and practical applications have been summarized for medical practitioners and practice nurses. In Chapter 4 we also look at developing skills in lifestyle medicine such as "shared medical Appointments," which take advantage of peer support by having individual medical consultations in front of a group of people with similar problems. While some health systems are not yet set up to accommodate the costs of this, new systems will need to be developed to account for what, essentially, is a more logical way of dealing with chronic diseases in primary care. Other processes and tools are also a work in progress as lifestyle medicine develops a structure and a pedagogy to deal with the problems of today. To illustrate this and the speed with which a new discipline is developing, extra chapters have been added to this third edition of the current text. No doubt the next edition (and there will be others) will add even more to the mix.

What Can Be Done in a Brief Consultation

- ☐ Ask what aspects of the patient's lifestyle s/he thinks could be improved.
- ☐ On the basis of the answers, enquire in more depth about the areas identified.
- ☐ Suggest/refer to an allied health specialist in this area or make another appointment to discuss these in more detail.
- ☐ If appropriate, carry out some basic measurements (e.g., waist circumference, weight, bioimpedance).
- ☐ Provide written material on the lifestyle areas identified.

Some Practice Principles for Lifestyle Medicine

While guidelines for the practice of lifestyle medicine have, as of 2016, not yet been fully endorsed, a number of draft practice principles were formulated by the Australian Society of Lifestyle Medicine. These may change with time. However, they offer a starting point for practice in the field and include the following:

1. Consider the lifestyle influences on chronic diseases.
2. Regard obesity (and other known risk factors) as signs for further investigation for chronic disease risk (but don't disregard those who are not overweight).
3. Engage individuals with chronic disease in some level of self-management.
4. Adopt a client-centered, counseling style, in either a 1:1 or shared format, focused on increasing motivation, health efficacy, and health literacy.
5. Provide a user-friendly practice environment, with team care involvement and understandable health information.
6. Consider nutrition, exercise, and stress management as the core of lifestyle medicine.
7. Do not ignore underlying social, economic, and environmental factors in chronic disease.
8. Consider lifestyle measures in addition to drug or surgical therapy in patients with later-stage lifestyle diseases.

SUMMARY

Lifestyle medicine represents a different approach to dealing with the significant proportion of patients with ailments caused predominantly by lifestyle and the environments driving such lifestyles, and now presenting to healthcare practitioners. Managing these problems changes the emphasis from conventional treatment to one where the patient needs to be more involved in his or her care, and this requires considerable knowledge about motivation and motivational skills on the part of the clinician as well as a different approach by the clinician to assist patients in managing their own chronic health problems. Developments in health funding, at least in Australia and some parts of Europe, involving allied health professionals as part of a healthcare team now make this not only financially feasible but necessary for dealing with the changes in disease etiology associated with our modern way of life.

REFERENCES

Goodyear-Smith, F., Arroll, B., Sullivan, S., et al., 2004. Lifestyle screening: development of an acceptable multi-item general practice tool. N. Z. Med. J. 117 (1205), U1146.

Greene, L.W., Kreuter, M.W., 1991. Health Promotion Planning: An Educational and Environmental Approach. Mayfield Publishing Co., Palo Alto, CA.

Knight, J.A., 2004. A Crisis Call for New Preventive Medicine: Emerging Effects of Lifestyle on Morbidity and Mortality. World Scientific Publishing Co., New Jersey.

Rippe, J. (Ed.), 2013. Lifestyle Medicine, second ed. Blackwell Publishing, NY, pp. 2–13.

FURTHER READING

(AIHW) Australian Institute of Health and Welfare, 2014. Australia's Health 2014. Cat. no. AUS 178. Canberra. Australia's Health Series No. 14.

(AIHW) Australian Institute of Health and Welfare, 2011. Health Determinants, the Key to Preventing Chronic Disease. Cat No. PHE 157 AIHW, Canberra. Available from: www.health.gov.au/internet/wcms/publishing.nsf/Content/health-pubhlth-strateg-lifescripts-index.htm.

Bodheimer, T., 2006. Primary care — will it survive? NEJM 355 (9), 861–864.

Sagner, et al., 2014. Lifestyle medicine potential for reversing a world of chronic disease epidemics: from cell to community. Int. J. Clin. Pract.

Professional Resources

A Lifestyle Screening Questionnaire

Patient instructions: For each question, give a score from 1 to 5 based on the following scale:

0 = Don't do
1 = Never do
2 = Rarely do
3 = Sometimes do
4 = Often do
5 = Very often do

And tick the column that best answers the question.

Question	Score	Would You Like Help With This? Yes	No	Later
Do you ever feel the need to cut down on your smoking?				
Do you ever feel the need to cut down on your alcohol intake?				
Do you ever feel the need to cut down on other drug use?				
During the past month, have you often been bothered by feeling down, depressed, or hopeless?				
During the past month, have you often been bothered by having little interest or pleasure in doing things?				
Have you been worrying a lot about everyday problems?				
Do you spend most days being physically inactive?				
Are you concerned about your current weight?				
Are you unhappy with the quality of your sleep?				
Do you eat less than about three servings of fruit and vegetables each day?				
Do you suffer side effects from any medications you take?				
Do you have concerns about the state of your skin/hair?				
Are you concerned in any way about your sexual performance?				
Do your teeth/gums cause you any problems?				
Do you regularly drink large amounts of soft drinks or fruit juices?				

Adapted from Goodyear-Smith, F., Arroll, B., Sullivan, S., et al., 2004. Lifestyle screening: development of an acceptable multi-item general practice tool. N. Z. Med. J. 117 (1205), U1146.

Chapter 2

The Epidemiology of Chronic Disease

Maximilian de Courten, Barbora de Courten, Garry Egger, Michael Sagner

For if medicine is really to accomplish its great task, it must intervene in political and social life. It must point out the hindrances that impede the normal social functioning of vital processes, and effect their removal.

Rudolf Ludwig Karl Virchow (1849)
(Quoted in P. Farmer Pathologies of Power; Farmer, 2004).

INTRODUCTION: A (VERY) SHORT HISTORY OF DISEASE

About five million years ago, a variation of a common ape began to walk out of the jungles of Central Africa. Not much is known about this antecedent of *Homo sapiens*, but we can speculate on a couple of things. First, it is highly unlikely that these creatures were obese. Although having the genetic capacity for storing fat (something that evolved in the early mammals as a means of getting through the inevitable lean times), the prospects of achieving this were almost certainly limited by a harsh environment and competition for scarce resources. Second, they were exposed to all manner of diseases, most of which were likely to be acute and contagious and for which protection would evolve through the development of disease-fighting antibodies. In the ensuing millennia, life would revolve around getting enough to eat, not being eaten, or killed in warfare, and battling the countless coevolving disease microorganisms, some of which still plague humanity today.

How are these things connected? Until the late 20th century this was not clear. Humans had progressively survived the "four horses of the apocalypse" to increase exponentially as a result of developments in disease prevention. Changes in public health and hygiene around the time of the industrial revolution of the late 19th century and the development of medical marvels, such as vaccinations and antibiotics, resulted in a massive blow to infectious diseases. In the euphoria of the 1960s, it appeared that the battle against disease had been all but won. Few would have predicted that the (apparent) demise of infectious

Lifestyle Medicine. http://dx.doi.org/10.1016/B978-0-12-810401-9.00002-4

diseases would be accompanied by a rise in diseases largely caused by the factors underlying this decrease in infections, namely, industrial development, "modernization," and economic growth. Lucretius may have been an exception when he wrote in 50 BC, "In primitive times, lack of food gave languishing bodies to death; now, on the other hand, it is abundance that buries them" (translation Latham) (Carus and Latham, 1993).

The emergence of chronic diseases at a population level began with the industrial revolution, which, for the first time, allowed machines to carry out much of the work otherwise done by humans. The increased production of food not only made what was once a relatively scarce resource into a common commodity but also increased the availability of processed foods, which are more energy dense with a greater fattening capacity. The technological revolution of the late 20th century then hastened the process. With greater access to mass-produced food pleasing to our primitive taste and obtainable with ever-decreasing effort, humans began to fulfill their genetic potential to store excessive fat. In fact, an epidemic of fatness of national and international scope began in the 1980s and, by the turn of the millennium, it involved at least half of the adult population and one-quarter of adolescents in industrialized countries. Because obesity has been shown to have a link to many diseases, this significantly changed the pattern of disease from those caused by an outside agent (e.g., bacterium, virus) to those primarily related to the way we live. This has not been helped by a return of acute and infectious problems, such as SARS, AIDS, and bird flu, which are also associated with living conditions (e.g., overcrowding, overpopulation, international travel, sexual behavior) but are less relevant to the discussion here, which we will confine to noninfectious, lifestyle-related chronic disease problems.

LIFESTYLE-RELATED CAUSES OF DISEASE

Epidemiologists have traditionally used mortality (death rates) from a particular disease as an indication of the severity of that disease. With chronic diseases, however, mortality tells only a part of the story. Morbidity (sickness statistics) is also limited in this respect. However, there is a new measure called "disability adjusted life years" (DALYs). DALYs for a disease are the sum of the years of life lost due to premature mortality and the years lost due to disability in a given population (World Health Organization, 2013). The DALY is a health gap measure that extends the concept of potential years of life lost due to premature death to include equivalent years of healthy life lost to states of less than full health (broadly termed disability). One DALY represents the loss of one year of equivalent full health.

In the United States, chronic diseases are estimated to account for 88% of total deaths (World Health Organization, 2016). The European Region is also strongly affected by chronic diseases, and their growth is startling. The impact of the major chronic diseases (diabetes, cardiovascular diseases, cancer, chronic respiratory diseases, and mental disorders) is equally alarming; taken together, these five conditions account for an estimated 86% of the deaths and 77% of the disease burden in the European Region (World Health Organization, 2016).

In Australia, for instance, the age-standardized mortality rate reduced from 440 in 2000 to 354 in 2011 (ASDR per 100,000; total deaths all causes, both sexes) whereas the World Health Organization DALY statistics for Australia show a significant increase from the 4,755,000 in the year 2000 to 5,220,000 for the year 2012 (World Health Organization, 2016). This demonstrates an increasing burden of diseases despite progress in mortality, driven largely by chronic diseases.

Table 2.5 then shows the 15 highest causes of DALYs in Australia in 1990 and how these changed to 2013. For over three decades, in Australia the top eight causes are dominated by noncommunicable diseases with all of them having a strong underlying lifestyle-related etiology.

Overall, it can be seen that a large burden of modern disease is currently being imposed by potentially preventable lifestyle causes (Tables 2.1–2.6).

Although smoking is still one of the main preventable causes of disease, this has now been overtaken by overweight as the major preventable cause of DALYs in the United States and Australia (Tables 2.2 and 2.6). This can probably be attributed to the significant successes achieved through systematic antitobacco campaigns since the 1970s. At the same time research is discovering more and more diseases linked to obesity that traditionally were not thought of as being related to that condition, such as several cancers, osteoarthritis, neurodegenerative diseases such as Alzheimer and Parkinson diseases, etc. It is also of note that the much-publicized benefit from drinking alcohol (in moderation) is, on a population level, more than canceled out by the harm caused by drinking.

Whilst the GBD-2010 study has fitted diseases and injuries into a hierarchy that incorporates 235 different causes of death (Murray et al., 2012), this is a necessary simplification when calculating these estimates (Byass et al., 2013). Globally, many individual episodes of disease and causes of death may not be diagnosed and documented with sufficient precision to be categorized, and underlying chronic diseases often get forgotten in death certificates listing the more immediate cause of death.

Nevertheless, four main causes—excess weight, poor diet, physical inactivity, and smoking—account for most of the mortality and morbidity of the major diseases of modern society, such as heart disease, stroke, colorectal cancer, depression, kidney disease, diabetes, osteoarthritis, osteoporosis, and neurodegenerative diseases.

ASSESSING RISK FACTORS

Risk factors are characteristics that are associated with an increased risk of developing a particular disease or condition. They can be demographic, behavioral,

TABLE 2.1 The Leading Causes of Disability Affected Life Years (DALYs) in the United States, 1990 and 2013

United States
Both sexes, All ages, Percent of total DALYs

1990 rank	2013 rank
1 Cardiovascular diseases	1 Cardiovascular diseases
2 Neoplasms	2 Neoplasms
3 Mental & substance use	3 Mental & substance use
4 Musculoskeletal disorders	4 Musculoskeletal disorders
5 Other non-communicable	5 Diabetes/urog/blood/endo
6 Diabetes/urog/blood/endo	6 Other non-communicable
7 Neurological disorders	7 Neurological disorders
8 Chronic respiratory	8 Chronic respiratory
9 Unintentional inj	9 Unintentional inj
10 Transport injuries	10 Self-harm & violence
11 Self-harm & violence	11 Transport injuries
12 Neonatal disorders	12 Neonatal disorders
13 Diarrhea/LRI/other	13 Diarrhea/LRI/other
14 HIV/AIDS & tuberculosis	14 Cirrhosis
15 Digestive diseases	15 Digestive diseases
16 Cirrhosis	16 Nutritional deficiencies
17 Nutritional deficiencies	17 HIV/AIDS & tuberculosis
18 Other group I	18 Other group I
19 Maternal disorders	19 Maternal disorders
20 NTDs & malaria	20 NTDs & malaria
21 War & disaster	21 War & disaster

Communicable, maternal, neonatal and nutritional diseases; Non-communicable diseases; Injuries

Institute for Health Metrics and Evaluation (IHME), 2015. GBD Compare. IHME, University of Washington, Seattle, WA. Available from: http://vizhub.healthdata.org/gbd-compare.

biomedical, genetic, environmental, social, or other factors, and they can act independently or in combination. Such a broad definition includes biomedical concepts of risk and behaviors considered as "risky" based on social values and cultural norms. A risk factor can therefore be understood as something lying within the realm of personal responsibility (something the individual can control) or something that either cannot be altered at all (e.g., genetic factors) or is very intangible and therefore hard to address (e.g., cultural traditions). To identify risk factors, then, does not necessarily lead to their swift reduction.

TABLE 2.2 Rank and Percentage of Disease Burden (Attributable DALYs as a Percentage of Total DALYs) Attributed to Selected Risk Factors of Health, United States 1990 and 2013

United States
Both sexes, All ages, Percent of total DALYs

1990 rank	2013 rank
1 Dietary risks	1 Dietary risks
2 Tobacco smoke	2 High body-mass index
3 High systolic blood pressure	3 Tobacco smoke
4 High body-mass index	4 High systolic blood pressure
5 Alcohol and drug use	5 Alcohol and drug use
6 High total cholesterol	6 High fasting plasma glucose
7 High fasting plasma glucose	7 High total cholesterol
8 Low physical activity	8 Low physical activity
9 Air pollution	9 Low glomerular filtration rate
10 Low glomerular filtration rate	10 Occupational risks
11 Occupational risks	11 Air pollution
12 Unsafe sex	12 Child and maternal malnutrition
13 Sexual abuse and violence	13 Low bone mineral density
14 Low bone mineral density	14 Sexual abuse and violence
15 Child and maternal malnutrition	15 Unsafe sex
16 Other environmental risks	16 Other environmental risks
17 Unsafe water, sanitation, and handwashing	17 Unsafe water, sanitation, and handwashing

Metabolic risks | Environmental/occupational risks | Behavioral risks

Institute for Health Metrics and Evaluation (IHME), 2015. GBD Compare. IHME, University of Washington, Seattle, WA. Available from: http://vizhub.healthdata.org/gbd-compare.

Biomedical risk factors (sometimes also called metabolic risk factors) are derived from body measurements, such as weight, blood pressure, and blood cholesterol. Because they are "within the body," biomedical risk factors carry comparatively direct and specific risks for health. However, not everything that can be measured counts.[1] Although biomedical risk factors often can be measured with great precision, their resulting statistical tight correlations with disease outcomes sometimes attributes them higher significance than other risk factors, such as stress, due to the fact that the latter are much harder to accurately assess in population surveys.

1. A variation of a quote wrongly attributed to Albert Einstein.

TABLE 2.3 The Leading Causes of Disability Affected Life Years (DALYs) in Europe, 1990 and 2013

Central Europe

Both sexes, All ages, Percent of total DALYs

1990 rank	2013 rank
1 Cardiovascular diseases	1 Cardiovascular diseases
2 Neoplasms	2 Neoplasms
3 Musculoskeketal disorders	3 Musculoskeketal disorders
4 Mental & substance use	4 Mental & substance use
5 Other non-communicable	5 Other non-communicable
6 Unintentional inj	6 Diabetes/urog/blood/endo
7 Diabetes/urog/blood/endo	7 Neurological disorders
8 Neurological disorders	8 Unintentional inj
9 Chronic respiratory	9 Chronic respiratory
10 Diarrhea/LRI/other	10 Self-harm & violence
11 Transport injuries	11 Cirrhosis
12 Neonatal disorders	12 Transport injuries
13 Self-harm & violence	13 Diarrhea/LRI/other
14 Cirrhosis	14 Digestive diseases
15 Digestive diseases	15 Nutritional deficiencies
16 Nutritional deficiencies	16 Neonatal disorders
17 HIV/AIDS & tuberculosis	17 HIV/AIDS & tuberculosis
18 Other group I	18 Other group I
19 Maternal disorders	19 War & disaster
20 NTDs & malaria	20 Maternal disorders
21 War & disaster	21 NTDs & malaria

Communicable, maternal, neonatal, and nutritional diseases

Non-communicable diseases

Injuries

Institute for Health Metrics and Evaluation (IHME), 2015. GBD Compare. IHME, University of Washington, Seattle, WA. Available from: http://vizhub.healthdata.org/gbd-compare.

Biomedical risk factors are often (but to varying degrees) a result of *behavioral factors* that are in turn influenced by *socioeconomic factors* (including environmental and occupational risk) and other even more distal determinants. Health behaviors such as eating healthy foods and being active tend to interact with each other and influence a variety of biomedical factors (e.g., body weight, blood pressure, blood levels of glucose and cholesterol). Each health behavior can exert this influence alone but act with greater effect when combined with

TABLE 2.4 Rank and Percentage of Disease Burden (Attributable DALYs as a Percentage of Total DALYs) Attributed to Selected Risk Factors of Health, Europe 1990 and 2013

Central Europe
Both sexes, All ages, Percent of total DALYs

1990 rank	2013 rank
1 Dietary risks	1 Dietary risks
2 High systolic blood pressure	2 High systolic blood pressure
3 Tobacco smoke	3 Tobacco smoke
4 High body-mass index	4 High body-mass index
5 Alcohol and drug use	5 Alcohol and drug use
6 High total cholesterol	6 High fasting plasma glucose
7 High fasting plasma glucose	7 High total cholesterol
8 Air pollution	8 Low glomerular filtration rate
9 Low glomerular filtration rate	9 Air pollution
10 Low physical activity	10 Low physical activity
11 Child and maternal malnutrition	11 Occupational risks
12 Occupational risks	12 Low bone mineral density
13 Low bone mineral density	13 Child and maternal malnutrition
14 Other environmental risks	14 Other environmental risks
15 Sexual abuse and violence	15 Unsafe sex
16 Unsafe sex	16 Sexual abuse and violence
17 Unsafe water, sanitation, and handwashing	17 Unsafe water, sanitation, and handwashing

Metabolic risks | Environmental/occupational risks | Behavioral risks

Institute for Health Metrics and Evaluation (IHME), 2015. GBD Compare. IHME, University of Washington, Seattle, WA. Available from: http://vizhub.healthdata.org/gbd-compare.

other behaviors. Behavioral and biomedical risk factors tend to amplify each other's effects when they occur together; for example, smoking tobacco and high blood pressure escalate the risk for stroke.

Assessing risk factors for disease is at the core of preventive clinical medicine. Risk factors in individuals can be measured by interview (e.g., questionnaires), physically (i.e., anthropometry), or biochemically (i.e., based on blood samples). Some of the common biomedical risk factor measures used are shown in Table 2.7 (the clinical relevance of some of these is discussed at the end of each of the relevant chapters dealing with their lifestyle-related links). Relevant risk factor questionnaires are discussed in some of the chapters to follow.

TABLE 2.5 The Leading Causes of Disability Affected Life Years (DALYs) in Australia, 1990 and 2013

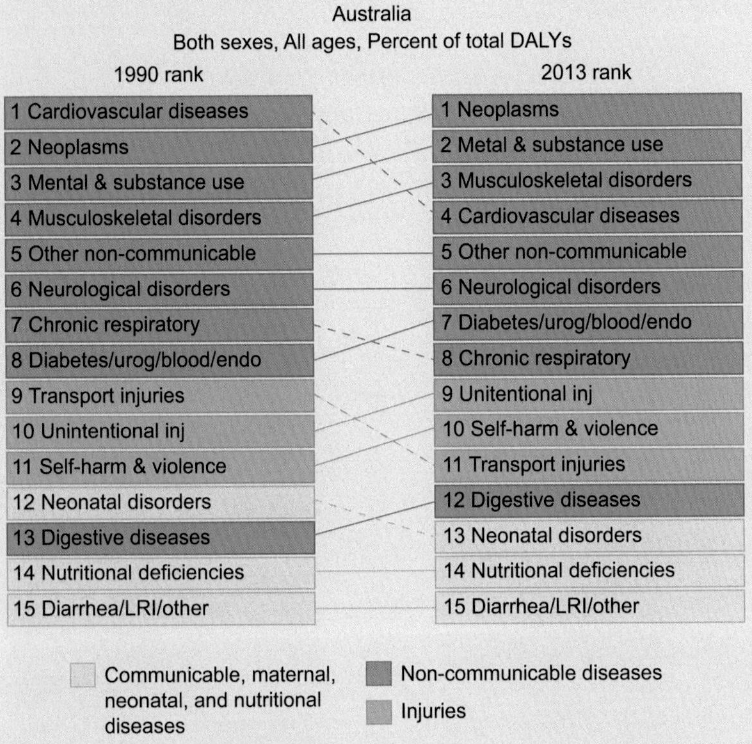

Australia
Both sexes, All ages, Percent of total DALYs

1990 rank	2013 rank
1 Cardiovascular diseases	1 Neoplasms
2 Neoplasms	2 Metal & substance use
3 Mental & substance use	3 Musculoskeletal disorders
4 Musculoskeletal disorders	4 Cardiovascular diseases
5 Other non-communicable	5 Other non-communicable
6 Neurological disorders	6 Neurological disorders
7 Chronic respiratory	7 Diabetes/urog/blood/endo
8 Diabetes/urog/blood/endo	8 Chronic respiratory
9 Transport injuries	9 Unintentional inj
10 Unintentional inj	10 Self-harm & violence
11 Self-harm & violence	11 Transport injuries
12 Neonatal disorders	12 Digestive diseases
13 Digestive diseases	13 Neonatal disorders
14 Nutritional deficiencies	14 Nutritional deficiencies
15 Diarrhea/LRI/other	15 Diarrhea/LRI/other

Communicable, maternal, neonatal, and nutritional diseases; Non-communicable diseases; Injuries

Institute for Health Metrics and Evaluation (IHME), 2015. GBD Compare. IHME, University of Washington, Seattle, WA. Available from: http://vizhub.healthdata.org/gbd-compare.

While many of the risk factor measures have been around for some time, new measures and markers are continually being assessed. Some of these extend existing measures, such as the inclusion of an ApoB/ApoA ratio, which further refines the risk of abnormal lipid levels for heart disease (Yusuf et al., 2004). Others, such as C-reactive protein (CRP) and homocysteine, take us into different areas of risk measurement (CRP namely reflecting chronic but subclinical inflammation, increasingly identified as underlying a range of chronic diseases). Still others increase the predictive potential of individual measures by combining more than one measure (Wang et al., 2006); the combination of a high triglyceride level and large waist circumference, for example, has been shown to be a reliable proxy measure of insulin resistance (Esmaillzadeh et al., 2006) and may be the first indication of insulin resistance.

TABLE 2.6 Rank and Percentage of Disease Burden (Attributable DALYs as a Percentage of Total DALYs) Attributed to Selected Risk Factors of Health, Australia 1990 and 2013

Australia
Both sexes, All ages, Percent of total DALYs

1990 rank	2013 rank
1 Dietary risks	1 High body-mass index
2 High systolic blood pressure	2 Dietary risks
3 Tobacco smoke	3 Tobacco smoke
4 Alcohol and drug use	4 High systolic blood pressure
5 High body-mass index	5 Alcohol and drug use
6 High total cholesterol	6 High fasting plasma glucose
7 High fasting plasma glucose	7 High total cholesterol
8 Low physical activity	8 Low glomerular filtration rate
9 Low glomerular filtration rate	9 Low physical activity
10 Occupational risks	10 Occupational risks
11 Sexual abuse and violence	11 Child and maternal malnutrition
12 Child and maternal malnutrition	12 Sexual abuse and violence
13 Low bone mineral density	13 Low bone mineral density
14 Air pollution	14 Unsafe sex
15 Unsafe sex	15 Other environmental risks
16 Other environmental risks	16 Air pollution
17 Unsafe water, sanitation, and handwashing	17 Unsafe water, sanitation, and handwashing

Metabolic risks | Environmental/occupational risks | Behavioral risks

Institute for Health Metrics and Evaluation (IHME), 2015. GBD Compare. IHME, University of Washington, Seattle, WA. Available from: http://vizhub.healthdata.org/gbd-compare.

However, identification of new risk factors for specific diseases is an enduring theme in medical research (Ware, 2006). This type of research often increases our understanding of the causes of diseases better than it actually translates into practice for treatment or prevention. Research done by Wang and colleagues is a good example; adding 10 novel biomedical risk factors for predicting the risk of cardiovascular events to the classic risk factors, such as blood pressure, cigarette smoking, and total cholesterol, resulted in only very small increases in the ability to classify risk (Wang et al., 2006). There is a real potential for contemporary medical risk factor research to get caught in Pareto's trap where 20% of the risk factors explain 80% of the disease but researchers try hard to

TABLE 2.7 Common Biomedical Risk Factor Markers for Lifestyle-Related Diseases

Physical Measures	Biochemical Measures
Blood pressure	Standard measures:
Fitness:	• Fasting lipids (TC, TG, HDL, LDL)
• Step test	• Fasting plasma glucose
• Submaximal test	• HBA_1C
• Stress test	• Uric acid
Weight/body fat/fat distribution:	• Liver function tests
• BMI (weight in kg/height in meters^2)	Other measures:
• Body fat (%)	• Glucose tolerance test
• Waist circumference	• C-reactive protein
• Waist-to-hip ratio	

chase the remaining 20% of disease with ever-increasing effort (Rose, 1991), even though there is already sufficient ground to act.

One should not underestimate the power of the classic risk factors at hand (abnormal lipids, smoking, hypertension, diabetes, abdominal obesity, psychosocial factors, inactivity, lack of consumption of fruits and vegetables, and alcohol), which accounted for more than 80% of the risk of myocardial infarction in a worldwide study (Yusuf et al., 2004). Such findings suggest that approaches to prevention can be based on similar principles worldwide and have the potential to prevent most premature cases of cardiovascular disease.

BEYOND SINGLE RISK FACTORS

Narrow attention on single risk factors and their association with disease led to clinical recommendations focused primarily on managing these individual risk factors, particularly seen in older guidelines for the control of raised blood pressure and cholesterol. Typically, separate guidelines were developed for each risk factor and treatment was recommended when that factor was above a specified level. Only since Rose's (1991) maxim, "All policy decisions should be based on absolute measures of risk; relative risk is strictly for researchers only," are more integrated recommendations based on absolute or total risk instead of the relative risk of individual factors emerging in guidelines and risk factor assessment applications. This orientation around the absolute risk should also apply to risk management in lifestyle medicine.

Thus, the newer clinical guidelines recommend that priority for treatment be given to patients with a high absolute risk of disease or death, defined as the probability of developing a particular endpoint over a specified period, instead of undue emphasis being placed on an individual risk factor. New and simple clinical tools for assessing absolute risk were developed and are being constantly refined and made available over the Internet.[2]

Clinical trial evidence clearly shows that the absolute benefits of treatment are directly proportional to the risk level before treatment. The absolute risk of a person's future cardiovascular disease is therefore strongly influenced by the combination of risk factors already present, particularly by the history of cardiovascular disease, age, sex, diabetes, smoking, blood pressure, and blood lipid concentrations. For example, a 50-year-old, nonsmoking woman with a blood pressure of 170/105 mm Hg, a total cholesterol of 7.0 mmol/L, and a high density lipoprotein (HDL) cholesterol of 1.0 mmol/L has about a 9% chance of suffering a major cardiovascular event in the next five years, if her risk is being calculated using an absolute cardiovascular disease risk calculator www.cvdcheck.org.au. However, a 60-year-old male smoker with a blood pressure of 160/95 mm Hg, a lower total blood cholesterol values (6.6 mmol/L), and slightly higher HDL value of 1.1 mmol/L has about a 26% risk.

Finally, the recent finding from a large mortality registration database looking at all-cause mortality during six years of follow-up among more than 200,000 Australians 45 years or older pointed out that some combinations of health risk behaviors may be more harmful than others (Ding et al., 2015). This suggests that some risk factor combinations have greater interactive (synergistic) effects on health outcomes that could help to guide the design of effective programs for the prevention of chronic diseases.

With antihypertensive or lipid-lowering drugs both these patients could reduce their risk of cardiovascular disease (CVD) by up to 25% over the next five years. The woman's absolute risk over the five years could fall from 9% to about 6.8% and the man's from 26% to about 19.5%. In other words, about 50 such women would require five years of treatment to prevent one cardiovascular event but only 15 such men. What is even more striking is the contribution of smoking to the high cardiovascular risk of our male example; if he gave up smoking, his five-year CVD risk would rapidly drop from the 26% to about 16%, a greater reduction than he could experience by adhering for five years to drug treatment of his high blood pressure or elevated lipids. This illustrates the importance of selecting lifestyle-related disease management based on the (evidence-based) impact any intervention could have (Tonkin et al., 2003).

---

2. For example, from the New Zealand Guidelines Group at www.nzgg.org.nz or the National Heart Foundation, Australia, at www.heartfoundation.com.au.

BEYOND RISK FACTORS ALTOGETHER

Risk factor assessment and monitoring leading to intervention is often at the center of clinical management but it is merely the start of diagnosis for chronic diseases. The task of the modern clinician is not only to recognize the importance of the immediate causes or markers of disease probability but also to address associated factors (e.g., mechanical and other problems such as joint problems associated with obesity, impaired vision with diabetes). Because these can often cause vicious cycles, they can hinder a lifestyle change being effectively undertaken or sustained by the patient (Swinburn and Egger, 2004). After all, a key characteristic of chronic diseases is that they are caused not by a single factor but by several factors. Consequently, merely correcting one risk factor (such as lowering blood glucose levels with medication) will often not suffice to cure the problem that caused blood sugars to rise in the first place (e.g., poor diet, inactivity).

LOOKING BEYOND IMMEDIATE CAUSES OF DISEASE

While Tables 2.5 and 2.6 give a general picture of "causality," this needs to be expanded. As can be seen in Fig. 2.1, there is a hierarchy of determinants[3] in modern diseases that needs to be discussed in any considered analysis. In the first place, there are the measurable risk factors discussed herein. These are a result of more proximal determinants such as smoking, obesity, or alcohol use, which are often not considered further. However, these proximal determinants in turn have more medial and, ultimately, distal determinants that present the true basis of disease.

Tracking disease determinants back like this helps to explain why a disease like type 2 diabetes is closely associated with affluence, industrialization and "modernity," and hence is a consequence or fault of modern society. At face value, diabetes appears to be caused by inactivity and poor nutrition, and so could be regarded as the fault of the individual (Chakravarthy and Booth, 2004). And while a clinician could be expected to have little effect on the more distal determinants, these need to be recognized in a comprehensive public health approach and any sophisticated epidemiological analysis. Each level in this hierarchy and its measures and implications for the clinician are discussed following.

PROXIMAL DETERMINANTS

Immediately recognized determinants of disease usually represent the reach of medical investigation into disease causality at the practice level, beyond

3. "Causality" is often problematic in chronic disease. Hence the term "determinants" is used herein (see Chapter 3).

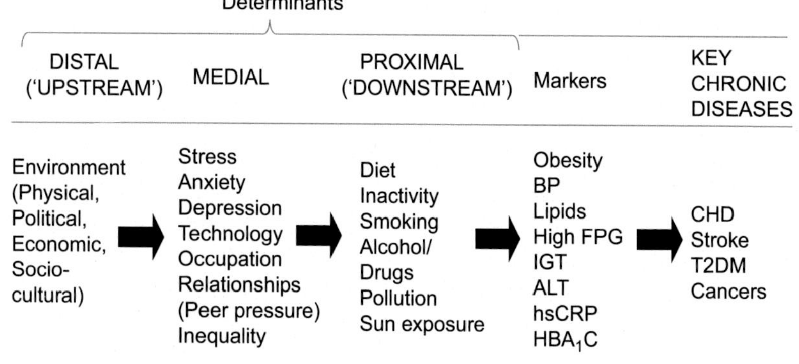

FIGURE 2.1 A hierarchy of chronic disease determinants. *ALT*, alanine aminotransferase; *BP*, blood pressure; *CHD*, coronary heart disease; *FPG*, fasting plasma glucose; *HBA_1C*, hemoglobin A_1C; *hsCRP*, high sensitive C-reactive protein; *IGT*, impaired glucose tolerance; *T2DM*, Type 2 diabetes mellitus.

measuring and monitoring the biomedical risk factors. Smoking, drinking habits, inactivity, obesity, drug use, unsafe sex, and poor diet are often pinpointed as the primary etiologies the treating physician will focus on. However, as argued later, these are at least only secondary factors. Proximal determinants are often seen as the endpoint of the causal chain when in fact they are not. This may be a consequence of the busy time schedules GPs face—rarely is there enough time to explore and address the underlying psychosocial experiences that resulted in disturbed body perception and eating patterns that in turn caused obesity. For example, "Eat healthier and move more" is often the only advice dispensed—well-meant but hardly appropriate.

Another glaring example is the failure to realize and act upon the significance of psychological factors such as stress and depression in the etiology of heart disease. The results of the Interheart study (Yusuf et al., 2004), which involved more than 11,000 cases of myocardial infarction in 52 countries and various ethnic groups, have now objectively indicated the need to extend the epidemiological analysis to include "causes of causes"—and stress in particular.

MEDIAL DETERMINANTS

Medial factors lead to the more obvious proximal determinants of disease. Tobacco smoking, for example, is prevalent because of peer pressure, social stress, anxiety, and other psychological factors. Regular smoking is not a behavior that occurs without a precipitating drive. Similarly, obesity is undoubtedly influenced by factors such as social pressures to eat, binge eating due to depression or anxiety, and genetic influences, among other things.

The value in identifying medial factors as driving forces behind the proximal determinants of disease is that any success in positively influencing these may lead to profound and long-lasting changes of a greater significance than treating the biomedical or proximal risk factor alone would produce. For example, mediating a bad relationship (be it at home or at work) can change eating and drinking habits and reinvigorate physical activity; the resulting joy in life in turn reduces overweight, blood pressure, and blood sugar probably more sustainably than a combination drug could have achieved. It is this potential widespread impact medial factors have on lifestyle-related diseases that makes recognizing and influencing them so important. And, unlike with distal determinants, the clinician has the potential to influence them—in fact, it is the hallmark of a good family doctor to address medial determinants while accompanying patients through their life.

What Can be Done In a Brief Consultation

- ☐ Discuss the lifestyle basis of much modern illness
- ☐ Explain the concept of a hierarchy of risk factors
- ☐ Measure anthropometric risk factors that can be done on the spot (e.g., weight, waist circumference, bioimpedance), and blood pressure
- ☐ Refer for appropriate tests of further biomedical risk factors (see text)

DISTAL DETERMINANTS

It is hard to escape the conclusion that many modern diseases have deep-seated environmental and economic grounds beneath their more apparent causes (Marmot, 2006). In the words of one writer, "We have dallied too long at the banquet of natural resources, only to discover that the only way out is past the cashier . . . the general consensus seems to be that with the aid of a little fast technological tap dancing most of us may make our escape without paying the full price. Not only would this involve a drastic and immediate reduction in the daily rate at which we gobble up the world's energy resources and dump our wastes, but we would have to sacrifice two of western civilisation's most sacred cows—Growth and Progress—to do it" (Morrison, 2003).

Growth economics has served humanity well over the past century, but there is little doubt that exponential development—of both population growth and resource use—cannot continue unabated (Egger, 2009). The debate on global warming suggests that carbon emissions are the problem, yet, seen more "distally," carbon emissions are influenced by population size, industrialization, and the economic growth imperative. On the one hand, growth economics has led to prosperity and the defeat of many health scourges; on the other, it is responsible for some of the world's most prominent current problems, including obesity and its associated diseases (Egger et al., 2012).

The answers to these problems are unlikely to be simple, because they go to the heart of how we (currently) live. Cynics argue that a failure to act now may lead to the irreversible depletion of resources such as oil and clean water and bring on a major world conflict or economic crisis. The influence of technology on health can be seen in the rate at which cars, computers, and effort-saving machines have reduced the need for humans to be physically active. These are macroenvironmental issues that the clinician may not be able to influence but that need to be recognized and that call for a fundamental shift in thinking. Nevertheless, clinicians may be able to influence microenvironmental issues, particularly around the house or neighborhood.

EXPANDING THE CONCEPT OF DISEASE AND INTERVENTION

Although we have discussed the hierarchy of disease etiology from endpoint disease to its distal determinants, a modern perspective calls for a more integrated understanding of disease outcomes and risk factors. Often a disease itself leads to a reduction in the capacity to change the lifestyle that is causing it. Damaged joints that result from being overweight, for example, lead to reduced activity, which, in turn, can increase body weight in a vicious circle (Swinburn and Egger, 2004). Consequently, increasing levels of effort and motivation are needed to make the required lifestyle changes to reduce illness.

In addition to this negative feedback between outcomes and preventive factors, there is also an incremental relationship between the motivation and effort needed to successfully implement lifestyle change, the complexity of the intervention required, and the barriers encountered along the scale of predominant disease determinants from proximal to distal. Fig. 2.2 shows that the complexity of the intervention recommended and the amount of motivation needed or barriers experienced when managing a specific chronic disease are dependent on its predominant underlying determinant. For instance, a clinician may focus on the biomedical risk factor of an elevated LDL, which almost dictates the intervention chosen (e.g., lipid-lowering drugs) and the amount of motivation needed or barriers to be overcome (e.g., obtaining the pills and taking them regularly). Looking at the middle of the spectrum, where more medial factors (psychosocial and cultural) are at the core of the lifestyle-related disease problem, much larger amounts of motivation are needed—not only by the "patient" but also by the family and society, who are called on to provide support—to counter the massive barriers complex interventions face. Such interventions are likely neither to bring quick and easy gains nor to be popular, but again if it is true that distal factors are the root of the problem, there is no easy way out.

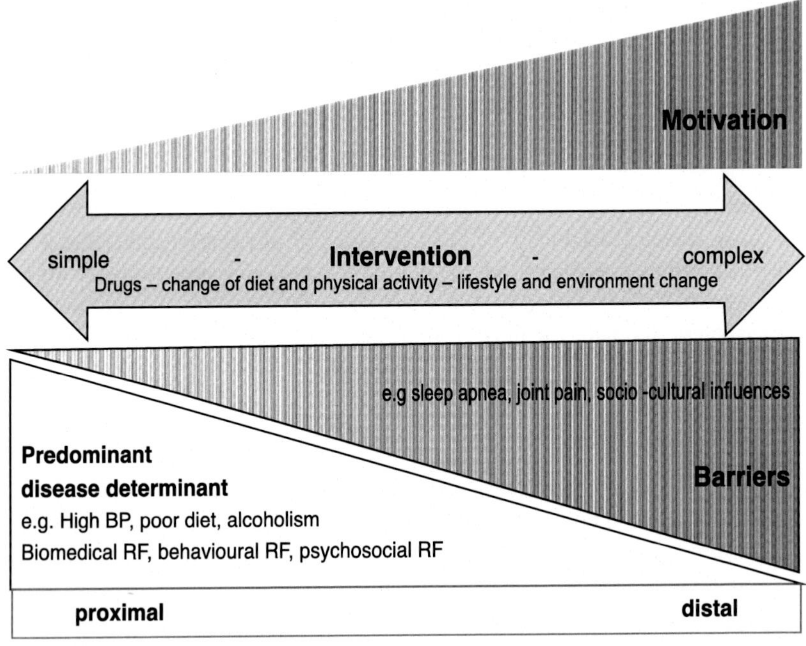

FIGURE 2.2 The motivation/intervention/barriers (MIB) model of disease.

THE CLINICIAN'S ROLE IN MANAGING LIFESTYLE-RELATED CHRONIC DISEASES

Understanding the etiology of disease is a significant part of primary health care. Yet in the area of chronic multifactorial diseases, our knowledge about causality is often less than in other medical areas. Research provides only probabilities, not straight causal relationships; a simple measurement of risk can give a probability estimate but is rather limited as a curative guide.

At the outset, the clinician should identify the problem and measure the obvious risk factors and assess overall risk. A practice nurse can usually undertake much of this. Immediate treatment of risks is often necessary and involves, especially at heightened risk levels, pharmaceutical management by a GP or specialist. Identification of the more distal causes and possible management of lifestyle-related disease, however, is vital for a holistic treatment of the patient. Identification may be part of a GP's diagnosis, but management is likely to take more time and may need to be left to the allied health team in collaboration with the GP. Detailed lifestyle management will require general skills in motivation (discussed in Chapter 5) as well as knowledge about the specific aspects of lifestyle associated with the measured risks.

In lifestyle interventions, motivation is crucial because a change in complex behavior requires a greater involvement than a relatively simple action such as taking medication. However, it should be apparent that governments also need to play a part by creating conducive and supportive environments through public health measures. Highlighting this, our experience with running quit smoking programs has shown that in the absence of a nonsmoking social and physical environment, those programs had little effect on smoking rates. Only when cigarette advertising was banned from the media in the 1980s and smoking became unacceptable in public places were significant inroads made into the smoking epidemic, shown by the decrease in smoking rates in Australia from about 40% during the 1950s to less than 15% today (Chapman, 2011).

Practice Tips

Managing Lifestyle-Related Chronic Disease

	Medical Practitioner	**Practice Nurse/Allied Health Professional**
Assess	• medical and family histories, focusing on lifestyle factors; refer for laboratory tests and treat appropriate risk factors • distal levels of causality (e.g., stress, depression, social factors) • whether lifestyle change has been a consideration • motivational factors	• blood pressure, weight, and waist circumference, routinely • smoking and alcohol consumption habits routinely • diabetes risk with diabetes risk assessment tests
Assist	• identification of the lifestyle changes required • risk reduction with medication where appropriate	• understanding of the proximal causes of risk factors • by reinforcing motivational factors • discussion of possible lifestyle changes • negotiation of reduction of barriers to change • development of a strategy for putting this into action
Arrange	• a list of other health professionals who may be involved and to develop a care plan • referral to a specialist where indicated	• a care plan if appropriate • contact with and coordination of other health professionals

Key Points

Clinical Management

- While treating proximal causes of disease, do not ignore the medial and distal factors.
- Check for comorbidities, such as depression or anxiety, with manifest disease.
- Use adequate measures for standard risk factor analysis.
- Use waist circumference in all standard consultations.
- Consider mechanical and other risks and their effects on compliance through reduced motivation.
- Be aware of the multiplicity of metabolic and mechanical risk factors of obesity and their impact on each other.

Professional Resources

Risk Factor Measures

Patients having or suspected of having lifestyle-related chronic disease can be tested or screened with a number of readily available standard measures. The following is a list of common measures used. More specific measures are covered in the individual chapters dealing with each lifestyle-related cause or outcome.

Nonintrusive Measures

- tobacco consumption—see Chapter 23
- alcohol consumption and dependency—see Chapter 23
- Diabetes Risk Assessment Tick Test
- resting pulse—a nonspecific indicator of possible health problems and potential lack of fitness; tachycardia may indicate lack of fitness (see Chapter 11) or an alcohol problem (Chapter 23)
- blood pressure—approximately 50% of hypertension is lifestyle based (Whelton et al., 2002)

Blood Assays

- fasting lipids
- total cholesterol—not a sensitive measure alone, except as an indicator for more detailed testing when very high
- LDL cholesterol—regarded as the "bad" form of cholesterol
- HDL—the "good" form of cholesterol
- triglycerides—an important dyslipidemic risk factor and often associated with a low HDL and increased girth
- blood sugars and glucose tolerance—see Chapter 5

Other

- thyroxine stimulating hormone—a sensitive test for thyroid dysfunction, although much less often associated with obesity than is often thought; usually only used if there are other symptoms
- measures of body fatness—see Chapter 4
- measures of fitness—see Chapter 6

REFERENCES

Byass, P., de Courten, M., Graham, W.J., Laflamme, L., McCaw-Binns, A., Sankoh, O.A., Tollman, S.M., Zaba, B., 2013. Reflections on the global burden of disease 2010 estimates. PLoS Med. 10 (7), e1001477.

Carus, T.L., Latham, R.E., 1993. On the Nature of the Universe, Monograph Collection (Matt - Pseudo).

Chakravarthy, M.V., Booth, F.W., 2004. Eating, exercise, and "thrifty" genotypes: connecting the dots toward an evolutionary understanding of modern chronic diseases. J. Appl. Physiol. (1985) 96 (1), 3–10.

Chapman, S., 2011. Lessons for the UK in how the Australians have reduced smoking. BMJ 343, d6797.

Ding, D., Rogers, K., van der Ploeg, H., Stamatakis, E., Bauman, A.E., 2015. Traditional and emerging lifestyle risk behaviors and all-cause mortality in middle-aged and older adults: evidence from a large population-based Australian Cohort. PLoS Med. 12 (12), e1001917.

Egger, G., 2009. Health, "illth," and economic growth: medicine, environment, and economics at the crossroads. Am. J. Prev. Med. 37 (1), 78–83.

Egger, G., Swinburn, B., Islam, F.M., 2012. Economic growth and obesity: an interesting relationship with world-wide implications. Econ. Hum. Biol. 10 (2), 147–153.

Esmaillzadeh, A., Mirmiran, P., Azizi, F., 2006. Clustering of metabolic abnormalities in adolescents with the hypertriglyceridemic waist phenotype. Am. J. Clin. Nutr. 83 (1), 36–46 quiz 183–184.

Farmer, P., 2004. Pathologies of Power: Health, Human Rights, and the New War on the Poor. Univ of California Press.

Marmot, M., 2006. Health in an unequal world: social circumstances, biology and disease. Clin. Med. (Lond.) 6 (6), 559–572.

Morrison, R., 2003. Plague Species: Is It in Our Genes? New Holland.

Murray, C.J., Ezzati, M., Flaxman, A.D., Lim, S., Lozano, R., Michaud, C., Naghavi, M., Salomon, J.A., Shibuya, K., Vos, T., Wikler, D., Lopez, A.D., 2012. GBD 2010: design, definitions, and metrics. Lancet 380 (9859), 2063–2066.

Rose, G., 1991. Environmental health: problems and prospects. J. R. Coll. Phys. Lond. 25 (1), 48–52.

Swinburn, B., Egger, G., 2004. The runaway weight gain train: too many accelerators, not enough brakes. BMJ 329 (7468), 736–739.

Tonkin, A.M., Lim, S.S., Schirmer, H., 2003. Cardiovascular risk factors: when should we treat? Med. J. Aust. 178 (3), 101–102.

Wang, T.J., Gona, P., Larson, M.G., Tofler, G.H., Levy, D., Newton-Cheh, C., Jacques, P.F., Rifai, N., Selhub, J., Robins, S.J., Benjamin, E.J., D'Agostino, R.B., Vasan, R.S., 2006. Multiple biomarkers for the prediction of first major cardiovascular events and death. N. Engl. J. Med. 355 (25), 2631–2639.

Ware, J.H., 2006. The limitations of risk factors as prognostic tools. N. Engl. J. Med. 355 (25), 2615–2617.

Whelton, P.K., He, J., Appel, L.J., Cutler, J.A., Havas, S., Kotchen, T.A., Roccella, E.J., Stout, R., Vallbona, C., Winston, M.C., Karimbakas, J., National High Blood Pressure Education Program Coordinating Committee, 2002. Primary prevention of hypertension: clinical and public health advisory from The National High Blood Pressure Education Program. JAMA 288 (15), 1882–1888.

World Health Organization, 2013. WHO Methods and Data Sources for Global Burden of Disease Estimates 2000–2011. Global Health Estimates Technical Paper. WHO, Geneva.

World Health Organization, 2016. Global Health Estimates: Deaths, Disability-adjusted Life Year (DALYs), Years of Life Lost (YLL) and Years Lost Due to Disability (YLD) by Cause, Age and Sex, 2000–2012. Retrieved February 17, 2016.

Yusuf, S., Hawken, S., Ounpuu, S., Dans, T., Avezum, A., Lanas, F., McQueen, M., Budaj, A., Pais, P., Varigos, J., Lisheng, L., Investigators, I.S., 2004. Effect of potentially modifiable risk factors associated with myocardial infarction in 52 countries (the INTERHEART study): case-control study. Lancet 364 (9438), 937–952.

Chapter 3

A "Germ Theory" Equivalent Approach for Lifestyle Medicine[1]

Garry Egger

a plague upon both your houses

William Shakespeare (*Romeo and Juliet*)

INTRODUCTION

In 1593, the closure of the theaters of London due to "unforeseen circumstances" set back a young playwright's career—fortunately, just temporarily. The circumstance was the great bubonic plague, a highly virulent disease that killed up to 20% of the population of the time. The writer was William Shakespeare.

The plague was just one of myriad infectious pestilences that have stalked mankind throughout evolution, keeping the average lifespan down to around 40 years of age through most of the time of *Homo sapiens*, and allowing little opportunity for the development of widespread degenerative disease.

But all that changed in the 19th and 20th centuries because of economic development and improvements initiated largely by the Industrial Revolution—public health and hygiene, the advent of antibiotics and vaccinations, and, driving these, the consolidation of the germ theory of disease (Harris, 2004). Since then, the lifespan of those in developed countries has almost doubled (Riley, 2001). It seemed that man's battle against disease had been all but won.

Mid-20th century optimism, however, was dampened by the reality of an epidemiological transition (Sanders et al., 2008) that occurs with economic development. In this transition, chronic diseases and conditions (e.g., heart disease, cancer, diabetes, chronic respiratory problems)—originally called the "diseases of civilization" (Bjorntorp, 1993)—replaced infections as the major source of disease. The epidemiological transition occurred in the latter half of the 20th century for many developed countries; approximately 70% of diseases

1. Modified from Egger B., 2012. In search of a "germ theory" for chronic disease. Prev. Chronic Dis. 9 (11), 1–7.

Lifestyle Medicine. http://dx.doi.org/10.1016/B978-0-12-810401-9.00003-6

now result from chronic conditions (Anderson, 2011). The same transition is occurring now in rapidly developing countries, such as China and Mexico, and is predicted in late-developing countries, such as India and Bangladesh (Harris, 2004).

The germ theory developed over several centuries, and represented the coming together of ideas that certain (predominantly infectious or "communicable") diseases are caused by microorganisms (bacteria, fungi, viruses, etc.) extant to the body, which together can be grouped under one generic term—"germs." This allowed a monocausal focus in disease that began with immunization and improvements in public health and hygiene, and culminated in the development of the antibiotics, sulfonomides, and other drugs that had a major impact on this class of diseases.

No equivalent of the germ theory has provided a unifying understanding of chronic disease etiology. The aging of the population and the dysmetabolism associated with aging have affected the prevalence of chronic disease; however, the increase in the prevalence of chronic diseases and associated risk factors and behaviors among all age groups limits aging as a sole explanation. Genetic influences and gene–age interactions are also incomplete explanations, in light of the sudden increase in, and other known causes of, chronic diseases. Many behaviors and environmental factors have been implicated, but a unifying theoretical underpinning has not been identified.

The discovery of a form of otherwise unrecognized inflammation in the early 1990s (Hotamisligil et al., 1993) and its widespread presence in many chronic diseases (Libby, 2007) led to the suggestion that many, if not all, such diseases may have this type of inflammatory basis (Scrivo et al., 2011). If so, and if a unifying cause could be identified to explain what is essentially a "multicausal enterprise" (Anderson, 2011), the implications for the management of chronic conditions could be significant, possibly reflecting the influence of the germ theory on changes in infectious disease prevention, diagnosis, treatment, and control.

INFLAMMATION AND DISEASE

Inflammation is a reaction by the immune system to restrict the spread of a potential disease, when the body suffers injury, or the invasion of a microbial organism (or "germ"), such as bacteria or a virus. As explained by renowned immunologist Dr. John Dwyer in his 1988 book *The Body at War* (Dwyer, 1988), the mammalian immune system is like an army, with specialist "warriors" designed to recognize, surround, delay, destroy, and then mop up the detritus of anything "foreign" (called an "antigen") entering the body with the potential to disturb its normally healthy equilibrium. Inflammation occurs in the early stages of invasion, and in classical infection is usually a large and acute response designed to keep the attackers "hemmed in" until help arrives. Scientists have only recently been able to decipher some of the "cross-talk" that goes on between chemical "warriors" in the bloodstream signaling such a response to the next line of defense.

Cytokines, or chemical messengers with abbreviated names like CRP, IL-6, IL-1β, and TNF alpha, among dozens of others, appear to play a proinflammatory role, signaling the buildup of a defensive "wall." These can rise by 100 or 200 times the normal level following an infection such as pneumonia. Other molecules, such as adiponectin and leptin, HDL cholesterol, and other members of the interleukin family such as IL-10 and IL-16, are antiinflammatory and help break down the defensive wall once a threat has passed (Libby, 2007) and restore the body to equilibrium.

For more than 2000 years, classical inflammation has been recognized by the symptoms identified by the Roman physician Aurelius Celsus as pain (dolor), redness (rubor), heat (calor), and swelling (tumor), with the more recent addition of loss of function (torpor). This form of classical inflammation is typically a short-term response to infection and injury, aimed at removing the infective stimulus and allowing repair of the damaged tissue, ultimately resulting in healing and a return to homeostasis. However, in 1993, researchers discovered a different type of prolonged, dysregulated, and maladaptive inflammatory response associated with obesity, which they suggested may explain the disease-causing effects of excessive weight gain (Hotamisligil et al., 1993). "Metaflammation" (Hotamisligil, 2006), as it was later called because of its link with the metabolic system, differs from classical inflammation in that it:

1. is low grade, causing only a small rise (i.e., a four- to sixfold increase vs. a several hundred-fold increase) in immune system markers such as CRP, IL-6, TNF alpha, and others;
2. is persistent and results in chronic, rather than acute, allostasis;
3. has systemic rather than local effects;
4. has antigens that are less apparent as foreign agents or microbial pathogens and hence have been referred to as "inducers";
5. appears to perpetuate, rather than resolve, disease; and
6. is associated with a reduced, rather than increased, metabolic rate.

In essence, although classical inflammation has a healing role in acute disease, metaflammation, because of its persistence, may have a mediating role, helping to aggravate and perpetuate chronic disease. The difference between these two forms of inflammation is illustrated in Fig. 3.1.

Metaflammation has been associated with many chronic diseases, conditions, and risk factors, ranging from type 2 diabetes and depression to heart disease, many forms of cancer, and even dementia (Libby, 2007). Possible reasons for these associations include facilitation of atherosclerosis (Mizumo et al., 2011), development of insulin resistance (Fernández-Real and Ricart, 2003), endoplasmic reticular stress (Hotamisligil, 2010), and changes in gut microbiota (Backhead, 2010). Metaflammation may be part of a causal cascade, including endoplasmic reticular stress and insulin resistance, or simply a

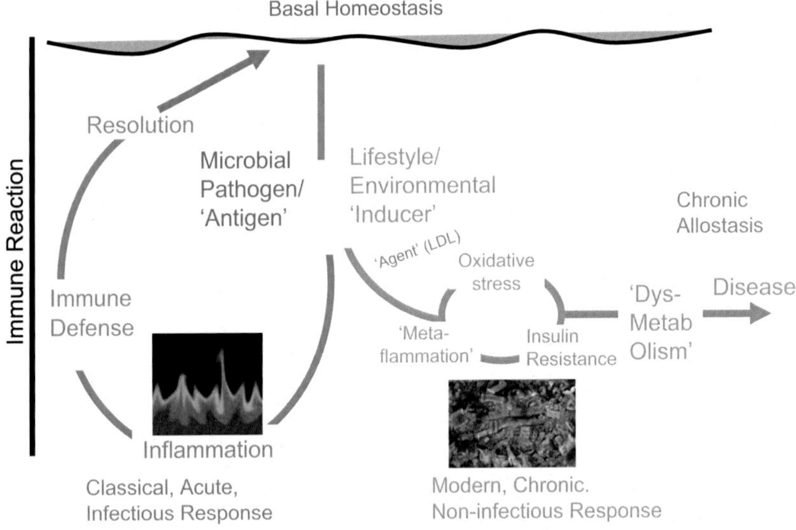

FIGURE 3.1 **Differences between classical inflammation and metaflammation.** A representation of the difference between classical inflammation (illustrated as raging fire), initiated by a microbial antigen or injury, and metaflammation (illustrated as smoldering fire), caused by inorganic "anthropogens." The scale of difference of immune reaction between the two forms (i.e., approximately 100-fold) is not shown. *LDL*, Low-density lipoprotein. *Adapted from Egger, G., Dixon, J., 2009. Should obesity be the main game? Or do we need an environmental makeover to combat the inflammatory and chronic disease epidemics? Obes. Rev. 10 (2), 237–249.*

defensive reaction to persistent stimuli that induce chronic disease. However, mounting evidence suggests metaflammation may develop as an intermediate immune system response to certain inducers, which, if maintained, can lead to the development and maintenance of dysmetabolic conditions and chronic disease (Libby, 2007; Hotamisligil, 2006). A commonly expressed view is that obesity is a prerequisite for metaflammation to occur or that metaflammatory inducers are necessarily either nutritional overload or nutrient based. Epidemiologic data suggest, however, that although several metaflammatory inducers are indeed associated with obesity, obesity is not a necessary precursor (Egger and Dixon, 2009; Warnberg et al., 2007); it may simply represent a "canary in the coal mine," warning of bigger problems in the broader environment (Egger and Dixon, 2009).

METAFLAMMATION AND "ANTHROPOGENS"

An increasing number of pro- or antiinflammatory biomarkers for measuring metaflammation have now been identified (Ouchi et al., 2011), and several avenues of research have sought to identify their nutritional, behavioral, and environmental inducers (Egger and Dixon, 2009; Warnberg et al., 2007). Many inducers have been identified (Table 3.1). These include not only behaviors linked to modern

TABLE 3.1 Lifestyle and Environmentally Related Metaflammatory "Inducers"

Proinflammatory	Antiinflammatory
A. Lifestyle	
Exercise • Too little (inactivity) • Too much Nutrition • Alcohol (excessive) • Excessive energy intake • "Fast food"/Western-style diet • Fat Saturated Trans fatty acids High fat diet High N6:N3 ratio • Fiber (low intake) • Fructose • Glucose High glucose/glycemic index foods Glycemic load Glycemic status • Meat (domesticated) • Salt • Sugar-sweetened drinks • Starvation Obesity/weight gain Smoking Sleep deprivation Stress/anxiety/depression/"burnout" "Unhealthy" lifestyle	Exercise/physical activity/fitness "Healthy" obesity Intensive lifestyle change Nutrition • Alcohol • Capsaicin • Cocoa/chocolate (dark) • Dairy calcium • Eggs • Energy intake (restricted) • Fish/fish oils • Fiber (high intake) • Garlic • Grapes/raisins • Herbs and spices • Lean game meats • Low glycemic index foods • Low N6:N3 ratio • Mediterranean diet • Fruits/vegetables • Monounsaturated fats • Nuts • Olive oil • Soy protein • Tea/green tea • Vinegar Weight loss Smoking cessation
B. Environment	
Age Air pollution • Indoor/outdoorAtmospheric CO_2 Perceived organizational justice (low) "Sick building syndrome" Second-hand smoke Status epilepticus (low)	

There is evidence that each of the inducers listed has a proinflammatory or neutral (in the case where there are similar proinflammatory inducers) response in humans, often without obesity. For a detailed list of discussions see references and Chapter 4.

lifestyles facilitated by postindustrial environments—for example, poor nutrition, inactivity, inadequate sleep, and stress (Willett et al., 2006)—but also components of these environments themselves, such as particulate matter (Mazzoli-Roch et al., 2010) and traffic-related air pollution (Alexeeff et al., 2011); endocrine-disrupting chemicals (EDCs), called obesogens because of their possible link to obesity (Neel and Sargis, 2011); social and economic conditions that create inequality and economic insecurity, including perceived organizational injustice or prejudice in the workplace (Elovainio et al., 2010); and the links between inequality and race/ethnicity as proposed in the "weathering hypothesis," which suggests that childbirth outcomes are more deleterious in older mothers of certain disadvantaged racial/ethnic groups (Geronimus, 1992). Such lifestyles and environments make up a category of inducers that can be labeled "anthropogens" because of their man-made origins and potential influence on health. Anthropogens are defined here as "man-made environments, their by-products, and/or lifestyles encouraged by those environments, some of which have biological effects that may be detrimental to human health." It is apparent that they have arisen in a chronological epoch labeled the Anthropocene (Zalasiewicz et al., 2010), signaling a geological period beginning when humans had a significant impact on Earth's ecosystems.

The underlying factors distinguishing proinflammatory and antiinflammatory inducers (or neutral, in the case where there are similar proinflammatory inducers) appear to be both temporal and of human origin (Table 3.1). All antiinflammatory or neutral inducers are those with which humans have had experience over hundreds or thousands of years, most of which are natural (e.g., fruits, nuts) or minimally modified (e.g., wine, beer). Proinflammatory inducers are recent and man-made (e.g., EDCs), are modified (e.g., processed food) or induced (e.g., inactive lifestyle), or are outcome effects (e.g., income/social inequality) of a man-made environment. Aging is an interesting proinflammatory inducer. The typical dysregulatory and metaflammatory effects of aging (also called "inflammaging") (Franceschi, 2007) are reduced by healthy lifestyles. Hence aging, although immutable, can be considered in the scope of an anthropogen, as the increase in longevity in modern populations can be ascribed to positive anthropogenic factors (e.g., medicine, immunization), leading to the decline of the infections.

Pro- and antiinflammatory inducers are related to major changes in human evolution (Fig. 3.2). The split between antiinflammatory or neutral inducers and proinflammatory inducers (anthropogens) is based on time and the amount of human involvement in developing such inducers (e.g., food processing, time-saving machinery).

CHRONIC DISEASE AND THE GERM THEORY

Given our modern understanding of the immune system, the ways in which anthropogens affect it could be revealing. Human immune responses are either innate or developed through exposure to unfamiliar stimuli over an extended time. Nontoxic antigens with which humans have evolved over thousands of

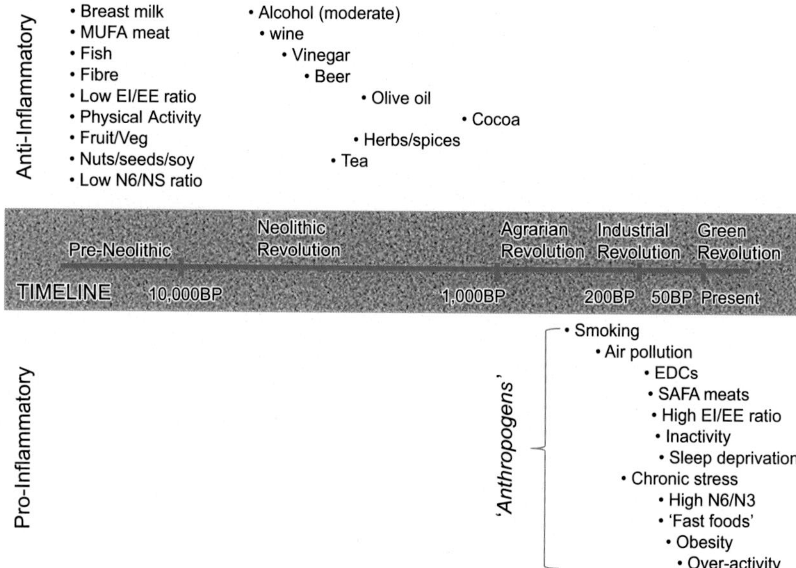

FIGURE 3.2 **The pro- or antiinflammatory effects of various inducers and their approximate (not to scale) introduction into the human environment.** The bullets associated with each inducer in the time frame indicated suggest the approximate time of introduction to the human environment. *BP*, Before present; *EDCs*, endocrine-disrupting chemicals; *EE*, energy expenditure; *EI*, energy intake; *MUFA*, monounsaturated fatty acid; *N3*, omega-3 fatty acid; *N6*, omega-6 fatty acid; *SAFA*, saturated fatty acid.

years ("good germs") are likely to cause little immune response, whereas anthropogens—man-made, novel, and recently introduced—are more likely to cause a response, albeit a low-level one, in a not-immediately-life-threatening situation. If exposure persists, the response may become chronic. A discussion of the possible mechanisms for this response is beyond the scope of this chapter, but the mechanisms may be related to genetic or epigenetic influences on chemical receptors, such as through nutritional factors (Choi and Friso, 2010). Because such a response is undifferentiated, it is likely to be systemic rather than localized.

What does the immune response tell us about the inducers identified in this chapter? It seems the human immune system is reacting to new, man-made environments and lifestyles (anthropogens), to which it has been exposed for a brief period (i.e., since the beginning of the Industrial Revolution, approximately CE 1800) and to which the immune system has not had time to develop. To characterize this reaction as lifestyle based offers only part of the explanation. For the last 30–40 years, in particular, we have experienced the emergence of a sedentary, nutritionally engineered, consumption-driven, security-conscious, screen-focused environment, the likes of which have never before been experienced by humans. The fact that most chronic diseases have also risen to prominence during this time (albeit along with an aging population and its genetic propensities) would point to the modern industrial and postindustrial environment as a distal "cause of the cause" of modern chronic diseases.

These ideas are not meant to imply that such a modern-day environment has not benefited human health. As discussed earlier, it has been primarily responsible for the decline in infectious diseases during the last one or two centuries. The rise in chronic diseases is a recent aberration and probably reflects an inevitable point of diminishing returns in a system of exponential economic growth, which is being reflected in our physiologic systems. Anthropogen-induced metaflammation could offer a generic etiology for chronic diseases of similar import, albeit slightly different in scope, to that of the germ theory for infectious diseases.

RELEVANCE FOR PUBLIC HEALTH AND LIFESTYLE MEDICINE

The germ theory had immediate implications for clinical medicine. Its relevance was extended to public health through the concept of the epidemiological triad, which stressed the need for attention to all three corners of a triangle: host, vector (and agent), and environment, for effective control of disease outbreaks. A focus on germs (i.e., agents) alone was insufficient. John Snow's removal of the handle of the Broad Street pump during the London cholera epidemic of the 1850s showed the relevance of both vector and environment to infection control.

Although less frequently applied, the triad has also proved robust in its application to chronic disease. The concept of anthropogens, as proposed here, offers a broader base for managing chronic disease than the germ theory did for infectious diseases. It offers a single concept for two corners of the triad, environment and vector (and agent). Although agents can be inducers of metaflammation—either exogenous or endogenous (e.g., EDCs, excessive cortisol levels)—these agents, plus the vectors through which they are delivered (e.g., polluted air/water, socioeconomic stress) and environments in which they exist (e.g., industrial output, inequality), are included under the single concept of disease promoting anthropogens. Public health and Lifestyle Medicine approaches in this model thus become generic, rather than fragmented (e.g., dietary intakes rather than specific nutrients); part of a "system" rather than a linear model (the interactions of diet, inactivity, stress, etc.), and distal, rather than proximal (e.g., industrialization/economic growth rather than fast-food marketing). Most important, such approaches need to involve disciplines often otherwise not considered in discussions of epidemic illness—such as macroeconomics, geography, ecology, and even business—in considering big-picture causality.

MANAGING ANTHROPOGENS

Obviously, not all anthropogens are unhealthy [although there are many others, not listed here, that possibly are (e.g., asbestos in building materials, shift work, increased atmospheric carbon dioxide)]. Any suggestion of halting human progress or returning to Paleolithic conditions would be akin to trying to wipe out all germs—good and bad—to manage infections. Acknowledgment of unhealthy anthropogens, however, would shift the focus from organic and multifocal causes of chronic diseases to vectors and environments and their distal economic, social, and physical drivers, just as the germ theory focused on microbial pathogens. This approach would also require a lifestyle and environmental approach, but also deflect attention from a simple lifestyle-based explanation alone. A focus only on lifestyle could inadvertently deflect criticism from systemic causes and put blame on the individual, whereas the recognition of an anthropogen-based etiology would cast a wider net.

The applications of this approach reside in the refocusing of testable hypotheses associated with chronic disease causality, as well as the recognition of a wider scope of influences on modern health and well-being that may incorporate a spectrum of current world issues, such as population growth and climate change (Egger, 2008). Pharmaceutical, medical, and surgical treatments will undoubtedly remain crucial in the treatment of chronic disease, as they are with infections, but an anthropogen-based etiology suggests the need for a bigger perspective on disease prevention and management. In the next chapter, we look at the relevance of this for how a new discipline of Lifestyle Medicine might be structured.

SUMMARY

Metaflammation, or low-grade, systemic inflammation, underlies most, if not all, modern chronic disease, and is manifest probably through a vicious cycle of cardiometabolic abnormalities including insulin resistance, disturbed blood fats, and high blood pressure. Although differing from classical inflammation in a number of ways, metaflammation, or "silent inflammation," is promoted by lifestyle and environmental "inducers" characteristic of the modern, post-Industrial Revolution environment called "anthropogens," because of their man-made origins in an epoch of the Athropocene. Metaflammation may or may not be mediated by obesity, which could be more of a "canary in the mineshaft" warning of bigger problems in our modern, Western way of life. Hence while it is imperative to attempt to reduce obesity and the chronic diseases associated with this, it should be kept in mind that a lifestyle medicine approach to dealing with these problems will involve an acknowledgment of broad economic and environmental issues, and not just palliative management of the disease.

Practice Tips

Metaflammation

- Look for potential signs of metaflammation in lifestyle
- Ask lifestyle questions of the lean as well as the obese
- Do not ignore low-level measures of inflammatory markers such as C-reactive protein
- Be on the lookout for loss of function as a standard inflammatory reaction

REFERENCES

Alexeeff, S.E., Coull, B.A., Gryparis, A., Suh, H., Sparrow, D., Vokonas, P.S., et al., 2011. Medium-term exposure to traffic-related air pollution and markers of inflammation and endothelial function. Environ. Health Perspect. 119 (4), 481–486.

Anderson, H., 2011. History and Philosophy of Modern Epidemiology. http://philsciarchive.pitt.edu/id/eprint/4159.

Bäckhed, F., 2010. 99th Dahlem Conference on Infection, Inflammation and Chronic Inflammatory Disorders: the normal gut microbiota in health and disease. Clin. Exp. Immunol. 160 (1), 80–84.

Björntorp, P., 1993. Visceral obesity: a "civilization syndrome." Obes. Res. 1 (3), 206–222.

Choi, S.W., Friso, S., 2010. Epigenetics: a new bridge between nutrition and health. Adv. Nutr. 1 (1), 8–12.

Dwyer, J., 1988. The Body at War. Allen and Unwin, Sydney.

Egger, G., 2008. Dousing our inflammatory environment(s): is personal carbon trading an option for reducing obesity and climate change? Obes. Rev. 9 (5), 456–463.

Egger, G., Dixon, J., 2009. Should obesity be the main game? Or do we need an environmental makeover to combat the inflammatory and chronic disease epidemics? Obes. Rev. 10 (2), 237–249.

Elovainio, M., Ferrie, J.E., Gimeno, D., Devogli, R., Shipley, M., Vahtera, J., et al., 2010. Organizational justice and markers of inflammation: the Whitehall II study. Occup. Environ. Med. 67 (2), 78–83.

Fernández-Real, J.M., Ricart, W., 2003. Insulin resistance and chronic cardiovascular inflammatory syndrome. Endocr. Rev. 24 (3), 278–301.

Franceschi, C., 2007. Inflammaging as a major characteristic of old people: can it be prevented or cured? Nutr. Rev. 65 (12 Pt 2), S173–S176.

Geronimus, A.T., 1992. The weathering hypothesis and the health of African-American women and infants: evidence and speculations. Ethn. Dis. 2 (3), 207–221.

Harris, B., 2004. Public health, nutrition and the decline of mortality: the McKeown thesis revisited. Soc. Hist. Med. 17, 379–407.

Hotamisligil, G.S., Shargill, N.S., Spiegelman, B.M., 1993. Adipose expression of tumor necrosis factor-alpha: direct role in obesity-linked insulin resistance. Science 259 (5091), 87–91.

Hotamisligil, G.S., 2006. Inflammation and metabolic disease. Nature 444 (7121), 860–867.

Hotamisligil, G.S., 2010. Endoplasmic reticulum stress and the inflammatory basis of metabolic disease. Cell 140 (6), 900–917.

Libby, P., 2007. Inflammation and disease. Nutr. Rev. 65 (12 Pt 2), S140–S146.

Mizuno, Y., Jacob, R.F., Mason, R.P., 2011. Inflammation and the development of atherosclerosis. J. Atheroscler. Thromb. 18 (5), 351–358.

Mazzoli-Rocha, F., Fernandes, S., Einicker-Lamas, M., Zin, W.A., 2010. Roles of oxidative stress in signalling and inflammation induced by particulate matter. Cell Biol. Toxicol. 26 (5), 481–498.

Neel, B.A., Sargis, R.M., 2011. The paradox of progress: environmental disruption of metabolism and the diabetes epidemic. Diabetes 60 (7), 1838–1848.

Ouchi, N., Parker, J.L., Lugus, J.J., Walsh, K., 2011. Adipokines in inflammation and metabolic disease. Nat. Rev. Immunol. 11 (2), 85–97.

Riley, J.C., 2001. Rising life expectancy: a global history. Cambridge University Press, New York.

Sanders, J.W., Fuhrer, G.S., Johnson, M.D., Riddle, M.S., 2008. The epidemiological transition: the current status of infectious diseases in the developed world versus the developing world. Sci. Prog. 91 (Pt 1), 1–37.

Scrivo, R., Vasile, M., Bartosiewicz, I., Valesini, G., 2011. Inflammation as "common soil" of the multifactorial diseases. Autoimmun. Rev. 10 (7), 369–374.

Wärnberg, J., Nova, E., Romeo, J., Moreno, L.A., Sjostrom, M., Marcos, A., 2007. Lifestyle-related determinants of inflammation in adolescence. Br. J. Nutr. 98 (Suppl. 1), S116–S120.

Willett, W.C., Koplan, J.P., Nugent, R., Dusenbury, C., Puska, P., Gaziano, T.A., 2006. Prevention of chronic disease by means of diet and lifestyle changes. In: Jamison, D.T., Breman, J.G., Measham, A.R., Alleyne, G., Claeson, M., Evans, D.B., et al. (Eds.), Disease Control Priorities in Developing Countries, second ed. World Bank, Washington, DC.

Zalasiewicz, J., Williams, M., Steffen, W., Crutzen, P., 2010. The new world of the Anthropocene. Env Sci. Tech. 44 (7), 2228–2231.

Chapter 4

A Structure for Lifestyle Medicine

Garry Egger, Michael Sagner, Hamish Meldrum, David Katz, Rob Lawson

Health is a state of complete physical, mental and social well-being and not merely the absence of disease or infirmity.

World Health Organization

INTRODUCTION: THE FIELD OF LIFESTYLE MEDICINE

Lifestyle Medicine (LM), as an application of clinical practice, is at the intersection of public health and other conventional clinical specialties. It arose in the first decade of the new millennium to provide new evidence-based treatment and prevention tools in response to the modern chronic disease epidemic (Egger et al., 2009). Since then, professional associations in LM have arisen in the United States, Europe, and Australasia (ESLM, 2015). Postgraduate courses are currently offered in a number of universities, and a growing number of texts are now available (Rippe, 2013; Guilliams, 2012; Lenz, 2013; Mechanick and Kushner, 2016; Egger et al., 2012). As the contributions of lifestyle and environmental factors to disease are now widely recognized; the field of LM is developing its own structure and pedagogy.

In this chapter we consider the current state of LM under four headings: (1) knowledge (the science), (2) skills (the art), (3) tools (the materials), and (4) procedures (the actions).

DEVELOPING THE KNOWLEDGE BASE: THE SCIENCE OF LIFESTYLE-RELATED DISEASES AND CONDITIONS

With infectious diseases, "causality" can usually be ascribed to biological causes ("germs"), using classical principles such as Koch's postulates (Koch, 1884). With chronic diseases and conditions, establishing causality is more problematic, as the true biological "cause" is more difficult to define (Bradfield-Hill, 1965), although the traditional "epidemiological triad" (host, vector, environment) does

Lifestyle Medicine. http://dx.doi.org/10.1016/B978-0-12-810401-9.00004-8

TABLE 4.1 Chronic Disease Categories With Lifestyle/Environmental Determinants

1. Cardio- and cerebrovascular disease
2. Cancers with lifestyle component
3. Endocrine/metabolic disorders
4. Gastrointestinal diseases
5. Kidney disease
6. Mental/CNS health
7. Musculoskeletal disorders
8. Respiratory diseases
9. Reproductive disorders
10. Dermatological disorders

offer some insight for chronic as well as infectious diseases here (Egger et al., 2003; Egger and Dixon, 2014). The closest we can often get is in describing *determinants* of disease at various levels as shown in Fig. 2.1 (Chapter 2).

The discovery of "metaflammation" as described in Chapter 3 gives some hints as to how this may be done. The proinflammatory inducers, or "anthropogens," shown in Table 3.1 (Chapter 3), for example, while not specifically *causes*, are the *determinants* of many, if not all, major classes of modern chronic disease (Egger and Dixon, 2010). In almost all cases these determinants are linked with metaflammation. The concept of anthropogens provides a single focus for the chronic disease categories shown in Table 4.1, like the "germ theory" afforded infectious diseases in the 19th and early 20th centuries (Egger, 2012).

Identifying Anthropogens

In most discussions of chronic disease etiology, smoking, poor nutrition, inactivity, excess weight, and alcohol use stand out as the dominant preventable determinants (Katz, 2014). However, research has expanded this considerably to take account of social, cultural, occupational, environmental, and other factors that interact with each other in a "systems"-type model (Hamed, 2009). These were listed in Table 3.1. In Table 4.2, they are restructured in a format using the acronym NASTIE MAL ODOURS. Most are covered in more detail elsewhere (Egger and Dixon, 2011; O'Keefe et al., 2008; Calder et al., 2009), and in the remaining chapters of this book. Hence they will be considered only briefly here, together with evidence of an inflammatory link, where this exists.

Nutrition

Inadequate and/or overnutrition account for a significant proportion of risk for chronic conditions like vascular disease, type 2 diabetes, and certain cancers (Calder et al., 2009). Studies have reported increased risk from excessive amounts of total energy, sugars, salt, alcohol, and (saturated and trans) fats,

TABLE 4.2 Anthropogens in Chronic Disease in Advanced Societies

Determinants	Decreases Disease Risk	Increases Disease Risk
Nutrition 1,2,3,4,5,6	Fruit/vegetables; dietary fiber; natural foods; food variety; healthy eating patterns; fish; low dose alcohol	High total energy; high energy density; excess processed foods; high glycemic index foods; saturated/trans fats; sugars, salt; excess alcohol; processed meats; obesity
(In) Activity 1,2,3,6,7,8,9	Aerobic/resistance exercise; flexibility; stability training	Sitting; sedentary work; excessive exercise; obesity
Stress, anxiety, depression 1,3,4,9	Perceived control; resilience; self-efficacy; coping skills; exercise/fitness; healthy diet	Overload; "learned helplessness"; early trauma; boredom; caffeine/drug use; excess alcohol use
Technopathology 7,10	Selective technology use; preventive care; limiting exposure	Machinery use; TV/small screens; repetitive actions; excessive noise; weapons of war
Inadequate sleep 1,3,6,10	Sleep hygiene; healthy diet; exercise/fitness	Shiftwork; excessive entertainment; sleep disorders; interactive media in room; obesity; drugs/alcohol; stress; activity before sleep
Environment 2,3,6,9,10	Political/economic structure; "nature therapy"; infrastructure for exercise; reduced chemical use	Political/economic structure; passive influences; second-hand smoke; particle pollution; endocrine disrupting chemicals; drug immunity (such as antibiotics)
Meaninglessness 1,2,6	"Something to do; someone to love; something to look forward to"; proactive self-balance	Unemployment; displacement; aging and loss of responsibility; depression; negative affect; early experiences
Alienation 1,2,6	Family relationships; improved parenting; increased competencies	Discrimination; early experiences; poor parental support; feelings of isolation; illness; emotional distress; social rejection
Loss of culture/identity 1,2,6	Cultural acceptance and support; conflict resolution; cultural pride/training	Warfare; domination by invading culture; displacement

Continued

TABLE 4.2 Anthropogens in Chronic Disease in Advanced Societies—cont'd

Determinants	Decreases Disease Risk	Increases Disease Risk
Occupation 1,2,8,10	Employment; social justice; work equality; economic security	Shiftwork; stress; hazard exposure; conflict; unhealthy interests/habits
Drugs, smoking, alcohol 1–10	Social support; relationships; resilience; employment	Stress, anxiety, depression; peer/social pressure; addiction; social disadvantage; relationships; social environment
Over- and **U**nderexposure 1,2,3	Sunlight (adequate); light; general stimulation	Climate; sunlight (excess); sunlight (inadequate); excessive darkness; low humidity; radiation; asbestos
Relationships 1,3,6	Companionship; peer support; maternal support in childhood; "love"	Loneliness; interpersonal conflict; lack of support; economic insecurity
Social inequity 1–10	Socioeconomic status; education; trust; economic security; support	Inequality; poverty; lack of welfare support

Numbers in column one refer to chronic disease categories (from Table 4.1) for which there is supporting evidence.

as well as inadequate levels of fiber, fruit, vegetables, and certain nutrients (Chapter 8). Levels of processing have been proposed as a general indication of risk (Monteiro, 2009), and there appears to be a clear postprandial "metaflammatory" trail from natural, whole foods to ultraprocessed foods and fluids (Egger and Dixon, 2011). Fluids in nutrition are covered in Chapter 9 and nutritional behaviors leading to nutrition problems in Chapter 10.

(In)Activity

The lack of regular, extended, physical activity is a significant driver of chronic disease in modern societies, with links to numerous common chronic conditions (Booth et al., 2012). Not only is this perpetrated through inadequate exercise and leisure and work-related movement (Chapters 11–12), but independently through sedentary activities such as excessive sitting (Dunstan et al., 2012). Weight gain is often a consequence of inactivity and overnutrition, but inflammatory processes can occur without obesity (Ross and Janiszewski, 2008; Bullo et al., 2007) suggesting these are independent determinants of disease.

Stress, Anxiety, and Depression

While no doubt existing in all human societies, stress, anxiety, and depression (Chapters 13–16) appear to have risen to epidemic proportions in modern Western cultures. Chronic psychological stress has been shown to trigger pathological pathways such as metaflammation. The link with heart disease and other chronic ailments is increasingly appreciated (Hajja and Gott, 2013) and interventions aimed at reducing or managing stress as a precursor are high on the list of desired lifestyle and behavioral prescriptions at the primary care level.

Technology-Induced Pathology

Changes in society invariably lead to changes in the types of diseases in those societies. Although not (yet) a widespread or accepted term, technology-induced pathology is a way of explaining the ill-health effects of certain types of modern technology (Chapter 17). It explains trauma from hi-tech warfare weaponry and motor vehicles or machines at one extreme, to endocrine disrupting chemicals and biological changes induced by extended small screen use at the other (Weng et al., 2013). As shown later ("Tools"), technology also has the potential to be part of the solution in chronic disease management.

Inadequate Sleep

Poor sleep may be one of the most underrecognized lifestyle determinants of modern disease epidemiology (Slarz et al., 2012). The practice of going to sleep and waking up at "unnatural" times as well as getting inadequate sleep has been described as "social jetlag" (Roenneberg, 2013). This can result from sleep disorders, but more commonly is associated with entertainment, drug use, and other aspects of the modern lifestyle, such as interactive screen use (Chapter 18). It also links closely with other determinants such as poor diet, fatigue, and inactivity and as a stressor leading to anxiety and depression.

Environment

Both macro- and microenvironments exert their effects on health and/or behavior through indirect mechanisms (Prüss-Ustun and Corvala, 2006), such as the passive endorsement of activities (overeating, inactivity) leading to obesity, or through more direct mechanisms such as pollution or exposure (Genuis, 2012). Within environmental sizes (micro, macro), there are environmental types (physical, economic, political, and sociocultural—Chapter 19). And while environmental influences are generally considered under public health, there are also implications for LM, such as in modifying the home or local environment to reduce disease determinants (Lyte, 2013).

Meaninglessness, Alienation, and Loss of Culture or Identity

Not unexpectedly *meaninglessness*, learned helplessness, or lack of purpose in life is associated with persistent elevated cortisol levels, raised inflammatory markers, and cardiovascular disease outcomes (Kim et al., 2014). *Alienation*, or estrangement, can result from many factors—discrimination, social isolation, rejection, or other adverse childhood experiences (ACEs). Mediating factors such as emotional distress and loss of control may even play a part in the recognized link between alienation and adverse cardiac events (Ketterer et al., 2011). *Loss of culture and/or identity* are significant factors in displaced indigenous populations, such as First Nation North America Indians and Aboriginal Australians. It can also result from confused identity around sexual orientation or family disruption or as a result of warfare or natural disasters (Shehab et al., 2008) where cultures are either wiped out or relocated (Chapter 20) (Doocy et al., 2013).

Occupation

Taken broadly to include "ways of occupying one's time" such as through habits, hobbies, and interests, "occupation" influences health directly through injury (such as sports/recreational, etc.) and toxicological or repeated exposures, but also through less direct processes such as shiftwork effects on physiological function and sleep, or "burnout" and economic insecurity relating to occupational status (Chapters 21–22). This interacts with other anthropogens discussed here such as "stress," "meaninglessness," and "relationships," and can have a link with metaflammation through various factors (Puttonen et al., 2011).

Drugs, Smoking, and Alcohol

Drugs, both licit and illicit, prescribed and nonprescribed, are responsible for a significant and increasing degree of morbidity and mortality in modern societies (Chapter 23). This includes iatrogenic reactions from adverse drug experiences, for which "deprescription" might be required as an LM intervention. Cigarette smoke is a proven toxicant with over 300 chemicals and links with cancers, heart disease, and respiratory problems. Tobacco control in most modern Western countries, however, has been one of the big success stories of health promotion and LM (Vic Health, 2015). Alcohol is a little more complicated with possible benefits coming from low doses but chronic problems, such as liver disease, and behavioral effects, such as violence from excessive and binge drinking.

Over- and Underexposure

The complex interactions and nonlinearity in biology and lifestyle-related chronic diseases often hinder understanding and lead to misconceptions in risk and disease management. A certain amount of physical activity or sleep, for example, is considered healthy (and reduces the risk for chronic diseases), whereas "overdosing" or "underdosing" quickly increase chronic disease risk. Exposure to ultraviolet radiation (UVR) from sunlight also follows a "U" or "tick-shaped" (nonlinear) relationship with health. UVR is classified as a carcinogen and a major determinant for several forms of skin disorders. Overexposure to heat and dryness (low humidity) is also thought to have adverse effects on the skin (Chapter 28). Underexposure to the same sunlight that can cause adverse effects from overexposure on the other hand can lead to vitamin D deficiency, and in some cases more extreme effects such as seasonal affective disorders.

Relationships

The quality of personal and social relationships is clearly linked to chronic disease outcomes including heart disease, stroke, some cancers, and all-cause mortality (Holt-Lunstad et al., 2010—Chapter 25). ACEs fit within this category as well as within other environmental exposures and alienation discussed earlier (Lanius and Vermetten, 2009). As yet, the pathways for this are unclear, but metainflammatory processes have been associated with poor social relations (Kiecolt-Glaser et al., 2010), and social support can alleviate the inflammation associated with childhood adversities (Runsten et al., 2013). Improving awareness of the importance of social support and assisting in finding such support should be integral to chronic disease management.

Social Inequity

Gaps between the rich and poor in a country or community add to the effects of poor interpersonal relationships on chronic disease as illustrated by epidemiological studies using relative income differentials within and between countries (Wilkinson and Picket, 2009). The mechanisms remain unclear but metaflammation appears again to play a mediating role in the chronic diseases that have been associated with such inequality (Friedman and Herd, 2010—Chapter 25).

SUMMARY

Chronic disease determinants, labeled "anthropogens" here, have varying levels of potency depending on a range of factors such as genes, environment, and exposure. Although each may impact independently in the development of chronic disease, findings have suggested it is more realistic to think of interactions as an in vivo "systems" model both within and between determinants in contrast to that which is characterized by a simple linear approach (Hamed, 2009). Isolating nutrients from foods, for example, ignores the interactive relationship of nutrients found in whole meal patterns, just as considering inactivity, sleep, or social factors in the absence of nutrition provides only part of the etiological answer to disease manifestation. The systems model interactions between determinants may be hidden below the surface, like an iceberg, but is a most important aspect of a modern Lifestyle Medicine approach to chronic diseases and conditions.

UTILIZING SKILLS: THE ART OF LIFESTYLE MEDICINE

Elaborating the determinants discussed earlier is relatively straightforward. Developing the skills required to change unhealthy lifestyles and (at least the "micro") environments driving these is more complicated. Clinical skills are similar to those required in conventional health care: *diagnosis*, *prescription*, and *counseling* in particular. The processes, however, are different and include the art of coaching, coaxing, and "nudging" patients from extrinsic to intrinsic motivation to facilitate change to healthy lifestyle behaviors.

Diagnosis in LM is focused on the underlying mechanisms and determinants of disease already discussed earlier, rather than the disease itself. Most chronic diseases share common dysfunctions/mechanisms (such as metaflammation) and signs in clinical practice (such as elevated inflammatory markers). Addressing these mechanisms can improve risks for several diseases. For example, obesity is a function of energy balance. But energy intake (food and drink) and energy expenditure (metabolism, physical activity) can be influenced by a range of other, less obvious factors, which need to be considered in any "systems model" approach. Stress, for example, can influence (positively or negatively) energy intake and metabolism, as well as activity levels. Inadequate sleep can lead to low activity levels during the day, which can then impact on diet and relationships, which can ultimately effect body weight outcomes.

Prescription involves both pharmaceutical and nonpharmaceutical interventions in the case of LM, although medication use is often an adjunct to a therapeutic lifestyle intervention (TLI, behavior change), more than as the primary treatment. This comes about with the recognition that medication is primarily aimed at disease or risk modification, whereas a more upstream lifestyle change is targeted for long-term management. An important tool in involving the patient in risk modification and decision making is an understanding of numbers needed to treat/benefit and numbers needed to harm.

Deprescription is a response to the overuse of medications (polypharmacy) that has become commonplace in modern affluent societies, particularly in the elderly (Chapter 24). Achieving the complex balance between managing disease and avoiding medicine-related problems may require discontinuing prescribed medicines. Deprescribing techniques have thus arisen to assist patients stopping medications that not only may cause harm but are also unlikely to result in benefit. While only in their infancy, such skills are likely to become a standard procedure in managing chronic diseases in the future (Pillans, 2015).

While all the usual *counseling* skills such as motivational interviewing, health coaching, interpersonal relationship training, cognitive behavior therapy, and self-management training are *necessary* in LM, they may not be *sufficient* for dealing with behaviors and the environments driving chronic diseases. Management may thus require a lateral shift, leading to different procedures, such as group education and shared medical appointments (SMAs), as discussed later.

EXPANDING THE RANGE OF TOOLS: THE MATERIALS FOR LIFESTYLE MEDICINE

LM tools are centered mainly around the concept of the "quantified self." This means evolving the role of the patient from a minimally informed recipient to an active collaborator in the patient–provider relationship. Many new devices can be seen as LM tools because they help to track and monitor signs and determinants related to lifestyle such as blood pressure, blood glucose, and physical activity levels. Digital tools can monitor and prompt action related to lifestyle, such as reducing sitting time, by alerting the user. Improvements in technology have given rise to new devices and developments called "mHealth," or health care and public health practices supported by mobile devices and other advances in telemetry (Table 4.3).

Limited evaluations of single mHealth devices have appeared in the literature since 2003, shadowing the recognition of chronic ailments as a rising category of disease. Two systematic reviews of such devices to date have highlighted the potential of these as a new set of tools for chronic disease management (Swan, 2009; Hamine et al., 2015).

Another review notes that: "Increasing adherence may have a greater effect on health than improvements in specific medical therapy" (Brown and Bussell, 2011). Well-controlled studies comparing adherence with mHealth devices compared to prescriptive advice controls typically show a 50% improvement in adherence from the former, more than justifying a serious look at these for chronic disease management. For example, adherence to smartphone food tracking for weight loss is higher than in traditional food diaries (Carter et al., 2013. Burke et al., 2013).

Mobile connective technology holds promise as a scalable mechanism for augmenting the effect of physician-directed weight loss treatment (Spring et al., 2013). SMS messages are the most popular current mHealth devices, used for medication reminders, education, or information about disease management. Simple SMS reminders or information about new programs or

TABLE 4.3 LM "Tools"

Clinical Application	Device	Measure
"mHealth" • Messages/education/marketing • Phone + feedback/Bluetooth	Mobile phone	Health promotion; reminders BP; HR; PG; temp; O_2 sats, COPD (level of pollution)
Telemetric monitoring • Movement • Asthma **Ambulatory monitoring** • Sleep • Blood quality • Heart measures • Neurological sensing **Implantable devices** • Cardiac rhythm • Resynchronization devices	Accelerometers; movement sensors "health buddy" Sensor devices Breathing sensors Meter Electrocardiograph (ECG) vest Inertial sensor Implantable cardiac devices ECG Defibrillator	Activity levels; sleep patterns BP; glucose; temp Breathing Apnoeas; sleep quality; brain wave monitoring; menstrual-cycle tracking; INR Sinus rhythm Gait, paralysis; early Parkinson's Atrial fibrillation Correct arrhythmias
Self-monitoring • Body composition • Strength • Lung function • Hypertension	BIA scales Dynamometer PICO6 (spirometer) BP machine	Body fat; lean tissue mass Grip strength FEV1/FV6; lung age BP; pulse
Brief assessments • Dietary questionnaire • DAB-Q • Physical fitness	Short diet questionnaire Diet/activity questionnaire SPAL questionnaire	Basic diet quality Diet/exercise (Est) VO_2 max
Internet connections • Prevention/rehabilitation	Internet user groups; bibliotherapy	Depression; anxiety; asthma; arthritis, etc.

BIA, bioimpedance analysis; *BP*, blood pressure; *COPD*, chronic obstructive pulmonary disease; *FEV1/FEV*, forced expiratory volume in 1 s or 6 s; *HR*, heart rate; *INR*, international normalized ratio; *PG*, plasma glucose; *SPAL*, subjective physical activity level.

treatments are not only effective, but cost-effective. As an example, one of us (GE) involved in the 1990s development of a men's waist loss program ("GutBusters": Egger and Dobson, 2000) found weekly advertising costs of $AUS10,000 for recruiting men through mainstream media crippling, leading to the early retirement of the program. Operating out of medical centers we have now found a personal SMS invitation to 10 times the desired number of men to fill an SMA group (see later) of 10–12, identified through medical records systems as falling within the required audience (BMI > 35; metabolic syndrome), is not only successful, but virtually cost free. SMS is also used for follow-up weekly tips. Fine targeting combined with a personalized invitation from the patient's GP could hold the key to better long-term chronic disease management in a number of disease areas (de Jongh et al., 2012).

Other mHealth devices include mobile phones plus software or applications, specific medical telemetry devices, or phone plus wireless or Bluetooth-compatible devices. Between them, these not only deliver education and reminders, but also monitor functions such as blood pressure, heart rate/rhythm, and blood sugars for patients and providers.

Multiple outcome measures were used in the most recent review, including usability, feasibility, and acceptability of the mHealth tools studied as well as adherence and disease-specific outcomes. Examples of improved management included reduced HbA_1C, reduction in hyperglycemic events, improved lung function, reduced blood pressure, better use of nebulizers, improved fitness levels, etc. mHealth tools were also found to increase self-care awareness and knowledge, improve patient confidence to monitor symptoms and signs of chronic diseases, and decrease anxiety about disease. Improvements were noted across all age and socioeconomic categories. However, as might be expected, take-up and use by adolescents was shown to be particularly effective.

Significantly, an mHealth system between the patient and provider was less burdensome and judgmental compared to face-to-face contact, making such tools likely to be even more effective in an SMA context or with individuals who are averse to the "scary" doctor–patient environment in a closed setting, such as Indigenous individuals or people of low health literacy.

The future for the management of chronic long-term conditions, and the potential for lifestyle-related disease management through mHealth, is encouraging. There are now telemetry tools such as movement sensors, portable sleep monitors, bioimpedance analysis scales, grip strength dynamometers, pulse measures, and other "tools" for self-monitoring. There are also skills for motivation, brief assessments, self-care measures (primary prevention), and self-management measures (secondary/tertiary prevention) that expand the prescriptive environment developed for acute disease. Instantly accessible internet assistance, self-help groups, and virtual games provide further assistance.

ADOPTING PROCEDURES: DEVELOPING THE ACTIONS FOR LIFESTYLE MEDICINE MANAGEMENT

A procedure in clinical practice is a sequence of steps to be followed in establishing a course of action. Examples of evidence-based procedures specific to an LM approach are weight loss programs, quit smoking programs, utilizing drug or patch therapy, meditation classes, and art therapy for ACEs (see Chapter 20). Other examples include: sleep hygiene; light therapy; chronic pain management programs; occupational health and safety; and self-help programs such as Alcoholics Anonymous and Narcotics Anonymous. Possible future processes include fecal microbial transplants (Chapter 19) or diet therapy for recolonization of the gut microbiome (Carvalho and Saad, 2013). Other, new or innovative procedures currently on the fringe of LM practice include the following.

Self-Management Training

Self-management is an important part of chronic disease management. Training is at different levels, from individual assistance by a health professional (Lawn et al., 2009) to group learning processes (Lorig and Holman, 2003). Self-management training as a procedure in LM is discussed in more detail in Chapter 6.

Shared Medical Appointments

At a conventional level, it has been assumed that the clinical process of one-on-one (1:1) consulting (despite the lack of data supporting this over other forms of clinical engagement) is set in stone and would continue to play a part in chronic disease management, which hitherto it has done in the past for acute diseases and injury.

Yet chronic diseases and conditions have distinct requirements over acute care. Consequently, short consultations, as covered by health systems determined in an acute disease era, are unlikely to be wholly appropriate for chronic disease. Health education in groups with an experienced leader (such as a diabetes educator, dietitian, exercise physiologist, or other health care professional) arose to help overcome these problems. But group education lacks medical input. Individual 1:1 consulting, on the other hand, has medical input, but lacks the educational component, time, and peer support associated with group education.

SMAs are "… consecutive individual medical visits in a supportive group setting where all can listen, interact, and learn" (Noffsinger, 2012). These have been used as an adjunct clinical approach in several countries (Noffsinger, 2012; Egger et al., 2015). They provide more time with the doctor, faster access to care, increased peer support, and greater opportunity for self-management. SMAs sit between clinical 1:1 care and group education, as shown in Fig. 4.1. In the future they are likely to become a standard procedure in clinical LM.

Clinical Care	**Shared Medical Appointments**	**Group Education**
(1:1)____________	!____________	(1:15)
I doc: 1 patient	1 doc; 1 Facilitator 6-12 patients	1 educator: up to 15-20 patients

FIGURE 4.1 Where shared medical appointments (SMAs) fit.

Total Telephone Triaging

Telephone triaging is a system used in an increasing number of centers in the United Kingdom to speed up the process of consulting with a doctor and utilize the doctor's time to better manage the 21st century tsunami of chronic diseases in primary care. Because this requires more than just a cursory 10-min consultation and prescription, which generally has been appropriate for most acute infectious disease episodes, screening by phone enables others in the medical center (initially a receptionist, then a nurse or allied health professional) to triage cases into different levels of need. It also allows for assessment of medical issues before utilizing valuable resources. This means patients suffering long-term conditions, which require much more attention and self-management opportunities, can get the required level of help while managing limited resources.

"Eduventures"

Eduventures are a new concept in experiential learning developed in Australia to help bring health professionals and patients together over an extended (2–7-day) period of interaction combined with Continuing Professional Development training during an extended physical activity. These include trail and beach walking, kayaking, skiing, sight-seeing, and cycling, generally from point A to point B as a form of "pilgrimage." Adventures and interactions between providers and patients are carried out during the day, and lectures and group discussions are held at night and in the mornings, sometimes around a campfire, in cabins, or at a meeting place along the way. Eduventures offer the unique opportunity of "walking the talk" in relation to lifestyle behaviors and health Fig. 4 2.

SUMMARY

Although not a departure from conventional medicine, LM knowledge, skills, tools, and procedures provide an adjunct approach to managing lifestyle and environmental determinants of much modern chronic diseases. Although still a work in progress, LM fits a role between clinical medicine and public health, enticing clinicians to consider both the underlying biological mechanisms and more distal environmental determinants of chronic disease than merely risk factors and behaviors within their area of interest. Self-management, SMAs, total telephone triaging, and educational adventures ("Eduventures") provide adjunct procedures for conducting LM consultations. New "tools," such as mHealth, are

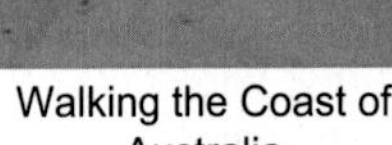

Walking the Coast of Australia

Kayaking the Myall Lakes

Skiing New Zealand

FIGURE 4.2 Lifestyle Medicine (LM) "Eduventures" for patients and providers.

Practice Tips

- Consider the underlying mechanisms in the patient's disease such as chronic inflammation;
- Use synergistic therapeutic lifestyle interventional: TLIs and pharmacotherapy (prescribing and deprescribing);
- Consider the potential *anthropogens* associated with chronic disease and the level of determinants (proximal, medial, and distal) associated with each of these;
- Although focusing on the primary determinants of chronic disease, consider the "systems" nature of lifestyle and environmentally related *anthropogens*;
- Utilize all levels of counseling skills as appropriate for managing lifestyle and environmentally related determinants of chronic disease;
- Use modern mHealth tools as appropriate to assist patient self-monitoring and self-management;
- Introduce SMAs as an adjunct consultation model to amalgamate the best features of each of the current consultation/learning models.

useful to capitalize on modern technological developments and increase treatment options. Undoubtedly the field will expand while further research assessing these ideas bears fruit.

REFERENCES

Booth, F., Roberts, C., Late, M., 2012. Lack of exercise is a major cause of chronic disease. Compr. Physiol. 2 (2), 1143–1211.

Bradfield-Hill, A., 1965. The environment and disease: association or causation? Proc. R. Soc. Med. 58 (5), 295–300.

Brown, M.T., Bussell, J.K., 2011. Medication adherence: WHO cares? Mayo Clin. Proc. 86 (4), 304–314.

Bulló, M., Casas-Agustench, P., Amigó-Correig, P., et al., 2007. Inflammation, obesity and comorbidities: the role of diet. Public Health Nutr. 2007 (10), 1164–1172.

Burke, L.E., Styn, M.A., Sereika, S.M., et al., 2013. Using mHealth technology to enhance self-monitoring for weight loss: a randomized trial. Am. J. Prev. Med. 43 (1), 20–26.

Calder, P., et al., 2009. Inflammatory disease processes and interactions with nutrition. Brit. J. Nutr. 101 (Suppl. 1).

Carter, M.C.1, Burley, V.J., Nykjaer, C., Cade, J.E., 2013. Adherence to a smartphone application for weight loss compared to website and paper diary: pilot randomized controlled trial. J. Med. Internet Res. 15 (4), e32.

Carvalho, B., Saad, M., 2013. Influence of gut microbiota on subclinical inflammation and insulin resistance. Med. Inflamm.:986734.

de Jongh, T., Gurol-Urganci, I., Vodopivec-Jamsek, V., Car, J., Atun, R., December 12, 2012. Mobile phone messaging for facilitating self-management of long-term illnesses. Cochrane Database Syst. Rev. 12, CD007459. http://dx.doi.org/10.1002/14651858.CD007459.pub2. Review 23235644.

Doocy, S., Sirios, A., Tileva, M., Story, J.D., Burnham, G., 2013. Chronic disease and disability among Iraqi populations displaced in Jordan and Syria. Int. J. Health Plann. Manage. 28 (1), e1–e12.

Dunstan, D., Howard, H., Healy, G.N., Owen, N., 2012. Too much sitting – a health hazard. Diabates Res. Clin. Pract. 97 (3), 368–376.

Egger, G., 2012. In search of a "germ" theory equivalent for chronic disease. Prev. Chronic Dis. 9 (11), 1–7.

Egger, G., Binns, A., Rossner, S., 2012. Lifestyle Medicine: Understanding the Diseases of the 21st Century, second ed. McGraw-Hill, Sydney.

Egger, G., Dixon, J., Meldrum, H., Binns, A., Cole, M.-A., Ewald, D., Stephens, J., 2015. Patient and provider satisfaction with shared medical appointments. Aust. Fam. Phys. 44 (9), 674–679.

Egger, G., Dixon, J., 2014. Beyond obesity and lifestyle: a review of 21st Century chronic disease determinants. Biomed. Res. Int. :731685. http://dx.doi.org/10.1155/2014/731685.

Egger, G., Dixon, J., 2010. Inflammatory effects of nutritional stimuli: further support for the need for a big picture approach to tackling obesity and chronic disease. Obes. Rev. 11 (2), 137–145.

Egger, G., Dixon, J., 2011. Non-nutrient inducers of low grade, systemic inflammation: support for a "canary in the mineshaft" view of obesity in chronic disease. Obes. Rev. 12 (5), 329–345.

Egger, G., Dobson, A., 2000. Clinical measures of obesity and weight loss in men. Int. J. Obes. 24 (3), 354–357.

Egger, G., Swinburn, B., Rossner, S., 2003. Dusting off the epidemiological triad: could it apply to obesity. Obes. Rev. 4 (2), 115–120.

Egger, G., Binns, A., Rossner, S., 2009. The emergence of a clinical process: "Lifestyle medicine" as a structured approach to the management of chronic disease. Med. J. Aust. 190, 143–145.

ESLM, 2015. European Society for Lifestyle Medicine (www.eslm.eu) American College of Lifestyle Medicine (www.aclm.net.au); Australian Society of Lifestyle Medicine (www.lifestylemedicine.com.au).

Friedman, E.M., Herd, P., 2010. Income, education, and inflammation: differential associations in a national probability sample (The MIDUS Study). Psychosom. Med. 72 (3), 290–300.

Genuis, S., 2012. What's out there making us sick. J. Environ. Public Health 2012:605137.

Guilliams, T.G., 2012. The Original Prescription: How the Latest Scientific Discoveries Can Help You Leverage the Power of Lifestyle Medicine. Point Institute.

Hajja, D.P., Gott, A.M., 2013. Biological relevance of inflammation and oxidative stress in the pathogenesis of arterial diseases. Am. J. Pathol 182 (5), 1474–1481.

Hamed, T., 2009. Thinking in Circles about Obesity. Springer, Monterey, CA.

Hamine, S., Gerth-Guyette, E., Fauix, D., Green, B.B., Ginsburg, A.S., 2015. "mAdherence" and chronic disease management: a systematic review. J. Med. Internet Res. http://dx.doi.org/10.2196/jmir.3951.

Holt-Lunstad, J., Smith, T.B., Layton, B., 2010. Social relationships and mortality risk: a meta-analytic. PLoS Med. 7 (7), e1000316.

Katz, D.L., 2014. Lifestyle is the medicine, culture is the spoon: the covariance of proposition and preposition. Am. J. Lifestyle Med. 8, 301.

Ketterer, M., Rose, B., Knysz, W., Farha, A., Deveshwar, S., Schairer, J., Keteyian, S.J., 2011. Is social isolation/alienation confounded with, and non-independent of, emotional distress in its association with early onset of coronary artery disease? Psychol. Health Med. 16 (2), 238–247.

Kiecolt-Glaser, J.K., Gouin, J.-P., Hantson, 2010. Close relationships, inflammation, and health. Neurosci. Biobehav. Rev. 35 (1), 33–38.

Kim, E.S., Strecher, V.J., Ryff, C.D., 2014. Purpose in life and use of preventive health care services. Proc. Natl. Acad. Sci U. S. A. http://dx.doi.org/10.1073/pnas.1414826111.

Koch, R., 1884. 2 Die Aetiologie der Tuberkulose. Mitt Kaiser Gesundh 1884, 1–88.

Lanius, R., Vermetten, E. (Eds.), 2009. The Hidden Epidemic: The Impact of Early Life Trauma on Health and Disease. Cambridge University Press, London.

Lawn, S., Battersby, M., Lindner, H., Mathews, R., Morris, S., Wells, L., Litt, J., Reed, R., 2009. What skills do primary health care professionals need to provide effective self-management support? Seeking consumer perspectives. Aust. J. Prim. Health 2009 (15), 37–44.

Lenz, T.L., 2013. Lifestyle Medicine for Chronic Disease. Prevention Publishing, Omaha, NE.

Lorig, K.R., Holman, H.R., 2003. Self-management education: history, definition, outcomes, and mechanisms. Ann. Behav. Med. 26 (1), 1–7.

Lyte, M., 2013. Microbial endocrinology in the microbiome-gut-brain Axis: how bacterial production and utilization of neurochemicals influence behavior. PLoS Pathog. 9 (11), e1003726.

Mechanick, J.I., Kushner, R.F., 2016. Lifestyle Medicine: a Manual for Clinical Practice. Springer, NJ 2.

Monteiro, C., 2009. Nutrition and health. The issue is not food, nor nutrients so much as processing. Public Health Nutr. 12 (5), 729–731.

Noffsinger, E., 2012. The ABC of Group Visits. Springer, London.

O'Keefe, J.H., Gheewala, N.M., O'Keefe, J.O., 2008. Dietary strategies for improving post-prandial glucose, lipids, inflammation, and cardiovascular health. J. Am. Coll. Cardiol. 51 (3), 249–255.

Pillans P., 2015. Medication Overload and Deprescribing. Paper presented to the Australian Disease Management Association (ADMA) Annual Conference, Brisbane, Sept., 2015.

Prüss-Ustun, Corvala, 2006. Preventing Disease through Healthy Environments: Towards an Estimate of the Environmental Burden of Disease (WHO Report).

Puttonen, S., Viitasalo, K., Härmä, M., 2011. Effect of shiftwork on systemic markers of inflammation. Chronobiol. Int. 28 (6), 528–535.

Rippe, J., 2013. Lifestyle Medicine, second ed. CJC Press, London.

Ross, R., Janiszewski, P.M., 2008. Is weight loss the optimal target for obesity-related cardiovascular disease risk. Can. J. Cardiol. 24 (Suppl. D), 25D–31D.

Runsten, S., Korkeila, K., Koskenvuo, M., Rautava, P., Vainio, O., Korkeila, K., 2013. Can social support alleviate inflammation associated with childhood adversities? Nord. J. Psychiatry 68 (2), 137–144.

Shehab, N., Anastario, M.P., Lawry, L., September–October, 2008. Access to care among displaced Mississippi residents in FEMA travel trailer parks two years after Katrina. Health Aff. (Millwood) 27 (5), w29–w416.

Slarz, D.E., Mullington, J.M., Meier-Ewart, H.K., 2012. Sleep, inflammation and cardiovascular disease. Front. Biosci. 4, 2490–2501.

Spring, B., Duncan, J.M., Janke, A., et al., 2013. Integrating technology into standard weight loss treatment: a randomized controlled trial. JAMA Intern. Med. 173 (2), 105–111.

Swan, M., 2009. Emerging patient-driven health care modes: an examination of health social networks, consumer personalized medicine and quantified self-tracking. Int. J. Environ. Res. Public Health 6, 492–525.

Vic Health, June, 2015. Witness Seminar on Smoking Control in Australia. Melbourne.

Weng, et al., 2013. Gray matter and white matter abnormalities in online game addiction. Eur. J. Radiol. 82 (8), 1308–1312.

Wilkinson, R., Pickett, K., 2009. The Spirit Level: Why Greater Equality Makes Societies Stronger. Bloomsbury Press, London.

Chapter 5

Everything You Wanted to Know About Motivation (But Weren't Intrinsically Motivated Enough to Ask)

John Litt, Rosanne Coutts, Garry Egger

What people really need is a good listening to.

Miller and Rollnick (2002)

INTRODUCTION: THE REQUIREMENTS FOR CHANGE

The principal patient requirement in lifestyle medicine is change. However, encouraging an individual to change requires the clinician to have an ability and/or a significant understanding of motivational interviewing principles. It is these principles that are at the core of lifestyle medicine and that differentiate it from conventional medicine.

Unlike a medication, which should have similar effects at different times across different individuals, motivation to act in a certain way will often depend on the context and differ both between individuals and within the same individual at different times. A significant challenge is to move from evanescent (or external) motivation to more internally derived motivation. A person who has little motivation to quit smoking can become extremely motivated by the death of a friend who smoked or an X-ray that suggests possible cancer; for others, these things may have little impact. It is for this reason that there is no motivational pill, no single solution that suits all. It is also wrong for clinicians to think that they alone can motivate a patient to act.

The verb *to motivate* means "to provide a prompt or incentive for a person to act in a certain way" (Macquarie Dictionary, 1997). Motivational interviewing (MI) is one way of doing this (Miller and Rose, 2009). Using the MI approach, the clinician acts as a facilitator by providing the prompt or incentive, framing an agenda for the patient, and exploring the costs and benefits associated with lifestyle-related behavior (Rollnick et al., 2008). In the end, the patient needs to

Lifestyle Medicine. http://dx.doi.org/10.1016/B978-0-12-810401-9.00005-X

be engaged and mobilized to identify internal sources and drivers of motivation to carry out the activity.

Also, prompting on its own may generate resistance in patients especially if:

- They feel there is a judgment involved;
- The prompt cannot easily be connected to their health complaints; and
- They perceive that the clinician does not really appreciate their difficulties or has not heard their concerns.

It is important to remember that the way the communication is provided often has just as big an impact as what is said (Epstein, 2006). Strategies that demonstrate a nonjudgmental approach and a sense of support help to encourage the patient to self-explore.

WHAT IS MOTIVATIONAL INTERVIEWING?

MI is a patient-centered clinical intervention intended to assist in strengthening motivation and readiness for action (Rollnick et al., 2008). An important goal of MI is to elicit and reinforce "change talk" from the patient (Apodaca and Longabaugh, 2009; Miller and Rose, 2009; Rollnick et al., 2008). In MI, the focus is on reflections and questions on topics that relate to ambivalence and action—what might promote action and what makes it difficult or inhibits it. The skillful MI counselor is attuned to change-relevant content in the patient's behavior and communication. Their thoughtful reflective listening statements help to facilitate action. At the same time, adopting the spirit of MI helps to explicitly affirm the person's autonomy and choice with respect to what, whether, and how to change.

A core component of the MI approach is the MI spirit—a mix of skillful counseling style blended with a clear patient-centered approach. Key elements of the MI spirit include (Miller and Rose, 2009):

- A **collaborative approach** rather than authoritarian. The counselor actively fosters and encourages power sharing in the interaction in such a way that the patient's ideas substantially influence the direction and outcome of the interview. Gaining a better understanding of the patient's ideas, concerns, expectations, and preferences through using the MI approach increases shared decision-making. Information is actively shared and the patient is supported to consider options, to achieve informed preferences.
- **Evocation.** The focus is on the patient's own motivation rather than trying to instill it. The counselor works proactively to evoke the patient's own reasons for action and ideas about how change should happen. All patients have goals, values, and aspirations. Part of the MI approach is to connect health-related behavior with the things that patients care about.
- Honoring and respecting the **patient's autonomy.** The MI process actively supports autonomy by building good relationships, respecting both individual

expertise/competence and interdependence on others. Patients can and do make choices and it is ultimately their right to choose what they wish to do, that is, patient self-determination is respected. Specifically, patients have the right to follow their own preferences and make their own decisions even if these are regarded as problematic by others.

MI is different from the transtheoretical model of behavior change. The latter is intended to provide a comprehensive conceptual model of how and why changes occur, whereas MI is a specific clinical method to enhance personal motivation for change (Miller and Rollnick, 2009).

WHY USE MOTIVATIONAL INTERVIEWING TECHNIQUES?

Motivational Interviewing Approaches Are Effective

Addressing motivation has been shown to be effective in changing a range of behaviors including alcohol use (Burke et al., 2003; VanBuskirk and Wetherell, 2014), smoking (Lindson-Hawley et al., 2015), diet and weight loss (Barnes and Ivezaj, 2015), exercise and physical activity (O'Halloran et al., 2014; Lundahl and Burke, 2009; Rubak et al., 2005). They also contribute to improved adherence in a range of chronic diseases and health problems (Alperstein and Sharpe, 2016).

Motivational Interviewing Approaches Are Patient Centered

MI strategies are respectful, empowering, and promote patient autonomy. These all contribute to their acceptability to patients.

Motivational Interviewing Approaches do Not Necessarily Take More Time

With practice, motivational interviewing can be used in routine consultations and do not take more time than traditional prescriptive methods. They are also a more efficient method of getting results.

The Impacts of Motivational Interviewing Approaches Are Durable

Evidence from the systematic review of MI techniques has demonstrated that their effects last up to 2 years.

MOTIVATIONAL GOALS AND STRATEGIES

Motivation is one of the key to lifestyle medicine. The ultimate goal of a motivator should be to move an individual from requiring extrinsic motivation through prompts or incentives (e.g., money, praise, support) to intrinsic motivation, or that which is internal and requires no outside prompting (although it may benefit from constant reinforcement). The process could be likened to switching on a motor and making sure it keeps running. To do this, it is necessary to do two things.

1. Make conscious (or create) a discrepancy between where an individual is and where he or she would like to be.

Motivation comes from desire—to have, to do, to maintain or to change. In lifestyle medicine, the desired form of motivation is one that would encourage the patient to change his or her unhealthy lifestyle behaviors. But for many people, the benefits of doing this, of moving from a current to a desired position, are often unclear or seen as too difficult to obtain. In some cases, the discrepancy is obvious (e.g., the desire to feel and look better) and needs little amplification. But for others it may be more obscure, as in the experience of someone with an existential dilemma, who has abused drugs or alcohol, being sober and having a greater enjoyment of family and friends instead of avoiding them.

Key assertions in MI are that counselors can influence the balance of change talk (pros) and sustain talk (cons) voiced by clients in an interview, and that this balance in turn predicts subsequent change (Miller and Rollnick, 2012). Substantial evidence has accumulated to support this causal chain (Miller and Rose, 2009; Moyers et al., 2009). In correlational transtheoretical model of behavior change (TTM) research, movement from ambivalence (contemplation) to preparation and action is marked by increased pros (relative to cons) of change. In psychotherapy sessions, the extent to which clients voice reasons for change (change talk) is directly related to the likelihood that change will occur (Gaume et al., 2010). Conversely, the more a client verbalizes arguments against change (sustain talk), the less likely change is to occur (Moyers et al., 2009). The change talk/sustain talk ratio is a good predictor of change in MI (Moyers et al., 2009).

As ambivalence is both common and normal part of the process of change, a useful strategy is to make this more visible, for example, using a decisional balance approach. In a nutshell, the client is asked to look at and balance the benefits and costs (likes and dislikes) associated with making a change such as quitting smoking or taking up exercise. The suggestions all need to come from the client and provide rich material to explore what is underpinning the patient's motivation. The ways of drawing this out are considered in the following discussion of motivational practices.

2. Help reduce the ambivalence about getting there.

If there is a discrepancy about where a person is and where he or she wants to be, it is because there is ambivalence about the two positions. Quitting smoking, for example, involves denying oneself a pleasure that may help relieve tension in order to get a perceived benefit of reducing the risk of a cancer that may not occur for decades.

A genuine desire or ability to change can only be increased by reducing the barriers and/or increasing the triggers associated with this ambivalence by discussing the costs, as raised by the patient. For example, a barrier to losing weight may be the perceived pain of having to exercise or the perceived difficulty of losing weight, reinforced by many previous unsuccessful attempts. Triggers could be the desire for fresh breath from quitting smoking or the desire to reduce health and/or sexual performance problems caused by an expanded waistline. The clinician can then use this knowledge to reduce clients' ambivalence and move them toward a greater state of readiness to change.

Focusing

Within a "helping" approach (Egan, 1998), three questions make the framework for focusing the individual. These are a starting point for encouraging new thinking and the gathering of information, so that interventions can be tailored to individuals within their current circumstances. The first question is, "What is going on?" This is where patients speak about themselves and describe aspects of their current lifestyle, often with suggestions for changes that they would like to make. The second question is, "What solutions would make sense for me?" This is where the clinician can guide and assist individuals to think about solutions that will support the changes. The third question is, "How can I make this happen?" This is about what individuals will actually do toward achieving the new goals.

Important findings by Landry and Solomon (2004) provide guidance on motivational strategy. They suggest that:

- Strategies that rely on coercion do not, in the long term, encourage lasting change (Wertheimer, 2013);
- The more reinforcing or encouraging forms of motivation will be more effective in encouraging change (Miller and Rose, 2009);
- Various rewards and threats have been found to be poor motivators when trying to encourage individuals to be more active or change their eating habits;
- Strategies reinforcing a more self-regulated approach (the individual organizing it themselves) are more likely to foster lasting change (Teixeira et al., 2015).

An understanding of how ready, willing, and able a patient is to make a change—as well as the barriers preventing and the triggers encouraging any particular change—will facilitate shifting the patient from requiring extrinsic motivation to being intrinsically motivated.

Being Ready, Willing, and Able

Miller and Rollnick (2002) use the terms "ready, willing, and able" to understand the processes inherent in a high level of motivation for change. These provide a format for discussing how to move patients closer to long-term change for a healthier lifestyle.

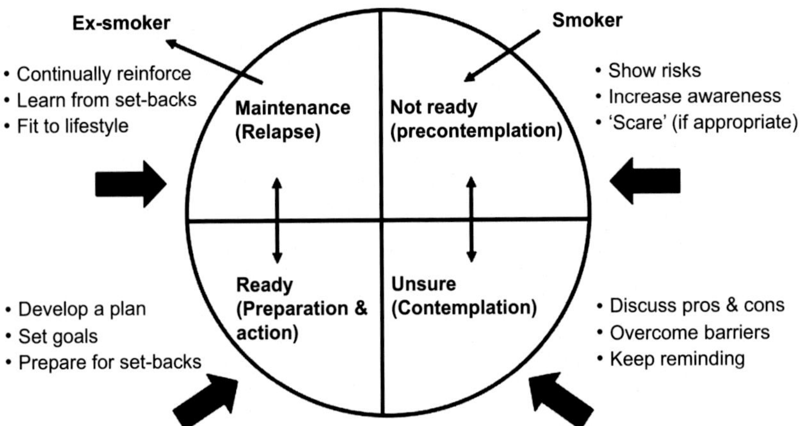

FIGURE 5.1 The stages-of-change model of readiness to act.

Being willing defines why a patient should change. Being able explains how. The only remaining question is when. This is where motivational instruction is probably most relevant, particularly the transtheoretical, stages-of-change model of readiness developed by US psychologists Prochaska and Velicer (1997). The stages-of-change model proposes that individuals go through about six stages in changing any behavior from the time that this is not even being considered to the adoption and maintenance of the new behavior as a part of their lives. The stages of change are shown in Fig. 5.1. The process is not necessarily a linear one and usually occurs in a more spiral form, with relapse and recovery over anything up to a two-year period until final adoption.

The TTM is commonly referred to as the "stages of change" model and has been used in many clinical settings to determine patient readiness for action. While the stages of change model can be useful, transition through the model is not usually linear. Further, external factors may have a large role in a subject's ability to change. It is preferable to maintain a degree of flexibility rather than adopt a rigid approach when choosing intervention strategies (West, 2005).

There is limited rigorous evidence of the effectiveness of the stages of change approach as the preferred counseling approach when assisting smokers to quit (Cahill et al., 2010). The evidence for the TTM model in helping patients lose weight or increase their physical activity is limited and equivocal (Hutchison et al., 2009; Spencer et al., 2006).

Interventions can be designed to encourage the individual to make forward moves from one stage to the next. Although the data are unclear on this, different interventions (processes of change) may work better in one stage than in another. Nevertheless, stages of change can help the doctor and the patient to understand the latter's interest or willingness to make changes. Using motivational interviewing principles as a guide, the doctor can explore the patient's motivation to elicit change talk. Increased change talk is associated with great likelihood of behavior change (Apodaca and Longabaugh, 2009; Copeland et al., 2015; Romano and Peters, 2014). Change talk is any expressed language that is an argument for change or a resolution of ambivalence in favor of change. The first three stages—precontemplation, contemplation, and preparation—are about thinking (cognitive processes). Action and maintenance are about actions and doing (behavioral processes). The Stages of Readiness to Change Scale allows the clinician to categorize an individual and then work on interventional approaches to assist with moving on. For example, for individuals in the contemplation stage, a discussion that may prove beneficial could be about how increasing physical activity, improving eating behaviors, or reducing smoking (whichever might apply) would enhance their lifestyle. For individuals in the action stage, assisting them with further social support can encourage or facilitate a move forward into the maintenance phase.

Of the six stages seen in Fig. 5.1, the first three are the main concerns of the clinician because during the stages beyond this—decision, maintenance, and transformation—the impetus will come more from the patient and the main clinical requirement is for support and encouragement. A description of the first three stages and consideration of strategies to elicit change talk is shown in Table 5.1.

An example of a readiness test using physical activity is included in Professional resources at the end of this chapter. This can be changed according to the behavior being tested. A positive response reflects the stage of readiness to act. The individual is asked to choose one statement. For example, people who choose option three (I currently am physically active, but not on a regular basis) are in the preparation stage. Stages of change can usually be determined early in a consultation. It is important to recognize that the stages can be quite fluid and vary considerably with time (and clinician!).

An alternative approach to questions that can elicit change talk is summed up under the acronym "DARN" as shown in Table 5.2.

Nine Strategies for Evoking Change

1. Ask evocative questions (e.g., open-ended questions).
2. Explore decisional balance.
3. Ask for elaboration.
4. Ask for examples.
5. Look back/look forward (e.g., "If you did change, what do you hope would be different in the future?" "Given what has happened so far, what might you expect to happen if you didn't make any changes?")
6. Query extremes (e.g., "What do you think is the worst thing that could happen?" "What do you think is the best thing that could happen if you made some changes?")
7. Use change rulers (i.e., rate readiness to change on a 1–10 scale).
8. Explore goals and values.
9. Come alongside.

TABLE 5.1 Stages of Change and Strategies to Shift

Stage	Description	Clinician's Goal	Strategies to Shift
Precontemplation	Not considering change (not ready)	Get the patient to consider the advantages of change.	Ask, "have you thought about…" check "importance" and "confidence." Find where the patient "would like to be" and work on barriers and triggers to getting there. Discuss past failures and "reframe" these as learning experiences. Systematically explore perceived benefits and costs associated with a behavior. If this doesn't work heighten awareness (but with options for reducing fear). If this doesn't work: Wait.

TABLE 5.1 Stages of Change and Strategies to Shift—cont'd

Stage	Description	Clinician's Goal	Strategies to Shift
Contemplation (unsure)	Moving toward thinking about change	Get the patient to reduce their ambivalence about changing behavior.	Provide information. Discuss outcomes if nothing is done. Discuss alternatives (e.g., exercise, diet). Set a manageable short-term goal around change options for reducing this (e.g., steps/ day). If this does not work, keep reminding.
Preparation (ready)	Ready to make a change but has not yet done so	Move the patient into making a change and then maintaining this over the long term	Discuss options. ACT—set a time and date and just "do it." BELONG—ask about someone (e.g., partner, club) over the long term for ongoing commitment. COMMIT—set some short- and long-term goals (e.g., see weight goals in Chapter 4). If this does not work: Provide evidence that underpins your concern by showing links with health effects of illness.

TABLE 5.2 Questions to Ask

	Questions To Elicit Change Talk
Desire	How would you like for things to change?
	Tell me what you don't like about how things are now?
	What do you hope will be different?
Ability	What do you think you would be able to change?
	Of the options you have considered, what seems most possible?
Reason	Why do you think it is important to change?
	What impact/effect do you think that your [behavior] is having on your health?
Need	How important is it to your health to change your diet?
	Do you have any worries about your [health-related behavior]? In what way does this worry or concern you?
	What do you think has to change for you to consider cutting down on your drinking?
	How urgent is it that something needs to be done?

Responding to Change Talk

A range of strategies can be used to encourage change talk. They are summarized by O.A.R.S:

Open-Ended Questions

- Ask for more detail, examples, elaboration.

Affirmation

- Recognize and respect what the person is saying about change.

Reflection

- Simple reflection makes a reasonable guess/statement at what the patient is saying. Statement is intended to mirror meaning. For example, "What I hear you saying is…It sounds like…"
- Complex adds additional or different meaning beyond what the patient has just said and posits a guess as to what the patient may have meant.
- It often includes a reflection of feeling that addresses the emotional content of the patient's statement.

Summary

- Focus on the change talk.
- Try and end with an open question.

Being Willing: How Important Is It to Change?

The extent to which a person wants to change is summed up in his or her willingness to do so. Willingness to change is in turn influenced by the discrepancy between the status quo and the goal (i.e., where he or she is and where he or she wants to be), the level of ambivalence this has created, and the perceived importance of wanting to move closer to the goal. If ambivalence is reduced by helping remove barriers and increase the triggers to action (either physically or psychologically), there should be greater willingness to move toward the goal.

Willingness can be determined by a simple question: On a scale of 1–10 (where 1 is not important at all and 10 is extremely important), how important is it for you to [reach the goal discussed]? If the answer score is high, the patient is willing to act and only his or her readiness and ability to do so need to be considered. If the answer is low, greater emphasis needs to be placed on increasing importance and hence reducing the ambivalence to act. If more time is available it is useful to use the nine strategies outlined herein to elicit change talk.

It is also worth checking what underpins the motivation or importance when the patient offers a relatively high number by asking, "What contributes to your score being an 8 (say) rather than a 2?" This will provide insight into what the patient believes is underpinning his or her motivation. This is useful information because you, as a facilitator, can reinforce the information and further explore it to see how robust it is. Some patients may have not thought it through and therefore will be more vulnerable to failure or lack of commitment. Others may be offering a high level of importance to please you (or get you "off their back"). An example would be an excessive drinker who sees no problem, physical or psychological, in continuing to drink excessively. In this case, the potential problems (e.g., liver damage, family breakdown) need to be highlighted to increase the discrepancy between status quo and goal and increase willingness to do something about this.

A further strategy to facilitate interest in quitting is to estimate how much their smoking is costing per year and ask how long before they plan to quit. A person smoking 20 cigarettes a day who thinks they may quit in 10 years will spend $AUS31,500 at today's prices for a packet of cigarettes. While the health effects may be hard to identify in some smokers, all will be affected by the price.

Being Able: How Confident Is the Patient About Changing?

A person can be extremely willing but lack the confidence or ability to change. This does not come just from a lack of knowledge (although in some cases it may do so) but from a lack of self-efficacy or feeling of one's ability to succeed in reaching a specific or general goal. Weight loss for men, for example, is often hindered by their lack of knowledge. Receipt of the accurate knowledge about nutrition and exercise is often a first step to increase ability to change. For many women, on the other hand, biological factors may make successful weight loss more difficult and previous failures, even with the right knowledge, contribute to low self-efficacy and make success less likely.

As with importance, this can be checked with a simple question: On a scale of 1–10 (where 1 is not confident at all and 10 is extremely confident), "How confident are you that you can [reach the goal discussed]?"

Again, explore the reasons underpinning a high level of confidence to check that patients are realistic. Many smokers and those with hazardous drinking habits often report high levels of confidence to quit or cut down. This belief may represent a rationalization of the need to change. If the score is low, ask, "What would need to happen to move this score from a 2 to a 3 or 4?" This provides an opportunity to focus the discussion on what the barriers are that you and the patient need to overcome in order to improve motivation. For example, many younger smokers state that they will quit when there is evidence of health effects. If the clinician can show how the patient's health is already being affected, for example, by reduced pulmonary function (common in smokers over the age of 35–40 years), this may alter the motivation to quit.

If the score is high for both confidence and importance, there may be little for the clinician to do except discuss when this will happen (readiness). If the ability score is low, one of two aspects of confidence need to be targeted.

- Increasing task efficacy means ensuring that there is confidence in the process to be used to affect change. Can the patient feel confident, for example, that a particular weight loss or quit smoking medication will actually work, given that so many such programs may have failed in the past? In this case it is up to the clinician to convince the patient that this time it is different and to provide specific reasons that are within the patient's control.
- Increasing self-efficacy means working on the patient's confidence in being able to follow and stick with the task at hand. Typical reactions may include self-blame, catastrophizing, and self-doubt. In these cases, modifying thinking patterns becomes the issue and there are a variety of established ways of doing this—for example, rational emotive therapy or cognitive behavior therapy—some of which are discussed in Chapter 11 (also see Elder et al., 1999).

Case Examples

Some typical responses to both the importance and confidence questions are:

- A male alcohol abuser who scores low on the importance of giving up drinking but high on confidence to do this, if he so chooses. In this case, increasing using MI techniques to elicit any issues he or she may see about their drinking, such as a damaged liver or the outcome of cirrhosis, may increase the perceived discrepancy between the status quo and goal and stimulate this to be acted on.

- A woman wanting to lose weight whose rating of importance is high but confidence low because of previous failures. The goal of the clinician here is to raise the patient's confidence about being able to make the required change or in the process. Some ideas and questions for raising importance and confidence in a patient are shown in Table 5.3.

Behavior change is not just about motivation. C.O.M.P.L.I.A.N.C.E is a useful acronym to think about the range of factors that influence lifestyle-related behavior (Table 5.5).

BARRIERS AND TRIGGERS TO CHANGE

At this stage it is important to reiterate that the goal of a motivator is to shift a patient from requiring extrinsic motivation to being intrinsically motivated. Intrinsic motivation comes from the benefits achieved from change. However, in the early stages these benefits are unlikely to be felt. Hence it is useful to raise extrinsic motivation by removing the barriers and increasing the triggers toward achieving this goal.

Barriers to change can be either physical or psychological (Table 5.3).

In some cases these may be obvious and easy to detect using any of the techniques discussed previously. In others they may be more deep seated and require extensive probing and discussion. Early sexual or physical abuse, for example, can be relatively common causes of obesity in women in order to gain the comforts of food and reduce attractiveness to future potential perpetrators. This requires disclosure and management through the development of trust and rapport with a clinician. Other causes are often more subtle. For example, a male patient disclosed after several weeks of treatment for obesity that he did not want to walk in public because of a fear of looking foolish to young onlookers. Providing patients with a supportive environment and encouraging them to be candid about the various triggers will help to identify these barriers.

Some common physical barriers to becoming active and some suggested solutions for these are outlined in Tables 5.4 and 5.6. Similar tables can be drawn up for barriers to other behavior change.

There are also motivational triggers that can help overcome these barriers and shift an individual toward being intrinsically motivated. For some, these might be financial; for others, physical (e.g., looks, healthy feeling) or impact on health; and for still others, a desire to please other people, such as a partner or children. These can assist a patient to get to the next level and be closer to being intrinsically motivated. Triggers generated by the client are more likely to be salient or important because they have some meaning or relevance to the patient. This does not preclude the clinician identifying some triggers but this should occur after the patient has exhausted the list of triggers that they can think of.

TABLE 5.3 Tips for Exploring Ambivalence, Enhancing Importance and Confidence

Exploring Ambivalence
What do you like about the behavior?
What don't you like about it? Does it concern you?
On a scale of 1–10 where 10 is very motivated to change, where would you rate yourself? If a high number, ask, *Why an 8 and not a 2? If a low number, ask, What would need to happen to make it an 8 or 9?*
Focus on discrepancies, e.g., *On the one hand, eating ice cream make you feel better, on the other hand you are concerned about your weight gain and its effect on your diabetes. What do you make of that?*
Enhancing Importance
Why did you score a 1 and not a 2 or 3 on this scale?
What would need to happen to move you from a [low] score of x to a score of 8 or 9?
What would encourage you to move (that is, from your low score to an 8 or 9 on this scale)?
How would your behavior change if you did rate a 10 on this?
What are the main barriers stopping you from getting a higher score?
Can you see a way to remove or reduce these barriers?
What are the pros and cons of trying and not making the desired changes?
Pick what seems to be important to the individual (e.g., impotence in men, efficiency in women) and concentrate on this.
Enhancing Confidence
Why did you score a (for example) 1 and not an 8 or 10 on this scale?
What do you think could help you score higher on this scale?
What have you done in the past that has been successful?
Why do you think you have failed in the past?
What has made it difficult to succeed?
What was the trigger for the slipup?
Reframe previous failures as "slips" or short-term inconveniences.
Stress that while the patient may have tried and failed before, this time it is a different approach and outline why.
Discuss the stage approach (see following) and how this involves some failures as well as successes before moving to the next stage.
Emphasize hope and build self-efficacy.

TABLE 5.4 General Barriers to Lifestyle Change (See Elder et al., 1999)

Physical	Psychological	Other
Mechanical problems	Social pressure	Lack of time
Fatigue	Peer pressure	Responsibilities
Lack of equipment	Laziness	Health literacy
Financial	Embarrassment	
Injury/adverse consequences	Sense of hopelessness	
Addiction	Denial	
	Perceived difficulty in changing	
	Lack of self-efficacy	

MOTIVATION AT THE PUBLIC HEALTH LEVEL

The previous approaches to motivation can be used by the clinician and other health care practitioners. There are also a range of public health approaches for lifestyle change that would help support clinical initiatives. Economic incentives, such as tax rebates or legislative changes that make healthier choices the easiest choices, can add incentives to the patient environment. A lower excise/tax on low-alcohol beer, legislation mandating seat belts or allowing random breath testing, and restriction of smoking environments are good working examples. These are usually beyond the ambit of the clinician but should be encouraged as part of a wider health promotion approach to lifestyle change. For example, an approach that financially rewarded patients and clinicians for reversing early stage disease, such as type 2 diabetes, would have long-term cost-saving consequences.

What Can be done in a Brief Consultation?

- ☐ Think about a "1-min" approach—what is realistic when time is very pressured?
- ☐ Assess the patient's readiness to change and target motivation to change.
- ☐ Check willingness to change based on a 10-point scale of importance.
- ☐ Check ability to change based on a 10-point scale of confidence.
- ☐ Work on increasing willingness or confidence, depending on which is deficient.
- ☐ Isolate and work on reducing barriers to change.
- ☐ Look for and increase triggers to change.

TABLE 5.5 Factors That Influence Lifestyle-Related Behavior: (C.O.M.P.L.I.A.N.C.E)

	Factor	Strategy
C	Context	Identify supports/networks and sources of influence; explore opportunities, address access issues understand antecedents and consequences and cultural issues Flexible, e.g., meal replacements, flexible exercise schedule
O	Outcomes (expectations, consequences, and rewards)	Explore beliefs/expectations about outcomes; identify rewards, minimize adverse events/effects; promote self-efficacy; active problem-solving
M	Monitoring	Goal setting; review and encourage self-monitoring
P	Preferences/values	Identify and acknowledge, use reframing, enhance trust, clarify roles
L	Literacy	Address mood, self-efficacy and knowledge needs, assess perception of risk, consider learning style, identify supports; foster skills, use decision aids, active follow up/continuity of care
I	Interest/Motivation	Avoid making judgments; reframe; use discrepancy, decision balance; information about impact
A	Addiction	Assess and manage, including impulse control
N	Need	Identity cues, rewards; explore concerns/needs
C	Coping style/coherence/resilience	Consider avoidance, denial, inertia/passivity/helplessness; promote resilience
E	Emotion	Address affect/mood, stress, fear/threat

SOME MOTIVATIONAL TACTICS

Set Short- and Long-term Goals

It can take a lot of work and dedication to make substantial lifestyle changes. Accomplishments can be realized through good planning of daily, weekly, and long-term personal goals. The clinician can guide a specific and measurable

TABLE 5.6 Typical Physical Barriers to Being Inactive

Identify the problem	
To what does the patient attribute the problem?	
Are there any medical reasons why there is a problem (that may not be directly related to the patient's behavior)?	
Barrier	*Possible solutions*
Chafing	Wear Lycra bike pants; use Vaseline
Leg pain	Change shoes; stretch; see a podiatrist
Back pain	Warm up; walk in water/soft sand; try another form of exercise
Joint pain	Use walking poles; try weight-supportive exercise; get medical assessment
Hypoglycemia (in diabetes)	Reduce medication accordingly; carry a carbohydrate drink
Shortness of breath	Eliminate/review any potential medical contributors, e.g., undertreated asthma, lung disease; revise activity or training schedule
Perceived lack of time	Outline all activities in a typical day

goal-setting exercise that captures the aims of the individual, both long and short term (O'Brien et al., 2015). The goals should be accepted or internalized by the individual. An effective goal is not one that is thought about and then forgotten. Getting the patient to write goals down and monitor how they are going is most important.

Goals should be short and long term and involve processes as well as outcomes. For example, a long-term goal may be a 10-kg weight loss but a short-term goal leading to this could be a 1% decrease in waist size per week over the first 6–8 weeks. A process goal in doing this may be to walk 70,000 steps a week and cut back soft-drink consumption.

Record Progress

Self-monitoring and recording of progress is one of the most effective strategies of behavior change. This can include things such as steps measured by a pedometer, days since last drug use, or an increase in fitness. Recordings should be kept in a prominent place where they can be easily and regularly seen; they may even be kept by significant others.

Use the Mate System

Where possible, involving a partner, spouse, relative, or friend, or working with a group of like-minded people, can help increase adherence to a program and ensure against relapse.

Plan Ahead

Planning what is going to be done in advance—for instance, on the night before—is always a good way of ensuring things are done. The busiest people in the world (e.g., politicians, businesspeople) have exercise built into their day because they have built it into their diary. Planning on the day, as a sort of second thought, increases the likelihood that things will be put off for more "important" matters that arise.

Encourage the Use of Fantasy and Imagination

It may not always work, but using fantasy during exercise (e.g., imagining winning the Olympic Games 10-km race) can often get the patient over the personal motivation line, if not the Olympic line.

Using imagery is associated with motivation and is a form of self-modeling. When we practice imagery, the neuromuscular patterns that are used are the same as when actually doing the activity (Smith and Collins, 2004). For example, with physical activity an individual can practice without moving a muscle. However, it must be emphasized that getting on and really doing the activity is still very important.

Imagery is helpful for gathering confidence, working through fears, exploring new skills, and generally supporting progress. Individuals who have been encouraged to use their imaginations can see themselves having increased energy, an improved appearance, or the skills to perform certain techniques, for example, dancing, gardening, walking, or playing a sport. Based on work by Vealey and Greenleaf (2001), a sample five-step program to enhance imagery ability is included next. This could be used to focus an individual on new tasks and goals.

1. Get comfortable in a quiet place, close your eyes, and breathe rhythmically to establish a relaxed state.
2. Visualize a colored circle that fills the visual field initially and then sinks to a dot and disappears. Repeat the process, making the circle a different color each time.
3. Select a variety of scenes and images and develop them in detail. Include scenes such as a swimming pool, a beautiful walking track, or a golf course next to the ocean. Enjoy the scenes.
4. Imagine yourself in a setting of your choice doing a skill or sport that you are keenly interested in. Project yourself into the image as if you were one of the performers. Imagine yourself successfully completing the task in the scene.
5. End the session by breathing deeply, opening your eyes, and slowly adjusting to the external environment.

Find a Motivational Soft Spot

Motivation is an intraindividual factor that varies between individuals and over time. Finding what motivates an individual at any particular time can help the clinician improve the prospects of long-term behavior change. Avoiding disease may motivate someone to want to lose weight while increased physical attractiveness may work for someone else. Finding the motivational "soft spot" of the moment is often the key to success.

Encourage Flexibility/Discourage Rigidity

Rigidity in thinking is the great stumbling block to behavior change. Perfectionism is often the cause of such rigid thinking. Flexibility allows for the odd mistake, or "falling off the wagon," which can easily be corrected, whereas associating success with just a single, unchangeable path is a recipe for failure. Adherence is more likely when the following happen:

- people have choices
- the intervention can be tailored to the needs of the individual
- the process is incremental
- flexibility is coupled with regular review

INTRINSIC MOTIVATION: THE END GAME

As pointed out at the start of this chapter, it is the goal of the motivator to shift the patient from needing external reward to being intrinsically motivated to maintain a lifestyle change. This is probably best seen in long-term exercisers, such as joggers, who become dedicated to their activity and need little outside prompting to carry it out, even in the presence of such negative conditions as pain, cold, or lack of time. Several books have been written over the years about the stages of this process. Essentially there is a shift from a discomfort stage to one where the rewards are physical and then to one where the rewards become psychological and where there is a feeling of oneness and well-being when carrying out the adopted behavior (Egger, 1981). Once the level has been achieved, the need for an outside person to motivate that individual to act becomes almost irrelevant. Anyone who has experienced this will be easier to motivate to seek that feeling again.

Knowledge is also a vital component of motivational strategy, particularly to encourage physical activity. Individuals most likely to develop a more physically active lifestyle have the following characteristics:

- confidence that they could be successful in the use of physical activity
- knowledge about what constitutes a healthy lifestyle
- knowledge about the importance and value of partaking in regular physical activity

- a perception that they did have a level of self-control
- good attitudes about the value and importance of regular exercise

SUMMARY

The goal of a clinician is to move patients from requiring extrinsic motivation to change their unhealthy behaviors to being intrinsically motivated to do so, and thus not need external support. This is done by assessing the patient's readiness, willingness and ability to act, and helping them see the gap between where they are and where they would prefer to be. Reducing the barriers and increasing the triggers to action will assist in shifting the patient toward the desired goal.

Practice Tips

Motivating a Patient to Change

	Medical Practitioner	**Practice Nurse/Allied Health Professional**
Assess	Lifestyle: • Need and readiness to change • Whether it is worth putting effort into trying to bring about change at this stage • Barriers to change	Lifestyle: • The specific behavior(s) to be changed • How ready, willing, and able the patient is to change • Importance and confidence, rated on a 10-point scale • Motivational "soft spots" • Barriers to change (see Professional resources for a test)
Assist	• By providing clear, concise, and non-judgmental advice to change • By providing encouragement • By identifying potential short- and long-term goals	• By using motivational tools to encourage change • By concentrating on weaknesses in importance and confidence • By using appropriate tactics to move through stages of readiness • By helping with time planning

Practice Tips—cont'd

Motivating a Patient to Change

	Medical Practitioner	Practice Nurse/Allied Health Professional
Arrange	• Support (e.g., family or significant others; include a motivational specialist, where appropriate) • List of other health professionals for a care plan • Other resources (e.g., QUIT line)	• Discussion of motivational tactics with other involved health professionals • Eferral to a personal trainer, nutritionist, or life coach, if appropriate • Ontact with self-help groups

Key Points

Clinical Management

- Consider and regularly review motivation before giving a lifestyle-oriented prescription.
- Stress to patients that change is necessary for disease correction/prevention.
- Review the barriers to change and work with patients to problem solve possible solutions.
- Keep your antennae tuned for any resistance. If you get ahead of the patient or push too hard they may manifest resistance, for example, "yes, but," agree with you but do nothing.
- Encourage patients to identify and review both short- and long-term benefits of making changes.
- Aim for intrinsic motivation to sustain behavior change.

REFERENCES

Alperstein, D., Sharpe, l, 2016. The efficacy of motivational interviewing in adults with chronic pain: a meta-analysis and systematic review. J. Pain17 (4), 393–403.

Apodaca, T.R., Longabaugh, R., 2009. Mechanisms of change in motivational interviewing: a review and preliminary evaluation of the evidence. Addiction 104, 705–715.

Barnes, R.D., Ivezaj, V., 2015. A systematic review of motivational interviewing for weight loss among adults in primary care. Obes. Rev. 16, 304–318.

Burke, B., Arkowitz, H., Menchola, M., 2003. The efficacy of motivational interviewing: a meta-analysis of controlled clinical trials. J. Consult. Clin. Psychol. 71, 843–861.

Cahill, K., Lancaster, T., Green, N., 2010. Stage-based interventions for smoking cessation. Cochrane Database Syst. Rev. 11, CD004492.

Copeland, L., McNamara, R., Kelson, M., Simpson, S., 2015. Mechanisms of change within motivational interviewing in relation to health behaviors outcomes: a systematic review. Patient Educ. Couns. 98, 401–411.

Egan, G., 1998. The Skilled Helper: A Problem-management Approach to Helping, sixth ed. Wadsworth Publishing Company, Pacific Grove.

Egger, G., 1981. The Sport Drug. Allen and Unwin, Sydney.

Elder, J., Ayala, G., Harris, S., 1999. Theories and intervention approaches to health-behavior change in primary care. Am. J. Prev. Med. 17, 275–284.

Epstein, R.M., 2006. Making communication research matter: what do patients notice, what do patients want, and what do patients need? Patient Educ. Couns. 60, 272–278.

Gaume, J., Bertholet, N., Faouzi, M., Gmel, G., Daeppen, J.B., 2010. Counselor motivational interviewing skills and young adult change talk articulation during brief motivational interventions. J. Subst. Abuse Treat 39 (3), 272–281. http://dx.doi.org/10.1016/j.jsat.2010.06.010. Epub 2010 August 13.

Hutchison, A.J., Breckon, J.D., Johnston, L.H., 2009. Physical activity behavior change interventions based on the transtheoretical model: a systematic review. Health Educ. Behav. 36, 829–845.

Landry, J.B., Solomon, M.A., 2004. African American women's self determination across changes of change for exercise. J. Sport Exerc. Psych. 11, 263–282.

Lindson-Hawley, N., Thompson, T.P., Begh, R., 2015. Motivational interviewing for smoking cessation. Cochrane Database Syst. Rev. 2 (3), CD006936. http://dx.doi.org/10.1002/14651858.CD006936.pub3.

Lundahl, B., Burke, B.L., 2009. The effectiveness and applicability of motivational interviewing: a practice-friendly review of four meta-analyses. J. Clin. Psychol. 65, 1232–1245.

Macquarie Dictionary, third ed.1997. Macquarie University: Macquarie Library Pty. Ltd.

Miller, W.R., Rollnick, S., 2002. Motivational Interviewing: Preparing People for Change, second ed. Guilford Press, NY.

Miller, W.R., Rollnick, S., 2009. Ten things that motivational interviewing is not. Behav. Cogn. Psychother.

Miller, W.R., Rollnick, S., 2012. Motivational Interviewing: Helping People Change, third ed. Guilford Publications, New York. ISBN 10 1609182278.

Miller, W.R., Rose, G.S., 2009. Toward a theory of motivational interviewing. Am. Psychol. 64, 527–537.

Moyers, T.B., Martin, T., Houck, J.M., Christopher, P.J., Tonigan, J.S., 2009. From in-session behaviors to drinking outcomes: a causal chain for motivational interviewing. J. Consult. Clin. Psychol. 77 (6), 1113–1124. http://dx.doi.org/10.1037/a0017189.

O'Brien, N., McDonald, S., Araujo-soares, V., Lara, J., Errington, L., Godfrey, A., Meyer, T.D., Rochester, L., Mathers, J.C., White, M., Sniehotta, F.F., 2015. The features of interventions associated with long-term effectiveness of physical activity interventions in adults aged 55 to 70 years: a systematic review and meta-analysis. Health Psychol. Rev. 1–29.

O'Halloran, P.D., Blackstock, F., Shields, N., Holland, A., Iles, R., Kingsley, M., Bernhardt, J., Lannin, N., Morris, M.E., Ttaylor, N.F., 2014. Motivational interviewing to increase physical activity in people with chronic health conditions: a systematic review and meta-analysis. Clin. Rehabil. 28, 1159–1171.

Prochaska, J., Velicer, W., 1997. The transtheoretical model of health behavior change. Am. J. Health Promot. 12, 38–48.

Rollnick, S., Miller, W., Butler, C., 2008. Motivational Interviewing in Health Care. Guildford Press, New York.

Romano, M., Peters, L., 2014. Understanding the process of motivational interviewing: a review of the relational and technical hypotheses. Psychother. Res. 1–21.

Rubak, S., Sandbaek, A., Lauritzen, T., Christensen, B., 2005. Motivational interviewing: a systematic review and meta-analysis. Br. J. Gen. Pract. 55, 305–312.

Smith, D., Collins, D., 2004. Mental practice, motor performance, and the late CNV. J. Sport Exerc. Psychol. 26, 412–426.

Spencer, L., Adams, T.B., Malone, S., Roy, L., Yost, E., 2006. Applying the transtheoretical model to exercise: a systematic and comprehensive review of the literature. Health Promot. Pract. 7, 428–443.

Teixeira, P.J., Carraca, E.V., Marques, M.M., Rutter, H., Oppert, J.M., de Bourdeaudhuij, I., Lakerveld, J., Brug, J., 2015. Successful behavior change in obesity interventions in adults: a systematic review of self-regulation mediators. BMC Med. 13, 84.

Vanbuskirk, K.A., Wetherell, J.L., 2014. Motivational interviewing with primary care populations: a systematic review and meta-analysis. J. Behav. Med. 37, 768–780.

Vealey, R.S., Greenleaf, C.A., 2001. Seeing is believing: understanding and using imagery in sport. In: Williams, J.M. (Ed.), Applied Sport Psychology: Personal Growth to Peak Performance. Mayfield Publishing Company, Mountain View, CA, pp. 247–272.

Wertheimer, A., 2013. Should "nudge" be salvaged? J. Med. Ethics 39, 498–499.

West, R., 2005. Time for a change: putting the transtheoretical (stages of change) model to rest. Addiction 100, 1036–1039.

Professional Resources

Measuring Motivation

The following are some simple tests to measure readiness, importance (willingness), and confidence (ability) to change a behavior (e.g., physical activity). The type of behavior can be changed as required (e.g., quitting smoking, reducing drinking, improving nutrition, losing weight).

Measuring Readiness to Change Physical Activity

Assess the stage of readiness to change

Current physical activity status—pick the description that applies to you.

1. I am currently not physically active and do not intend to start being physically active in the next six months. (Precontemplation)
2. I am currently not physically active, but I am thinking about becoming physically active in the next six months. (Contemplation)
3. I am currently physically active but not on a regular basis. (Preparation)
4. I am currently physically active regularly, but I have only begun doing so within the last six months. (Action)
5. I am currently physically active regularly and have been so for longer than six months. (Maintenance)

Continued

Professional Resources—cont'd

Assess Concern

How concerned are you about your current level of physical activity? (Circle the number)

1	2	3	4	5	6	7	8	9	10
Not at all					Very concerned				

Measure Willingness to Change Physical Activity (Importance)

How important is it for you to exercise to lose weight? (Circle the number)

1	2	3	4	5	6	7	8	9	10
Not important					Extremely important				

Measure Confidence in Changing Physical Activity (Ability)

How confident are you that you can exercise properly to lose weight if you try?

1	2	3	4	5	6	7	8	9	10
Not confident					Extremely confident				

Chapter 6

Self-Management in Lifestyle Medicine

Malcolm Battersby, Garry Egger, John Litt

Most people would like their own ways and other people's means.

Sydney Tremayne

INTRODUCTION: SELF-CARE VERSUS SELF-MANAGEMENT

A core principle of Lifestyle Medicine, as shown in Table 1.1, and in the revised definition of Lifestyle Medicine in Chapter 1, is for the individual to become active in his or her own health care, that is, to "self-manage." The term "self-management" refers to the knowledge, skills, and behaviors that apply across the spectrum from wellness to risk factor and chronic condition management (Lawn et al., 2009). It encompasses "self-care," a broad set of behaviors, which include daily living activities that impact on health, from lifestyle to medical management, when an individual has a chronic condition. It involves decisions and behaviors that patients with chronic illness engage in that affect their health. It can also be seen as an *intervention* used to bring about specific outcomes. It is not about the patient acting alone, but sharing care and responsibility with the health professional team.

Self-management has been defined in a number of ways, but is basically: "... active participation by people in their own care" (RACGP, 2008). The rationale behind this is explained by Holman and Lorig (2004): "With acute disease, the patient is inexperienced, the health professionals are knowledgeable, and they apply that knowledge to a passive patient. With chronic disease, those roles are no longer appropriate. The patient should be an active partner, applying his or her knowledge continuously to the care process. But initially, the patient is inexperienced in this new role, and must learn how to be an effective participant."

COMPONENTS OF SELF-MANAGEMENT

Chronic condition self-management is a process. It includes a broad set of attitudes, behaviors, and skills and is directed toward managing the impact of the disease or condition on all aspects of living by the individual with a chronic condition.

Lifestyle Medicine. http://dx.doi.org/10.1016/B978-0-12-810401-9.00006-1

SELF-CARE IS A NECESSARY BUT NOT SUFFICIENT COMPONENT

A number of factors contribute to the process:

- ☐ Having adequate health literacy that includes the skills and knowledge of the condition and/or its management (Liddy et al., 2014; Berkman et al., 2011);
- ☐ A systematic approach to planning care, with subsequent adoption of a self-management care plan agreed and negotiated in partnership with health professionals, significant others, and/or carers and other supporters;
- ☐ Coordinated care across health care providers and actively sharing in decision making with health professionals, significant others, and/or carers and other supporters (Smith et al., 2012b);
- ☐ Monitoring and managing signs and symptoms of the condition (McBain et al., 2015);
- ☐ Managing the impact of the condition on physical, emotional, occupational, and social functioning;
- ☐ Adopting lifestyles that address risk factors and promote health by focusing on prevention and early intervention (Schulman-Green et al., 2016);
- ☐ Having access to (Kruse et al., 2015) and confidence in (Liddy et al., 2014) the ability to use support services.

Ideally, it is a collaborative activity where patients can be supported by family and friends and the health professional team.

EVIDENCE OF EFFECTIVENESS

Evidence for the effectiveness of a self-management approach comes from a variety of areas, and includes, but is not limited to, improved clinical outcomes with ailments such as diabetes (Ricci-Cabello et al., 2014), asthma (Peytremann-Bridevaux et al., 2015), chronic pain (Richardson et al., 2014), and joint protection (Kroon et al., 2014). There is also evidence for better use of health services and reduced health care utilization, improvements in "self-efficacy" in dealing with chronic diseases, improved health-related behaviors such as exercise and nutrition, and improved service satisfaction.

FORMATS SUPPORTING SELF-MANAGEMENT

Self-management interventions contain principles and processes, and draw on a range of health behavior theories (including: the health belief model, the transtheoretical model/stages of change, social cognitive theory), which can be complementary and additive (Lawn and Schoo, 2009). While taking several formats, there have been four main approaches to self-management.

The Stanford Chronic Disease Self-Management Program

The Stanford Chronic Disease Self-Management Program, developed by Kate Lorig at Stanford University (Holman and Lorig, 2004), is primarily a peer-facilitated 6-week group course with a strong skills-based, goal-setting, and problem-solving process aimed at building individual confidence in the ability to manage health.

The Flinders Program

The Flinders Program is a patient-centered program developed by the Flinders University Human Behaviour and Health Research Unit in South Australia (Bettersby et al., 20008). In contrast to the group initiative, this is a one-on-one semistructured approach aimed at increasing collaborative care between clinicians and clients to bring about behavior change through the development of a care plan with regular monitoring and follow-up.

Health Coaching

Health coaching complements, rather than competes with, other self-management models, utilizing principles from psychology, counseling, and business, with training of health professionals typically carried out over 1–2 days in the context of health behavior change for disease (risk factor) prevention and/or chronic disease self-management. The approach is semistructured and flexible, and adds to approaches such as rehab programs and telephone coaching.

Motivational Interviewing

Motivational interviewing is not so much an independent process as a client-centered, nonjudgmental method for enhancing intrinsic motivation, which is also usually a component of most other self-management initiatives. Developed originally for alcohol- and smoking-related behavior change, it has been applied to a range of health behaviors (for more detail on motivational interviewing see Chapter 5).

What Influences Self-Management?

Clinician factors (Taylor et al., 2014; Smith et al., 2012b):

- Having knowledge of the condition and/or its management;
- Negotiating a self-management care plan agreed and negotiated in partnership with the patient, significant others, and/or carers and other supporters;
- Actively sharing the decision-making process with the patient, significant others, and/or carers and other supporters;
- Managing the impact of the condition on physical, emotional, occupational, and social functioning.

Patient factors (Coventry et al., 2015; Liddy et al., 2014; Smith et al., 2012a):

- Health literacy;
- The ability to self-monitor and managing signs and symptoms of the condition;
- Adopting a self-management care plan agreed and negotiated in partnership with health professionals, significant others, and/or carers and other supporters;
- Managing the impact of the condition on physical, emotional, occupational, and social functioning;
- Adopting lifestyles that address risk factors and promote health by focusing on prevention and early intervention;
- Access to, and confidence in, the ability to use support services;
- System support for self-management, for example, adequate time, training, resources, and support.

LIFESTYLE MEDICINE AND SELF-MANAGEMENT

Self-management is both a philosophy and a process within the broader field of Lifestyle Medicine. It has been predominantly developed to facilitate better outcomes in clients with chronic disease. In the absence of chronic disease (and/or symptoms), Lifestyle Medicine promotes self-care to address a range of health-related behaviors to reduce the likelihood of the development of chronic disease. Client adherence to health-promoting activities and self-care is different and often more problematic in the absence of symptoms or impairment (Hall and Rossi, 2008). As such, Lifestyle Medicine provides a vehicle for key components from each of the specific approaches to self-management discussed earlier.

Most lifestyle-related chronic disease is, at least in theory, amenable to, and thus likely to benefit from, at least some level of self-management by the client or patient. This requires that the recipient of care is actively involved in:

Monitoring—the condition(s);

Managing—the symptoms, treatment, and consequences of his or her condition(s);

Making—the necessary changes to adapt to the demands of the condition(s); and

Maintaining—a quality of life in the face of this or these condition(s).

The question is, how do you get patients to do it? Years of promise based on the "magic bullet" approach to infectious diseases have imbued the public with a sense of their right to effort-free treatment of all that ails. Self-management support is designed to get clinicians, patients, and practice teams working differently to change this way of thinking and acting.

APPROACH

IMPLICATIONS	A. Client Centred	B. Motivation Focused	C. Health literacy Oriented
1. Counseling	• Shared care	• Motivational interviewing	• Tutoring
2. Practice	• 'User friendly'	• Team arrangements	• 'hub-&-spoke' information' service
3. Outcomes	• *'Health efficacy'*	• *'Health determinancy'*	• *'Health literacy'*

FIGURE 6.1 Principles of self-management.

Government offerings in various countries have gone some way to help this, but the principles are often vague. In summary, it comes down to three things: (1) being client centered (Liddy et al., 2014), (2) being motivation focused (Coventry et al., 2015), and (3) being health literacy oriented (Liddy et al., 2014).

To make these changes effectively in the primary care environment (if they do not already exist) would involve a number of changes from the traditional model of health care delivery, such as:

1. A reorientation in counseling—to reflect greater patient involvement in care, particularly in shared decision making;
2. A restructuring of the practice environment—to become more "user-friendly" in line with the cooperative concept; and
3. A reordering of priorities—based on the client's priorities not the clinician's—particularly relating to outcomes or measures of "success."

The goal of clinicians in this respect is to increase the client's health:

- Efficacy (belief in his or her ability to do this);
- "Determinancy" (motivation to want to do something about the problem); and
- Health literacy (understanding of the problem and how to effectively reduce the consequences of this).

This is shown in matrix form in Fig. 6.1. Components of the matrix are discussed next.

PRINCIPLES OF SELF-MANAGEMENT

Being Client Centered

Few clinicians would ever admit to *not* having their patient's best interests at heart. However, much conventional practice is still problem, rather than client centered (Hibbard, 2003). The shift required is reminiscent of a parent's reaction to a child growing up. In the early stages (i.e., as with infectious diseases) this involves instruction, from the more knowledgeable to the least knowledgeable partner in the interaction. However, as the adolescent matures and is ready to move from home, a more interactive style is necessary. Counseling then often benefits from initially providing the patient with what is desired, rather than what is required, with anticipation that the latter change may then follow. An example is illustrated in the autobiography of famous Australian eye surgeon Fred Hollows, who expressed surprise going into his first Aboriginal community that no one jumped at his simple, revolutionary eye treatment. "The dogs have fleas," he was told by an Aboriginal elder, putting things into perspective and changing the modus operandi.

Being client centered involves the notion of "shared care" in counseling including:

- *Reflective listening*—Listening is a skill, that is, use the OARS approach of **O**pen-ended questions, **A**ffirmations, **R**eflective listening, and **S**ummarizing. Observe in a typical interview who is doing most of the talking? Most commonly, it should be the client. Observation includes being able to interpret body language of the client, as well as understanding one's own body language as a clinician.[1] Simple seating arrangements and the messages they portray, as shown in Fig. 6.2, are an example of this.

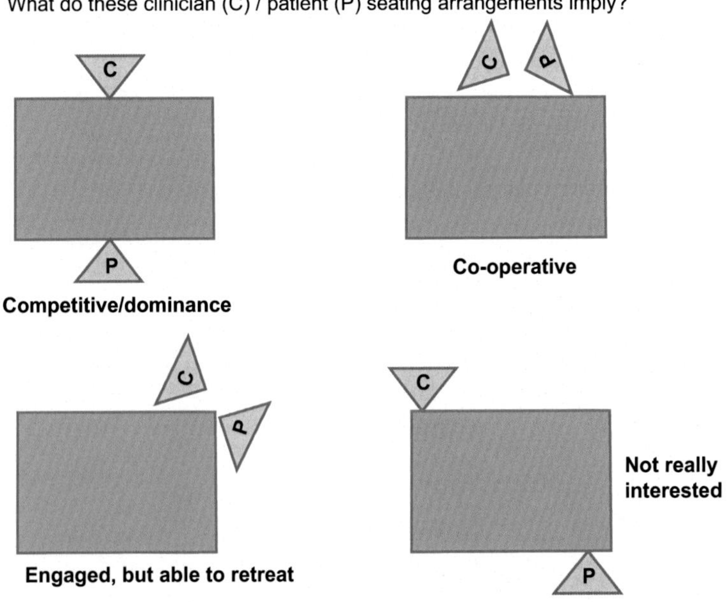

FIGURE 6.2 Seating arrangements and body language.

1. Some of the "art" of understanding body language is illustrated in the reference by Pease and Pease (2009) in the reference section.

- *Balancing the client's and clinician's goals and progress*—Goals should be SMART (**S**pecific, **M**easurable, **A**ction based, **R**ealistic, and **T**imely) set collaboratively between the clinician and the client. Priorities and progress measures need to be balanced between the needs of the client (e.g., the ability to fit into a suit for a special occasion) and the desires of the clinician (e.g., a decrease in blood pressure).

At the practice level this involves *having a "user-friendly" practice environment*—thinking about the design of the service experience for the client and carer and the staff, that is, the previsit, the visit, and the postvisit. Previsit includes bloods, clinical measures, self-management, and goal ratings. The visit includes a self-management-oriented waiting room with the philosophy clearly stated: "Our goal is to assist you with the self-management of your condition," and receptionists with readable health materials, internet education facilities, child friendly arrangements, etc.

At the outcome level this involves *improving patient "health efficacy."* In contrast to "self-efficacy," which has been defined as "…a belief about one's capability to produce designated levels of performance that exercise influence over events that affect their health," with health efficacy, the desired outcome of effective self-management counseling is "the belief in one's ability to influence one's own health." This can be increased in the patient in a number of ways, but in particular by:

- Building on small successes, and reducing the saliency of failures;
- Providing support/encouragement/reward;
- Having regular follow-ups;
- Teaching the patient to set SMART goals;
- Increasing health literacy—through the provision of appropriately targeted educational/self-help materials.

BEING MOTIVATION FOCUSED

At the counseling level, a motivational focus generally involves principles from the practice of motivational interviewing, as developed in particular by Miller and Rollnick (2002, 2009). The technique is underpinned by the patient-centered approach to care (Scholl et al., 2014). Motivational interviewing is centered around creating ambivalence in the client between his or her current situation and where s/he would like to be (see Chapter 5). This means accentuating the benefits and reducing the barriers to change and combining this with interpreting the client's readiness, willingness, and self-confidence in being able to change (Chapter 5).

At the practice level, it can become financially feasible to involve an allied health care team working with a GP as a coordinator to help patients manage, and hopefully over the long term self-manage, their chronic conditions. More detail on this is included in the Appendix to this chapter.

The outcome focus should be on increasing the "determinancy" or intrinsic motivation of the patient such that external reward is no longer an option for improving health behavior. Techniques for doing this are covered in more detail in Chapter 5.

BEING HEALTH LITERACY ORIENTED

Health literacy is one of the least understood and most overlooked factors in health management (Guzys et al., 2015). It is defined as: "The degree to which individuals have the capacity to obtain, process, and understand basic health information and services needed to make appropriate health decisions" (Sorensen et al., 2012). The level of an individual's health literacy impacts on tasks such as reading dosage instructions on a package of medicine, understanding prescription advice, whether s/he seeks screening or diagnostic tests, or making informed decisions such as choosing to proceed with treatment. It is a key factor in the appropriate use of any health care system and a determinant of one's ability to be able to effectively self-manage one's own ailment(s) (Keller et al., 2008; Osborne, 2013).

Studies show that up to 50% of the population may be functionally health illiterate (Guzys et al., 2015). This increases to ~65% in over 60 year olds, and, as might be expected, is highest in those with low education levels, those from non-English-speaking and indigenous backgrounds, and those from low socioeconomic status groups. On top of this, there is a significant proportion who have some level of dyslexia, which would handicap their understanding of information available. This is rarely communicated, however, because there is a level of embarrassment in admitting to a problem of understanding, and hence so much good advice can go begging.

Low functional literacy influences health through a number of different processes including more hospitalizations, higher health care costs, less knowledge about health problems, and generally poor health status (Guzys et al., 2015).

A number of behaviors may flag low literacy skills to the clinician (Hersh et al., 2015). These include:

- Asking staff for help;
- Bringing along someone who can read;
- Inability to keep appointments; making excuses ("I forgot my glasses");
- Noncompliance or poor adherence with medication and recommended interventions;
- Postponing decisions and mimicking the behavior of others.

Tests of self-literacy, such as the REALM (**R**apid **E**stimate of **A**dult **L**iteracy in **M**edicine; IOM, 2004), and Newest Vital Sign (Duell et al., 2015), which involves answering questions after reading a food label (see Appendix), are available for quantifying health literacy.

Enhancing Health Literacy

Ideally, increases in health literacy would come from improved communications of scientific findings. Currently, scientific knowledge about health is channeled through a number of sources, including the media, friends, associates, and acquaintances, not all of whom are reliable in their interpretation or explanation of known facts. Effective strategies to improve patient comprehension include conveying a few key points at each patient visit, jargon-free communication, use of pictures to clarify concepts, and confirmation of patient comprehension via the "show-me" or "teach-back" method (Berkman et al., 2011).

Hence it is left to the clinician to attempt to enhance understanding among patients with low literacy. Safeer and Keenan (2005) suggest a number of ways in which clinicians can "tutor" patients through their counseling. These should be preceded by the suggestion that it is usually better to under-, rather than overestimate the level of patient health literacy:

- Slow down and take time to assess the patient's health literacy;
- Use "living room" language instead of medical terminology;
- Show or draw pictures to enhance understanding and subsequent recall;
- Use a "teach-back" or "show-me" approach to confirm understanding;
- Be respectful, caring, and sensitive, thereby empowering patients to participate in their own health care.

At the practice level, an information service, with a wide range of materials in different forms specifically targeted at low health literacy, should be easily accessible for patients, without the embarrassment of having to ask. Preparation of patient materials, handouts, etc., should be focused at around a year 6 level of literacy. And while there is now extremely accessible and useful information available on the internet, it should not be assumed that all patients can easily access this. Those who can use the internet would appear to be unlikely to have low health literacy skills. Referral and assistance in accessing self-help groups is of value. This allows the practice to operate as a "hub-and-spoke" information service with a function to other services.

What Can Be Done in a Brief Consultation?

- ☐ Realistically assess the patient's level of health literacy;
- ☐ Listen and reflect on the patient's expected goals from the consult;
- ☐ Observe any "hidden" messages in body language;
- ☐ Show a genuine desire to work cooperatively with the patient;
- ☐ Explain the need for, and the concept of, self-management;
- ☐ Use the "teach-back" technique to make sure you are being fully understood.

Practice Tips

Increasing Involvement in Self-management

	Medical Practitioner	Practice Nurse/Allied Health Professional
Assess	• Health literacy • Self-management knowledge, behaviors, strengths, and barriers • Motivation • Chronic disease(s)	• Health literacy • Self-management knowledge, behaviors, strengths, and barriers • Interest in self-management • Risk factor assessment • Ability to self-manage
Assist	• Identify the problem • Identify goals • Support goal attainment • Discuss "health efficacy" • By "tutoring"	• Identify the problem • Identify goals • Work toward goals • Teach problem-solving skills • Identifying appropriate learning style and match with relevant educational approaches • Increase "health efficacy" • By monitoring outcomes
Arrange	• Chronic disease management • Referrals • Self-help groups	• Team care arrangement • "User-friendly" practice • Hub-and-spoke information

REFERENCES

Battersby, M., Lawn, S., Litt, J., Hood, C., Brennan, A., Ackland, A., Sudha, H., 2008. The Development of an Integrated Training Program for Prevention and Self-management of Chronic Conditions for the South Australian Workforce. Flinders Health Behaviour and Health Research Unit (FHBHRU), Flinders University, Adelaide.

Berkman, N.D., Sheridan, S.L., Donahue, K.E., Halpern, D.J., Viera, A., Crotty, K., Holland, A., Brasure, M., Lohr, K.N., Harden, E., Tant, E., Wallace, I., Viswanathan, M., 2011. Health literacy interventions and outcomes: an updated systematic review. Evid. Rep. Technol. Assess. (Full Rep.), 1–941.

Coventry, P.A., Small, N., Panagioti, M., Adeyemi, I., Bee, P., 2015. Living with complexity; marshalling resources: a systematic review and qualitative meta-synthesis of lived experience of mental and physical multimorbidity. BMC Fam. Pract. 16, 171.

Duell, P., Wright, D., Renzaho, A.M., Bhattacharya, D., 2015. Optimal health literacy measurement for the clinical setting: a systematic review. Patient Educ. Couns. 98, 1295–1307.

Guzys, D., Kenny, A., Dickson-Swift, V., Threlkeld, G., 2015. A critical review of population health literacy assessment. BMC Public Health 15, 215.

Hall, K.L., Rossi, J.S., 2008. Meta-analytic examination of the strong and weak principles across 48 health behaviors. Prev. Med. 46 (3), 266–274.

Hersh, L., Salzman, B., Snyderman, D., 2015. Health literacy in primary care practice. Am. Fam. Phys. 92, 118–124.

Hibbard, J., 2003. Engaging health care consumers to improve quality of care. Med. Care 41 (1 Suppl.), I61–I70.

Holman, H., Lorig, K., 2004. In: Public Health Reports, vol. 119, pp. 239–243.

IOM (Institute of Medicine), 2004. Health Literacy: A Prescription to End Confusion. www.iom.edu/report.asp?id=19723.

Keller, D.L., Wright, J., Pace, H.A., 2008. Impact of health literacy on health outcomes in ambulatory care patients: a systematic review. Ann. Pharmacother. 42 (9), 1272–1281.

Kroon, F.P., Van Der Burg, L.R., Buchbinder, R., Osborne, R.H., Johnston, R.V., Pitt, V., 2014. Self-management education programmes for osteoarthritis. Cochrane Database Syst. Rev. 1, Cd008963.

Kruse, C.S., Bolton, K., Freriks, G., 2015. The effect of patient portals on quality outcomes and its implications to meaningful use: a systematic review. J. Med. Internet Res. 17, e44.

Lawn, S., Schoo, A., 2009. Supporting self-management of chronic health conditions: common approaches. Patient Educ. Couns. 80 (2).

Lawn, S., Battersby, M., Lindner, H., et al., 2009. What skills do primary health care professionals need to provide effective self-management support?: A consumer perspective. Aust. J. Prim. Health 15 (1), 37–44.

Liddy, C., Blazkho, V., Mill, K., 2014. Challenges of self-management when living with multiple chronic conditions: systematic review of the qualitative literature. Canad. Fam. Phys. 60, 1123–1133.

McBain, H., Shipley, M., Newman, S., 2015. The impact of self-monitoring in chronic illness on healthcare utilisation: a systematic review of reviews. BMC Health Serv. Res. 15, 565.

Miller, W.R., Rollnick, S., 2002. Motivational Interviewing: Preparing People for Change. Guilford Press, London.

Miller, W.R., Rollnick, S., 2009. Ten things that motivational interviewing is not. Behav. Cogn. Psychother. 37, 129–140.

Osborne, H., 2013. Health Literacy: From A to Z. Jones and Bartlett Learning. Burlington, MA.

Pease, A., Pease, B., 2009. The Definitive Book of Body Language. Pease International, Hong Kong.

Peytremann-Bridevaux, I., Arditi, C., Gex, G., Bridevaux, P.O., Burnand, B., 2015. Chronic disease management programmes for adults with asthma. Cochrane Database Syst. Rev. 5 Cd007988.

RACGP, 2008. Putting Prevention into Practice. Guidelines for the Implementation of Prevention in the General Practice Setting, second ed. Royal Australian College of Physicians, Melbourne, p. 94.

Ricci-Cabello, L., Ruiz-Perez, I., Rojas-Garcia, A., Pastor, G., Rodriguez-Barranco, M., Goncalves, D.C., 2014. Characteristics and effectiveness of diabetes self-management educational programs targeted to racial/ethnic minority groups: a systematic review, meta-analysis and meta-regression. BMC Endocr. Disord. 14, 60.

Richardson, J., Loyola-Sanchez, A., Sinclair, S., Harris, J., Letts, L., Macintyre, N.J., Wilkins, S., Burgos-Martinez, G., Wishart, L., Mcbay, C., Martin Ginis, K., 2014. Self-management interventions for chronic disease: a systematic scoping review. Clin. Rehabil. 28, 1067–1077.

Safeer, R.S., Keenan, J., 2005. Health literacy: the gap between physicians and patients. Am. Fam. Phys. 72, 387–388.

Scholl, I., Zill, J.M., Harter, M., Dirmaier, J., 2014. An integrative model of patient-centeredness - a systematic review and concept analysis. PLoS One 9, e107828.

Schulman-Green, D., Jaser, S.S., Park, C., Whittemore, R., 2016. A metasynthesis of factors affecting self-management of chronic illness. J. Adv. Nurs. 72 (7).

Smith, S.M., Soubhi, H., Fortin, M., Hudon, C., O'Dowd, T., 2012a. Interventions for improving outcomes in patients with multimorbidity in primary care and community settings. Cochrane Database Syst. Rev. 4, CD006560.

Smith, S.M., Soubhi, H., Fortin, M., Hudon, C., O'Dowd, T., 2012b. Managing patients with multimorbidity: systematic review of interventions in primary care and community settings. BMJ 345, e5205.

Sorensen, K., Van Den Broucke, S., Fullam, J., Doyle, G., Pelikan, J., Slonska, Z., Brand, H., 2012. Health literacy and public health: a systematic review and integration of definitions and models. BMC Public Health 12, 80.

Taylor, S.J.C., Pinnock, H., Epiphaniou, E., Pearce, G., Parke, H.L., Schwappach, A., Purushotham, N., Jacob, S., Griffiths, C.J., Greenhalgh, T., Sheikh, A., 2014. A rapid synthesis of the evidence on interventions supporting self-management for people with long-term conditions: PRISMS - Practical systematic Review of Self-Management Support for long-term conditions. Health Serv. Deliv. Res. 2 (53) NIHR Journals Library, Southampton, UK.

Professional Resources

Self-management

A. Self-management audit

Used to test the clinician's readiness and ability to practice self-management.

How confident are you that you could:	Level of confidence										
	None					Some					Extensive
1. Teach patients to identify & self-manage their chronic disease(s)?	0	1	2	3	4	5	6	7	8	9	10
2. Assist the patient develop SMART goals & personalised action plans?	0	1	2	3	4	5	6	7	8	9	10
3. Use reflective listening skills (OARS)?	0	1	2	3	4	5	6	7	8	9	10
4. Determine a person's readiness, willingness and ability to change?	0	1	2	3	4	5	6	7	8	9	10
5. Match the appropriate intervention to the patient's stage of change?	0	1	2	3	4	5	6	7	8	9	10
6. Move the patient towards 'intrinsic' motivation for lifestyle change?	0	1	2	3	4	5	6	7	8	9	10
7. Use available computer-based and other tools during assessment?	0	1	2	3	4	5	6	7	8	9	10
8. Challenge psychological defenses against lifestyle change?	0	1	2	3	4	5	6	7	8	9	10
9. Understand body language?	0	1	2	3	4	5	6	7	8	9	10
10. Direct a patient to lifestyle related publications/groups/sources etc?	0	1	2	3	4	5	6	7	8	9	10

TOTAL SCORE = AVERAGE SCORE =

Continued

Professional Resources—cont'd

B. Assessing health literacy—Newest vital signs

Figure 1A. The newest vital sign – English.

Nutrition Facts

Serving Size	1/2 cup
Servings per container	4
Amount per serving	
Calories 250	Fat Cal 120
	%DV
Total Fat 13g	20%
Sat Fat 9g	40%
Cholesterol 28mg	12%
Sodium 55 mg	2%
Total Carbohydrate 30g	12%
Dietary Fiber 2g	
Sugars 23g	
Protein 4g	8%

* Percent Daily Values (DV) are based on a 2,000 calorie diet. Your daily values may be higher or lower depending on your calorie needs.

Ingredients: Cream, Skim Milk, Liquid Sugar, Water, Egg Yolks, Brown Sugar, Milkfat, Peanut Oil,Sugar, Butter, Salt, Carrageenan, Vanilla Extract.

Note: This single scenario is the final English version of the newest vital sign. The type size should be 14-point (as shown above) or larger. Patients are presented with the above scenario and asked the questions shown in Figure 1b.

Figure 1B. Questions and answers score sheet for the newest vital sign – English.

	ANSWER CORRECT? YES	NO
READ TO SUBJECT: This information is on the back of a container of a pint of ice cream.		
QUESTIONS		
1. If you eat the entire container, how many calories will you eat? **Answer** ❑ 1,000 is the only correct answer	___	___
2. If you are allowed to eat 60 g of carbohydrates as a snack, how much ice cream could you have? **Answer** Any of the following is correct: ❑ 1 cup (or any amount up to 1 cup) ❑ Half the container Note: If patient answers "2 servings," ask "How much ice cream would that be if you were to measure it into a bowl?"	___	___
3. Your doctor advises you to reduce the amount of saturated fat in your diet. You usually have 42 g of saturated fat each day, which includes 1 serving of ice cream. If you stop eating ice cream, how many grams of saturated fat would you be consuming each day? **Answer** 33 is the only correct answer	___	___
4. If you usually eat 2500 calories in a day, what percentage of your daily value of calories will you be eating if you eat one serving? **Answer** 10% is the only correct answer	___	___
Pretend that you are allergic to the following substances: Penicillin, peanuts, latex gloves, and bee stings.		
5. Is it safe for you to eat this ice cream? **Answer** ❑ No	___	___
6. (Ask only if the patient responds "no" to question 5): Why not? **Answer** Because it has peanut oil.	___	___
Total Correct	___	

Professional Resources—cont'd

C. Self-assessment—Inflatable tire

Scale and join points before and after self-management instruction. A "flat tire" should become an inflated tire.

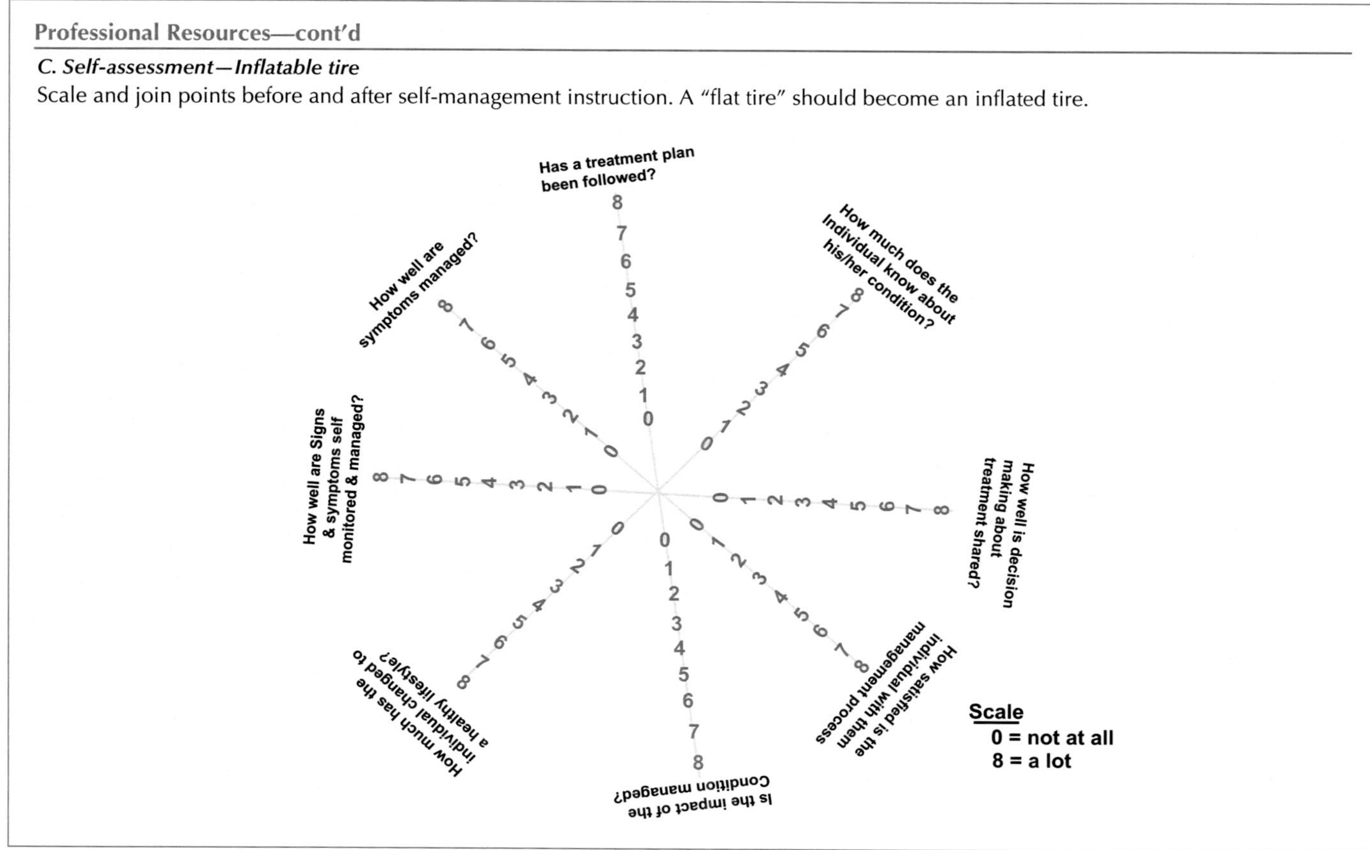

Chapter 7

Overweight and Obesity: The Epidemic's Underbelly

Stephan Rössner, Garry Egger, Andrew Binns, Michael Sagner

Any important disease whose causality is murky, and for which treatment is ineffectual, tends to be awash in significance.

Susan Sontag ("Illness as Metaphor")

INTRODUCTION: OBESITY AND ILL HEALTH

In 1970 a small group of dedicated scientists met in London at the First International Congress on Obesity to discuss what was then a distinctly "unsexy" topic. Although French endocrinologist Jean Vague had outlined the potential health problems associated with increased abdominal fat, or a "pot belly," as far back as 1947, it had taken this long for the modern obesity epidemic to germinate. By 2014 the numbers at the 12th International Congress of Obesity in Malaysia had grown to almost 3000, with widespread media interest in finding the answer to a problem now affecting up to 60% of people in many developed countries and an estimated 15% of the total world population (International Association for the Study of Obesity; World Health Organization, 2016).

Obesity is associated with a range of different diseases (e.g., diabetes, heart disease, cancers)[1] and a growing list of everyday ailments (e.g., sleep apnea, arthritis, musculoskeletal problems); it exists within a cascade of epidemics making it part of a "syndemic," or combination of epidemics; and it is one of the few health problems that feeds back on itself causing vicious cycles (Fig. 7.1). These factors make the situation even more serious (Swinburn and Egger, 2004).

FAT DISTRIBUTION: NOT ONLY *IF* YOU'RE FAT BUT *WHERE*

Research has confirmed the early suggestion of Vague (1947) of the importance of subcutaneous adipose tissue (SAT), or a "pot belly." Abdominally stored fat

1. Although obesity is "associated" with these diseases, it is not correct at this stage to say that it is a "cause." As discussed in Chapter 2, causality in chronic disease is difficult to determine.

Lifestyle Medicine. http://dx.doi.org/10.1016/B978-0-12-810401-9.00007-3

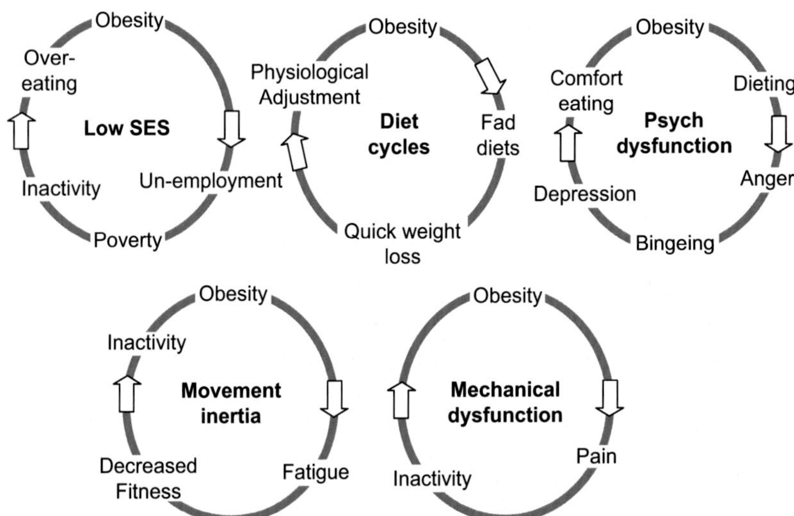

FIGURE 7.1 Vicious circles associated with obesity (SE = Socioeconomic status). *Reprinted with permission from Swinburn, B., Egger, G., 2004. The runaway weight gain train: too many accelerators, not enough brakes. Br. Med. J. 329 (7468), 736–739.*

(i.e., the android or "apple" shape), which is most characteristic of males and postmenopausal females, is more metabolically dangerous than fat stored in the gluteal region (i.e., the gynoid or "pear" shape) mainly around the hips and buttocks of premenopausal females. In the 1990s it was shown that abdominal visceral adipose tissue (AVAT) or internal fat around the organs of the trunk is even more associated with metabolic risk and endpoint disease than subcutaneous adipose tissue (SAT). Even more recently, a link has been shown between epicardial adipose tissue (EAT) and heart disease, and AVAT and metabolic syndrome (Wu et al., 2016). In all cases it appears that such risk-associated fat comes from ectopic supplies, or that which has "spilled over" from the fat cells, or adipocytes, in which it is normally stored with no apparent health consequences.

Seen diagrammatically, it appears that a normal adipocyte, such as that shown on the left-hand side of Fig. 7.2, has the potential to expand normally without implications for ill health, at least to a point (which is probably determined by genetics). Beyond this point, the cell (like a balloon filled to its maximum with water), is unable to store extra energy, which then "spills over" into ectopic fat supplies in the blood, liver, and muscle, where it becomes much more dangerous. In this respect, therefore, fat on the body is perfectly healthy, *until* it spills over, causing inflammation and oxidative stress as shown in Fig. 3.1 (Chapter 3) that may then lead to chronic allostasis and disease status. This may explain why a significant number of obese individuals (called "obese, but metabolically normal individuals") have no disease risk factors

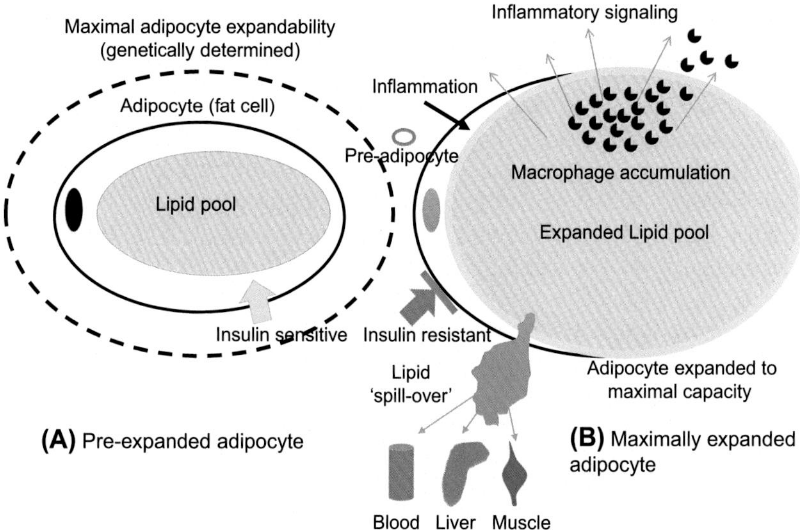

FIGURE 7.2 The fat "spillover" hypothesis.

and appear perfectly healthy, while others (called "not obese, but metabolically obese individuals") are lean but with all the risk factors expected of the obese (Phillips et al., 2013).

DEALING WITH OBESITY

Perhaps naturally, the ideal medical focus on dealing with obesity would be toward developing a pharmaceutical "cure," a quick fix, with little behavioral cost to the individual, or a surgical fix that can be carried out without risk and relatively cheaply. Many pharmaceuticals have shown promise but none has yet yielded widespread success over the long term, and the nature of obesity suggests this is unlikely to happen in the near future. Bariatric surgery is very effective, but expensive, complex, and with potential complications. It is also unlikely to be a solution at the population level. As with any chronic disease, it needs to be accepted that there is unlikely to be a single, simple, low-cost, long-term solution to obesity management.

Fat is stored in fat cells (adipocytes) under normal conditions as in (1) leading to expansion and contraction of the cell as shown by the dotted lines in Fig. 7.2. When maximum expansion is reached (probably determined genetically) as shown in (2), macrophage accumulation signals potential development of inflammation and lipid spillover occurs into ectopic supplies in blood, liver, and muscle, resulting in pathological fat stores.

In this chapter, therefore, we concentrate on the strategies that work best for weight loss within a system of triaging at the individual level, but stress that all (including surgery) are fully effective only in the context of a lifestyle change involving modification of energy intake (EI; i.e., food and drink) and/or energy expenditure (EE, i.e., exercise, metabolic rate, and thermogenesis). Both EI and EE will be dealt with in detail in subsequent chapters. Here we confine ourselves to a discussion of the energy balance equation and those factors influencing energy balance. We also look at developing trends in weight control management and summarize the available strategies for dealing with the problem, and we finish with a consideration of the environmental issues causing the problem at the population level. This is meant to complement the various national clinical guidelines from a range of countries (Table 7.1).

Before we turn to treatments, however, we need to look at the basic physics and bioenergetics of body weight.

TABLE 7.1 Clinical Guidelines for Weight Control and Obesity Management

Adults

1. Discuss weight and whether measurements (height, weight, BMI, WC) should be taken at this stage.
2. Assess and treat comorbidities associated with weight and determine the patient's need to lose weight.
3. Ascertain the patient's readiness and motivation to lose weight.
4. Assess *why* energy imbalance has occurred.
5. Assess *how* energy imbalance has occurred.
6. Determine the level of clinical intervention required.
7. Devise goals and treatment strategies with the patient.
8. Prescribe or refer for dietary and physical activity advice.
9. Prescribe medication, or where appropriate refer for obesity surgery and/or conduct or refer for behavior modification as appropriate.
10. Review and provide regular assistance for weight management and maintenance of weight change, and change program as required.

Children

1. Assess the extent of overweight or obesity in the child or adolescent in relation to other children at the same stage of development.
2. Assess comorbidities associated with weight and treat these independently where appropriate.
3. Assess *why* energy imbalance has occurred.
4. Assess *how* energy imbalance has occurred.
5. Determine the level of clinical intervention required.
6. Devise treatment strategy with the patient and family.
7. Devise treatment goals, including outcome indicators not related to weight.
8. Review and provide regular assistance for weight management and maintenance of weight change, and change program as required.

Adapted from National Institutes of Health, NHMRC, National Health and Medical Research Council, 2013. Clinical Practice Guidelines for the Management of Overweight and Obesity in Adults, Adolescents and Children in Australia. https://www.nhmrc.gov.au/guidelines-publications/n57. *BMI,* body mass index; *WC,* waist circumference.

IS A CALORIE REALLY A CALORIE?

Traditionally, the formula for body weight has been:

weight = energy in (from food and drink) – **energy out** (from exercise, thermogenesis, and metabolism)

This is a physics formula, with little relevance to real life. According to the equation, people eating an extra slice of toast and butter each day over the course of their lives would all be approximately 170 kg heavier than if they did not eat this. This is clearly not correct. The formula, which is used widely by the public and popular media, is based on the idea that energy balance is static, rather than dynamic (as it should be in a biological organism). Egger and Swinburn (1996) devised a more realistic formula (Fig. 7.3), that takes account of biological adaptation as well as other factors. But even this is simplistic. As shown by the Foresight study in the UK (Foresight, 2007) and Tarek Hamed's (2009) seminal book *Thinking in Circles About Obesity*, weight control, like chronic disease management in general, requires "systems model" thinking where everything (diet, exercise, sleep, stress, environment, etc.) interacts with everything else in a complex feedback system that requires consideration of all factors, not just diet and/or exercise, for successful long-term weight loss.

Using this conceptual approach, it is clear that energy balance is not only influenced by a range of environmental (e.g., food availability, energy-saving devices), biological (e.g., age, gender, ethnicity), and behavioral influences (e.g., early experiences, stress, depression) but also by physiological adjustments that occur to moderate energy imbalances and to interactions between aspects of energy intake and expenditure. In the case of the toast eater, for example, energy expenditure through heat loss, metabolism, or exercise may rise to ensure that overall there is only a minor net change in body weight as a result of a small change in energy intake. These adjustments can vary between and within individuals, according to metabolic rate, effects on hunger, and other physiological factors. In a practical sense, this means that weight loss and long-term maintenance of weight loss is much more complicated than is promoted in the popular

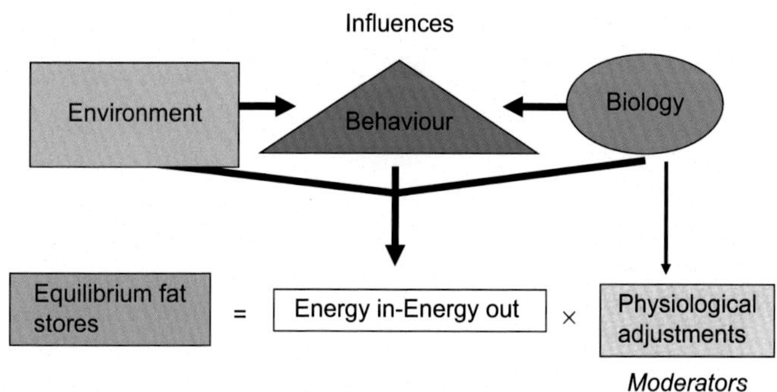

FIGURE 7.3 An ecological approach to obesity.

magazines that advertise the latest diet or exercise plan as a single solution to the problem. Ultimately sophisticated computer programs may be needed to properly devise individualistic programs. In the meantime, an attempt to individualize programs to similar groups of individuals (males vs. females; morbidly obese vs. overweight; active vs. inactive, etc.) as illustrated in programs such as that by Kushner and Kushner (2008) may be as close as we can get.

Measuring the components in energy balance is also a scientifically frustrating problem. Although everybody agrees that measurement of intake is complicated, it has remained in use, since nothing better is available. In 2015 a group of scientists challenged the assumption that "something is better than nothing" and argued that such research should not be published until more precise methods have been developed (Dhurandar et al., 2015).

CHANGING ENERGY BALANCE

Given the previous proviso, it is still important to consider energy balance, within lifestyle change, as the core of weight loss instruction (albeit with the knowledge of big possible variations around a mean weight loss response between individuals). This involves modifying food intake and exercise output. Although there are innumerable diets, products, and programs for doing this, all effective treatments essentially involve a change in the balance of calorific volume, or the amount of calories consumed, in relation to those expended. This is relatively simply defined in both cases as being the product of three factors, as shown in Fig. 7.4. Each of these is discussed in detail in the following chapters on exercise and nutrition.

TRENDS IN WEIGHT CONTROL MANAGEMENT

Optimal management in weight control is unlikely to come from unlocking a lock with a single key. The appropriate analogy is more likely to be that of a barrel lock, which will not open unless all locks on the barrel are lined up.

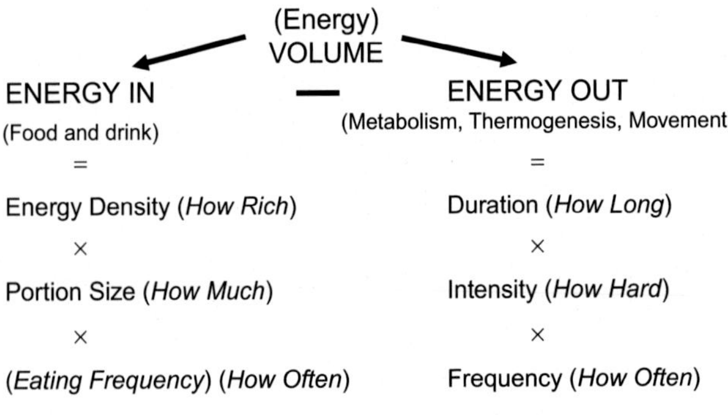

FIGURE 7.4 A calorific volume approach to energy balance.

Some of these strategies are discussed in following chapters. Meanwhile, there are a number of recent developments that can help facilitate the process. Among the more recent are putting the emphasis on lifestyle change, matching strategies to the patient, and combining treatment strategies.

Triaging Obesity Management

Excess body weight can range in medical seriousness from relatively benign to extremely morbid and severe, not always indicated by size alone. As discussed earlier, studies have shown that around one in three obese patients do not have the comorbid risk factors otherwise predicted by obesity, while around one in four lean patients can have these risks (Phillips and Pery, 2013). Triaging patients into three or four categories based on principles such as the Edmonton Staging Model (Sharma and Kushner, 2009) can help locate individuals in categories of response ranging from simple nutrition and exercise programs to corrective surgery or palliative care.

Putting the Emphasis on Lifestyle Change

With all the hype about new diets, drugs, surgery, and other weight loss treatments, it is sometimes forgotten that every treatment that works requires a lifestyle change involving the energy balance formula shown herein and that such a change should be the core of weight loss management. That this is often not the case is due to the fact that in a modern "obesogenic" environment, such a change is difficult to make. More detail on nutrition and physical activity is covered in the following chapters. A structured approach to this is covered in the Australian National Clinical Guidelines on weight control and obesity management (NHMRC, 2013) and in Egger and Binns (2001).

Matching the Strategies to the Patient

A one-suit-fits-all approach is no longer regarded as acceptable in weight loss management. For one thing, the strategies are complementary. Furthermore, they should be used dynamically such that the strategy or strategy mix implemented at the start of a program may be totally different to that used after early losses or for long-term maintenance. The skill of the clinician is in matching the appropriate strategies (and methods within strategies) to the patient.

Recognizing "Unhealthy" versus "Healthy" Body Fat

As discussed before, abdominal and particularly visceral fat are associated with greater metabolic risk than gluteal, or lower body, fat stores. Fat cells also have a maximum capacity to store fat, or lipids, which may occur at different levels of expansion in different individuals. Where fat storage reaches its maximum capacity, both through hyperplasia (the expansion of new fat cells) and hypertrophy (the expansion of the existing cell with lipid; see Fig. 7.2) and spillover occurs, excess energy is then stored ectopically (i.e., away from its normal place) in muscle, liver, or the bloodstream. Although it is generally associated with excess abdominal fat, spillover can occur in some individuals who do not appear overweight and not occur in others who are obese, hence possibly helping to explain the "healthy" and "sick" fat phenomenon at different levels of obesity. The difference can generally be detected by measuring fasting lipids, which may show raised triglycerides in association with an excess waist circumference, which together are highly predictive of insulin resistance (Arsenault et al., 2010).

Combining Treatment Strategies

Often in the past only one strategy (e.g., drugs, diet) has been applied to any or all individuals. This is now considered to be less effective than combining strategies, at least in the appropriate situation. Medication and surgery, for example, can be used together with meal replacements; preprepared meals can be used with a nutritional education and counseling program; and exercise should always be combined with a hypocaloric diet, which can consist of meal replacements, low energy meals, or a hunger-suppressant drug where indicated.

Making a Distinction Between Short- and Long-Term Loss

It is now well recognized that short-term weight loss is relatively easy. Long-term weight loss maintenance, on the other hand, is more difficult (and possibly dependent on the length of time being obese) but should be the main goal of any effective weight loss program. In the early stages, quick weight losses may be important for increasing motivation to proceed to longer-term benefits. It is much easier to do this at the outset by reducing calorific intake than by increasing exercise output. However, over the longer term, an increase in daily activity has been found to be one of the most significant factors in sustaining losses (Wing and Hill, 2001).

From Diets to a Lifestyle Eating Plan of Reduced Energy Eating

Because of the vagaries of physiological adjustment and the difficulty in trying to psychologically overcome a biological drive (hunger), diets (as in a structured reduction in energy intake) are unlikely to be able to be maintained over the long term. Because "going on" a diet usually results in "coming off" it sooner or later, weight regain is likely. And because of feedback mechanisms, this is likely to exceed that which is lost during the energy restriction phase, causing the common tick-shaped response (quick loss but overcompensated weight gain). Consequently, the modern shift is to a change in the pattern of eating, reducing (in particular) fats and other high energy-dense foods and increasing fiber and foods with low energy density.

Use of Clinically Supervised Meal Replacements

Although they have been around for some time, meal replacements in the form of powdered drinks or food bars have been generally frowned upon because they have not been nutritionally balanced and are often used over the short term only without clinical supervision. Advances in food technology and an increase in clinical knowledge have resulted in an increase in their use under certain conditions (Egger, 2006).

PRACTICAL IMPLICATIONS OF PHYSIOLOGICAL ADJUSTMENT

One of the most obvious implications of physiological adjustment to changes in energy balance is plateaus in weight loss. These are a source of frustration for patients and clinicians alike, many of whom often expect weight loss to occur in a perfect linear fashion. As the great Harvard nutritionist Jean Mayer once said, "Like a wise man will reduce spending when his income is cut, the body reduces the amount of energy it expends when energy intake (food) is reduced."

The difference between the wise person and the body is that energy use, in the form of metabolism, is dropped below that of energy intake in order to reduce the imbalance even further. In other words, a decrease of 10% in energy intake may lead to a decrease of 12–15% in energy expenditure because, like the spender without a loan, the body cannot afford to go into debt.

Plateaus occur as a result of the body's adaptation to the rate of energy intake in relation to energy expenditure. Unfortunately, nobody can say when and for how long a plateau will continue, as this is a highly individual process. It is dependent on a number of factors not yet identified but probably including the time spent being overweight, age, gender, and the actions taken to lose weight. The length of a plateau is difficult to determine because there has been little research. At this stage, and without an evidence base, there are two relevant factors. First, being on a plateau is far from a sign of failure. On the contrary, it indicates the body has adjusted to change and is readapting to these changes. As the average individual in advanced countries is gaining from 1 to 2 g a day, being on a plateau means a patient is winning. Second, as a plateau is a period of adjustment, it makes sense that getting off the plateau will involve change of some kind. Some possibilities are suggested in Table 7.2.

CURRENT STRATEGIES IN WEIGHT LOSS

A strategy is defined as a broad general approach to treatment. We have listed 11 such strategies used in the management of weight control (Table 7.3). Within these strategies there are a number of methods that are more specific ways of achieving successful weight loss. A strategy, for example, may be an exercise approach to weight loss; a method within this could be the use of resistance training. The level of evidence for these strategies is variable, so some will be treated with more detail than others. We have categorized them into strategies with a limited evidence base,

TABLE 7.2 Suggested Ways of Breaking Through a Weight Loss Plateau Through Change

General measures	Take a holiday.
	Go to bed earlier.
	Get up later.
	Go camping.
	Go hiking or bushwalking.
	Try a career change.
	Review and modify eating habits.
Changing energy intake	Change eating patterns.
	Try different drinks.
	Use meal replacements.
	Eat different foods.
	Go low carb (for a while).
	Monitor food intake.
Changing energy expenditure	Try different exercises.
	Monitor food intake.
	Add weights to exercise.
	Increase speed.
	Walk a different route.
	Walk backwards.
	Stand for longer.
	Walk instead of drive.
	Catch public transport.

Reprinted with permission King, N., Caudwell, P., Hopkins, M., et al. August 2006. Plateaus in body weight during weight loss interventions. Proceedings, 10th International Congress on Obesity, Sydney.

those with some evidence, and those with a sound evidence base. A base of lifestyle change is assumed as a prerequisite for success of all of the strategies listed.

Strategies With Limited or No Supporting Evidence

(Most) Over-the-counter Products

Over-the-counter (OTC) weight loss products are defined here as those ingestible substances usually (but not always) derived from herbal or natural

TABLE 7.3 Efficacy of Current Major Weight Loss Strategies

Clinically Proven Rx	~% Weight Loss @ 1 Year	Comment
Surgery	>24%	Often a last resort for the morbidly obese or those with major risks. Expensive but effective. Available within the health system in some countries.
Medication + CBT + MR	15%	Varies with type of medication. Psychological input good for ingrained habits and irrational or counterproductive thinking.
Very low energy diets	14.7%	Often as low as 800 kcal (3360 kJ)/d. Difficult to maintain over the long term. May be useful pre-op.
Very high energy exercise	?	Although high energy users are typically slim, there is little data showing the effects of very high intensity exercise alone on weight loss.
Combination drug (e.g., phentermine/ topiramate)	12.6%	Varies with drug combination. Adverse side effects sometimes shown after wide-scale use. Drug use has to be continuous.
Xenical + low fat diet & phentermine	8.6%	Although Xenical not used much today, the diet + drug combination can be useful for short-term motivation in particular.
Lifestyle alone (Diet + exercise)	8.1%	For better effects lifestyle changes need to underlie all other treatments and in combination with other factors, and not used in isolation.
Meal replacements alone	6%	Inexpensive, balanced way of reducing food input. Needs ongoing clinical support.
Diet alone	4.4%	Unlikely to be maintained over the long term unless built into a lifestyle eating pattern.

Continued

TABLE 7.3 Efficacy of Current Major Weight Loss Strategies—cont'd

Clinically Proven Rx	~% Weight Loss @ 1 Year	Comment
Exercise alone	0–2%	Possible confounding with compensatory eating, particularly if done in small amounts. Best effects over the long term for weight loss maintenance.
Commercial weight loss programs	0–4%	Variability depending on the program. Mainly targeted at women. Best results appear to be in social interaction groups.
Self-help/self-directed diets, books, etc.	?	Inexpensive alternative although outcomes are difficult to assess. Logic suggests that many people must use such an approach.
Pre-prepared meals	?	Can be combined with and assist other techniques. Best use is of nutritionally balanced low energy meals as recommended by food standards groups.
Herbal and other supplement OTC products	–	No proven long-term benefits. May have a placebo effect with lifestyle change over the short term.
Alternative treatments (e.g., hypnotherapy, acupressure, etc.)	–	May be an "add on" to lifestyle change in some people. No proven long-term effectiveness when used alone.

Based on Australian National Clinical Guidelines for Weight Control and Obesity Management in Adults. *CBT*, cognitive behavior therapy; *MR*, meal replacements; *OTC*, over-the-counter.

substances for which weight loss properties are claimed. American figures show that as many as 15% of the general population have used a weight-loss dietary supplement at some point in their lives, with more women reporting use (20.6%) than men (9.7%) (Blanck et al., 2007). Several reviews agree that, despite their widespread use, there is no convincing evidence of efficacy for the vast majority of OTC products (Egger and Thorburn, 2005; Pittler and Ernst, 2004). In addition, dietary supplements, often used for weight loss purposes, may have severe side effects. A recent analysis of the number of emergency care admissions due to adverse events related to dietary supplements indicated that in the United States more than 23,000 visits per year were caused by dietary supplements, most commonly being herbal products. Most adverse effects were cardiovascular, and most of the products causing admittance were for weight loss (Geller et al., 2015).

Despite all this, the category should not necessarily be dismissed totally and some OTC products may benefit from further research (National Institutes of Health, www.ods.od.nih.gov/factsheets/WeightLoss-HealthProfessional/#en12). Nevertheless, early meta-analyses and a current review of supplements have shown a lack of evidence and safety information for any of the available substances often promoted for weight loss (Esteghamati et al., 2015; Pittler and Ernst, 2004).

Alternative Noningestible Treatments

"Alternative" approaches to weight loss include topical treatments such as creams or soaps, body wrapping, aromatherapy, acupuncture/acupressure, electrical stimulation, hypnosis, passive exercise devices, and yoga and other Eastern meditation techniques (Egger and Thorburn, 2005). Minor weight loss benefits have been reported with treatments such as hypnosis, which could have adjunct benefits (even as a placebo) to other lifestyle change interventions. Other treatments such as body wrapping and passive exercise devices have no reliable supporting evidence.

Strategies With Some Supporting Evidence

Commercial Diet-oriented Weight Loss Programs

Data from some countries suggest that up to 13% of women and 5% of men participate in commercial weight loss programs (Latner, 2001). Because of commercial concerns, however, few results of these programs have been reported. In a recent review of 32 major commercial weight loss programs in the United States, researchers from the Johns Hopkins University School of Medicine found that only two—Weight Watchers and Jenny Craig—are backed by reliable data showing sustained weight loss. Other programs show promising results, but the report suggests that more studies are needed to look at long-term outcomes (Gudzune et al., 2015). Although most commercial programs are oriented toward women, limited data indicate that male programs may have a higher success rate than female programs (Egger et al., 1996), although commitment is a limiting factor.

Many claims about the effectiveness of commercial weight loss programs are not supported by adequate head-to-head randomized clinical trials. When such trials have been carried out it seems that Atkins, Weight Watchers, and Zone achieved rather modest and similar weight loss results over 12 months. Surprisingly enough, in spite of the huge amount of money spent on diets of this kind, long-term results are conflicting. In the absence of a superior program it has therefore been suggested that people who want to lose weight should adhere to the diet they best can tolerate over time (Atallah et al., 2014).

In summary, based on the available data, it seems that commercial weight loss programs can lead to modest weight losses and weight loss maintenance in some individuals. It would be speculative, but not unreasonable, to propose that existing commercial programs have most appeal to women who prefer a structured program with ongoing feedback through social or individual support.

Self-Help

For our purposes, self-help is considered to be those situations where no other structured person-to-person contact is made on an ongoing basis. By its nature, this is difficult to evaluate, but would include self-motivated lifestyle change, computer-assisted interventions, and packaged programs. Self-help has the advantages of increasing empowerment, self-efficacy, and self-esteem. These can effect positive change to self-reliance, rated by individuals as one of the most effective strategies for affecting obesity (Furnham, 1994). Using the American National Weight Control Registry, Latner (2001) claimed that up to 45% of those who have lost weight and kept it off may have done so through self-help. Because diet books are ubiquitous, it must be assumed that these are a common form of self-help, although success is difficult to assess.

Recently, attempts have been made to use the Internet as a self-help tool. Earlier programs have shown modest success but these have developed into more effective tools (Svensson et al., 2014). Such programs typically contain a self-monitoring section for detailing habits and physical activity, an interactive component used to design a personal program aimed at a reasonable weight loss over time, feedback, and coaching. Often, recipes, which can be adapted or modified, are included. Although the dropout rate in such programs is high, it is reasonable to assume that they provide a very cost-effective way to treat obesity. They are cheap, available 24 hours a day, require a minimum of technical knowledge, and are often seen as fun (Svensson et al., 2014). Swedish data (Svensson et al., 2014) suggest that those who chose an Internet program were 85% middle-aged women with an initial BMI of 29 kg/m^2 who lost about 3 kg over the first 3-month period. Older people did just as well as younger participants. Since more and more people have access to the Internet and can master the programs, the Internet has developed into a promising tool.

AT THE RETIRED SUPERHEROES CONVENTION.

STRATEGIES WITH GOOD SUPPORTING EVIDENCE

Exercise-Based Programs

Physical activity and a hypocaloric diet (+other lifestyle changes) form the basis of all effective weight loss programs and therefore must make up the generic component of any successful weight loss initiative. However, where exercise alone is the main consideration, the emphasis can vary according to requirement. There is little doubt that it is easier to modify energy balance in the short term by reducing EI than by increasing EE. A big person is more likely to be able to reduce EI [e.g., 4400 kJ (1,100 kcal) per day in a 13,200–17,600 kJ (3–4000 kcal) per day diet] than to increase EE by the same amount (e.g., by walking an extra 10–12 km/day). Also, as efficiency increases (as a result of the improvements in aerobic capacity and decreases in body weight with regular exercise), absolute EE for a set task will decline, thereby reducing the effectiveness of a set unit of physical exertion.

Despite sometimes limited effects on body mass index or weight loss, exercise is associated with beneficial changes in body composition and improvements in insulin sensitivity. Gains in muscle tissue (muscle tissue weighs more than fat tissue because of its higher water content) may lead to perceived "weight gain" or "weight stagnation" by the patient. Adults with excess body fat may benefit particularly from resistance exercise, which increases lean body mass in normal-fat participants and reduced fat mass in overweight and obese adults (Drenowatz et al., 2015).

In a study examining prospectively whether weight training, moderate to vigorous aerobic activity, and replacement of one activity for another were associated with favorable changes in waist circumference and body weight, weight training has the strongest association with less waist circumference increase (Mekary et al., 2015).

Increases in physical activity have been found to be of particular value in maintenance, but initial weight loss is probably the best predictor for success in obesity treatment (Elfhag and Rossner, 2010). Studies show that exercise prescription (with diet) for weight loss should be for a minimum of around 150 min (2.5 h) or 11,000 kJ (2600 kcals) per week (Jakicic et al., 2001). Using a pedometer or movement sensor, this equates to around 10,000 steps/day in an 80 kg person. Exercise alone is unlikely to be effective for weight loss until around 200–300 min (3–5 h) or 15,400 kJ (3500 kcals) per week (approximately 12,500 steps) is expended, with effects proportional to exercise volume. Data from the US National Weight Control Registry's 10-year follow-up suggests that exercise for weight loss maintenance in the post-obese may need to be at about 60–80 min/day of moderate activity, combined with a hypocaloric diet (Thomas et al., 2014) and that this aids in long-term weight loss maintenance.

In summary, it seems that while exercise is a necessary component of any weight loss program, it is usually not sufficient (without diet) unless carried out at levels that are unlikely to occur as part of normal modern daily life and in obese individuals. Exercise programs through fitness or health centers that lack a nutritional input and/or are at a level insufficient to compensate for the negative energy balance inherent in a modern obesogenic environment are unlikely to have a lasting impact on weight control.

Counseling and Behavioral Approaches

Counseling is the process of providing advice and guidance to a patient. This includes several behavioral and cognitive-behavioral approaches that, when integrated with dietary and exercise advice, have been shown to add to the long-term effectiveness of a program (Wing, 2007). These are based on the idea that eating and exercise behaviors are learned and therefore can be modified through such techniques as self-monitoring, stimulus control, problem solving, contingency management, cognitive restructuring, social support, and stress management. Improvements in these techniques in recent years have added to their value as part of an integrated weight loss program, and a recent review of the data suggests that behavioral therapy does add to the benefits of other forms of weight loss treatment (NHMRC, 2003) and thus should be a part of all weight loss interventions.

The classical counseling paradigm for weight management based on a 1:1 (counselor: patient) situation has recently been questioned (Noffsinger, 2012). A review of weight loss in this type of clinical situation, for example, suggests up to 26 sessions with a doctor may be required for a 3–7 kg average weight loss (Ard, 2015), a situation that is obviously not cost-effective. Trials of shared medical appointments, involving up to 15 people in a consultation with a doctor and health care facilitator, offer a potential new approach to this form of counseling, which may help overcome some of these problems (Egger et al., 2014).

Preprepared Low Energy Meals

There are a number of commercially available, preprepared, low energy meals available through retail outlets or direct. Preprepared meals, which are usually frozen for reheating, are energy controlled for a meal and nutritionally balanced, enabling users to objectify their energy intake. Such meals may be used as a total or partial substitute for ad lib eating. As such, they are useful not only for initial weight loss but also for use by a skilled clinician for "re-feeding" after more severe caloric restriction. A number of studies have shown positive benefits from preprepared meals (Metz et al., 2000), although their use in combination with other strategies has not yet been widely reported. They are most likely to appeal to individuals with time limitations or a lack of culinary skills or nutrition knowledge.

Meal Replacements (Very Low Energy Diets)

Meal replacements in the form of very low energy diets provide balanced nutrition in powdered form mixed as a drink or shake. These can be either low [e.g., fewer than 5280 kJ (1200 kcals) per day] or very low [e.g., fewer than 3520 kJ (800 kcals) per day] in energy and are usually used as a replacement for some, rather than all, meals or in combination with preprepared meals. Meal replacements can result in big, quick losses but can also be followed by rebound weight gains on their cessation, unless a maintenance program follows immediately afterward.

They have been discouraged for personal use but can be effective with close clinical supervision, particularly to increase motivation in some patients. Established recent research suggests that meal replacements are safe, can produce good losses, and can help reduce comorbidities (NHMRC, 2003). One review suggested that almost everyone in the modern environment could benefit from the regular use of a meal replacement, because of excessive daily food and energy intake (Egger, 2006).

What Can Be Done in a Brief Consultation?

- ☐ If appropriate, and agreed to by the patient, take anthropometric measures (waist circumference, weight, bioimpedance).
- ☐ Calculate and discuss ideal weight goals, using the formula given in professional resources.
- ☐ Ask whether the patient thinks diet, inactivity, or both are adding to weight gain.
- ☐ Ask about stress in the patient's life and the general reaction to this. For example, does she or he eat more or less or move more or less when stressed?
- ☐ Suggest the purchase or loan of a movement sensor (e.g., "Fitbit") for measuring baseline activity levels (steps) over a week.
- ☐ Provide information for the patient to complete the Diet, Activity and Behavior Questionnaire (DAB-Q) (see www.lifestylemedicine.com/DAB-Q).
- ☐ Explain the strategies available and the need for lifestyle modification in any weight loss.
- ☐ Explain the chronic nature of obesity, the long-term basis of any treatment, and the importance of maintenance in contrast to initial loss.

Medication

Although there is a considerable literature on prescribed weight loss medications (Jones and Bloom, 2015), these come and go and there are currently only a few options available, with many new versions currently being tested and some being available in some countries but not others. Drug treatment of obesity to date has not become a lasting success. Drugs that were initially considered effective have been withdrawn for safety reasons. New drugs are being developed but this process is slow. The situation differs between countries and drugs, for example, which are available in the United States have not been accepted in Europe. Thus it is difficult to assess what role pharmacotherapy has in obesity management at present. However, if and when such drugs are developed, the principle for their use remains unchanged and they should always be considered an adjunct to other basic strategies.

Orlistat (Xenical 120 mg × 3) is a gastric lipase inhibitor, which reduces energy intake through the partial elimination of dietary fats. For this reason it can be used as a "learning drug" in individuals who are unaware of the fat content of foods. About 30% of the fat consumed will be lost in the feces. If patients adhere to a low-fat diet, that is, less than approximately 67 g/day, this fat loss will not cause any clinical problems and the patient will lose weight, both by fat restriction and the modest excretion of fat consumed. With higher fat intake, steatorrhea develops, which is unpleasant and reinforces better dietary habits. In that sense, orlistat has been termed an "Antabuse" for fat overconsumption, because patients quickly learn how to adapt. Orlistat is also the active substance in Alli (60 mg × 3), which is sold as an OTC product in many countries. The weight loss effects are about 60–70% of those with the 120 mg × 3 dose.

Phentermine (Duromine) is an adrenergic that is only available for short-term (three months) use. Phentermine is especially useful where quick weight loss is desired to increase motivation or to help break through plateaus. There are side effects of dysphoria and tachycardia in a small number of people, but these occur immediately and can be eliminated by discontinuing the drug. Close clinical supervision is also advisable on withdrawal to prevent a weight rebound.

Lorcaserin (Belviq) was approved in the United States in 2012 for long-term weight management. Lorcaserin stimulates 5-HT (serotonin) receptors on anorectic POMC neurons. However, it was developed as a selective agonist of the 5-HT2C receptor to avoid 5-HT2B-mediated valvulopathies, which afflicted fenfluramine, an earlier agent that has since been taken off the market. In phase 3 trials, lorcaserin achieved average weight loss of 3.0–3.6% better than placebo, with 2.3 times as many patients losing at least 5% body weight in the treatment groups. It is not available in Europe, however, because of possible heart valve problems and psychiatric morbidity.

The combination of **Phentermine** and **Topiramate** (Qysmia) can have an appetite suppressing and energy stimulating effect with modest reductions in blood pressure and an impressive placebo-subtracted 6.6% weight loss at the rate of 7.5 mg phentermine/46 mg Topiramate. Similarly, the combination of **Buproprion** and **Naltrexone** (Contrave in the United States, Mysimba in Europe), both antidrug addictive medications, has shown potential as a hunger suppressant with 3.2–5.2% greater weight loss than placebo, over one year.

Liraglutide (Saxenda) is the first of the GLP-1 mimetics to be granted an obesity indication. GLP-1 is an endogenous incretin, released by intestinal L cells in response to nutrient ingestion, which enhances glucose-stimulated insulin release by pancreatic beta cells and acts on satiety pathways, including hypothalamic POMC neurons, to reduce food intake. Current studies show a 4–5.4% weight loss above placebo over a 12-month period.

Some antidepressant (SSRIs) medications have weight loss effects while others tend to cause weight gain, possibly as a result of the patient's consumptive reaction to depression (i.e., those who eat more when depressed will lose weight when on antidepressants and those who eat less may gain weight). Antidepressants may have a particular benefit in relapse prevention.

In summary, prescribed medication is useful but should not be considered as monotherapy in the absence of lifestyle modification. And while new drugs regularly cause excitement in the popular media, the cupboard remains relatively baer when it comes to effective, safe pharmaceutical preparations for weight loss.

Surgery

In terms of potential weight loss, obesity surgery is clearly the most effective strategy available (NHMRC, 2003). Recent data from the Swedish Obesity Study specify that both reduced cardiovascular mortality and a lower incidence of cardiovascular events are observed after a median follow-up of 15 years (Sjöström et al., 2012).

Most bariatric surgery has been performed laparoscopically, although this is now changing more to other forms of surgery. There are several types of bariatric surgery, the main forms of which (gastric banding, vertical-banded gastroscopy, gastric sleeve, and gastric bypass) involve the formation of a gastric pouch below the gastroesophageal junction. This was thought to reduce weight by restricting food ingested, but recent studies have shown that physical stimulation may induce hormonal hunger suppressant mediators. Laparoscopic banding ("lap banding") has been one of the most commonly used procedures, but is now less in favor due to its high postoperative requirements ("after sales service").

Because of its intrusive nature, expense, and possible (although now extremely slight) risks, obesity surgery has been seen as a "last resort" for intractable and life-threatening cases (NHMRC, 2003). However, improvements in operative techniques have made it a preferred option in some less-morbid cases. Postoperative treatment (and band adjustment in the case of lap banding) is the key to success, and, like all other treatments, surgery is only effective when combined with lifestyle changes.

Therapeutic Lifestyle Changes

Lifestyle modification remains the focal point of all effective weight control strategies. It involves changing energy intake and/or energy expenditure but may also involve looking at why an imbalance has occurred. For best effect, lifestyle modification usually involves some of the other strategies outlined here, but patients need to be made aware that permanently changing some aspect of their current lifestyle is a necessary precursor to effective long-term weight loss. Specific lifestyle modifications are discussed in more detail in subsequent chapters.

Combining Strategies

The practice of weight loss management requires creative interventions, as well as the selection of evidence-based strategies. Most effective treatments are likely to involve combining strategies and matching the right strategy to the characteristics of the patient. A literature review by Elfhag and Rossner (2005) produced a list of factors that influence weight loss and regain and these are shown in Table 7.4.

A key component of treatment is recognition of two main phases. The first is immediate weight loss and the second is long-term weight loss maintenance, and these invariably require different treatment emphases. The use of strategy mixes may also change between these phases. For example, a patient needing quick success for motivational purposes may be prescribed meal replacements or a prescription drug, then be gradually "re-fed" with preprepared meals or a balanced low energy diet with a graduated exercise program. Counseling and lifestyle modification would be continued throughout. A patient with less need for such a quick loss may be simply given an information-based program of diet and exercise. Packaged programs with clinical backup can reduce the repetitive nature of discussing details with the patient and the time required of the clinician.

Although long-term weight loss maintenance is difficult to achieve, particularly in an "obesogenic" environment, 10 year follow-up data from the US National Weight Control Registry (albeit using a self-selected group) show that it is possible. Mean weight loss at baseline was 31.3 kg (a loss of at least 13.5 kg was necessary to be included in the registry), 23.8 kg after 5 years and 23.1 kg after 10 years (Thomas et al., 2014). The main lifestyle habits associated with maintained loss were:

1. Eating breakfast
2. An ongoing pattern of daily exercise
3. Low energy diet
4. Effective stress management techniques
5. Regular self-monitoring (particularly weighing)

TABLE 7.4 Factors Associated With Weight Loss Maintenance and Regain After Intentional Weight Loss

Weight Maintenance	Weight Regain
An achieved weight loss goal	Attribution of obesity to medical factors
More initial weight loss	Perceiving barriers for weight loss behaviors
Physically active lifestyle	History of weight cycling
Regular meal rhythm	Sedentary lifestyle
Breakfast eating	Disinhibited eating
Less fat, more healthy foods	More hunger
Reduced frequency of snacks	Binge eating
Flexible control over eating	Eating in response to negative emotions/stress
Self-monitoring	Psychosocial stressors
Coping capacity	Lack of social support
Capacity to handle cravings	More passive reactions to problems
Self-efficacy	Poor coping strategies
Autonomy	Lack of self-confidence
"Healthy narcissism"	Psychopathology
More confidence as a motivator	Medical reasons as a motivator
Stability in life	Dichotomous thinking
Capacity for close reasoning	

Reprinted with permission from Elfhag, K., Rossner, S., 2005. Who succeeds in maintaining weight loss? A conceptual review of factors associated with weight loss maintenance and weight regain. Obes. Rev. 13 (6), 1070–1076.

Future Perspectives

While it's true that most weight loss treatments arise serendipitously from treatments for other ailments (particularly medications), there are some possible treatments on the horizon that may hold future promise. Fecal microbial transplants (FMT), based on restoring gut microbiosis, have now been used successfully for certain problems such as *Clostridium difficile*, a medical problem resulting from overuse of antibiotics in hospital situations. In limited animal studies, FMT using transplants from lean, healthy rats have been shown to reduce weight in obese transplant recipients, and vice versa (Duca et al., 2014). For ethical reasons, studies based on restoring the gut to a healthy, lean commensal have not yet been carried out in humans, but are already planned for the future. Using a similar rationale, prebiotics and probiotics may function like an FMT in restoring healthy gut function that could aid weight loss, although results to date do not strongly support this. Better targeted biotic supplements with a better understanding of the gut might help change this in the long run (Park and Bae, 2015). Deep brain stimulation (DBS) is a neuromodular process that has been used for treating Parkinson's disease, cluster headaches, and other problems and which has been found to have weight loss side effects. A number of trials are currently underway to test DBS as a specific treatment for obesity (Kumar et al., 2015).

OBESITY: THE "CANARY IN THE MINESHAFT"

Recent developments in obesity research suggest that while obesity is often associated with the metaflammatory processes referred to in Chapter 3, these may occur in the absence of obesity, but often in the presence of lifestyle and environmental factors that can cause obesity. Overnutrition, inactivity, inadequate sleep, smoking, stress, and environmental pollutants can all cause increases in inflammatory markers and hence chronic disease, sometimes in the presence of, but also in the absence of, obesity. This suggests that while obesity may lead to increased health risks in some people, this is neither a necessary nor a sufficient condition. Obesity may be just a "canary in the mineshaft." It could signal that something may be wrong in an individual, therefore requiring further investigation. It should not be assumed, however, that being lean implies a lack of metabolic health problems, and hence further questions need to be asked about the lifestyle factors that may independently lead to disease. When obesity is in epidemic proportions, as it is in most Western countries today, it suggests that something definitely is wrong in the broader environment, and that this should be looked at as part of a total approach to chronic disease prevention and treatment (i.e., see Chapter 30).

SUMMARY

Many medical problems (such as appendicitis or pneumonia) call for a single treatment option (appendectomy, antibiotics), which results in a definite cure of that particular medical problem. Chronic diseases with a lifestyle-based etiology do not have a single treatment option. Their cause is also often not as apparent as it may seem, with the broader environment playing a significant part. The skill of the clinician, therefore, is in matching the available, evidence-based options to the patient and being aware of the need for public health as well as clinical involvement. As chronic diseases now account for >60% of the world's deaths and 47% of the global burden of disease, clinical management is increasingly more likely to require this kind of "mixing and matching." Obesity treatment could benefit from a reconceptualization in this light.

Practice Tips

Weight Loss

	Medical Practitioner	Practice Nurse/Allied Health Professional
Assess	• Whether weight loss has been considered • Comorbidities • Family history of weight/comorbidities • Medication effects on weight • Ontraindications for treatment options • Whether surgery or medication may be appropriate (e.g., triage into levels of treatment)	• Body shape • Anthropometric measures • Stage of readiness to change • Goal weights/waist • Revious history of weight gain/loss • Smoking history • Diet and exercise (DAB-Q test; www.lifestylemedcine.com.au/DAB-Q) • Treatment options • Other factors (e.g., stress, depression, early experience) • Environmental influences
Assist	• By treating comorbidities • By identifying potential short- and long-term goals • By reviewing progress of drug therapy for weight loss • By reviewing ongoing special dietetic programs such as meal replacements • By using templates as appropriate	• By educating about eating • Increases in basic physical activity; use a pedometer for monitoring steps; use appropriate tactics to move through stages of readiness to change • Time planning • By using templates as appropriate
Arrange	• Referral for comorbidities where appropriate • List of other health professionals for a care plan	• Discussion of motivational tactics with other involved health professionals • Referral to a personal trainer, nutritionist, or life coach if appropriate • Contact with self-help groups if appropriate

Key Points

Clinical Management

- Do not automatically weigh and/or measure (at least at the start of a consultation) because this may impact negatively on some people.
- Try to determine the main contributing cause(s) to energy imbalance (e.g., food intake, energy expenditure).
- Within these, look for key factors (e.g., excess sweet or fat foods, portion size problems, fast foods only, emotional eating, night eating, sugar-sweetened fluids, excessive computer use, more than 15 hours of TV watching per week, weekend laziness).
- Consider/dismiss the impact of psychological factors such as stress, early abuse, or compulsive/binge eating.
- Use tools such as pedometers, questionnaires, or smartphone applications to encourage self-monitoring and feedback (see Chapter 4).
- Encourage regular visits (e.g., fortnightly/monthly), at least in the early stages, to assist with progress.
- Aim for long-term weight loss and maintenance but use short-term losses to motivate for ongoing lifestyle learning.

REFERENCES

Ard, J., 2015. Obesity in the US: what is the best role for primary care? Brit. Med. J. 350, g7846.

Arsenault, B.J., Lemieux, I., Despres, J.-P., Wareham, N.J., Kastelein, J.J.P., Khaw, K.-T., Boekholdt, S.M., 2010. The hypertriglyceridemic-waist phenotype and the risk of coronary artery disease: results from the EPIC-Norfolk Prospective Population Study. Can. Med. Ass. J. 182 (13), 1427–1432.

Atallah, R., et al., 2014. Circ. Cardiovasc. Qual. Outcomes 7.

Blanck, H.M., Serdula, M.K., Gillespie, C., Galuska, D.A., Sharpe, P.A., Conway, J.M., et al., 2007. Use of nonprescription dietary supplements for weight loss is common among Americans. J. Am. Diet. Assoc. 107, 441–447.

Dhurandar, N.V., Schoeller, D., Brown, A.W., Heymsfield, S.B., Thomas, D., Sørensen, T.I., Speakman, J.R., Jeansonne, M., Allison, D.B., 2015. Energy balance measurement: when something is not better than nothing. Int. J. Obes. 39 (7), 1109–1113.

Drenowatz, C., Hand, G.A., Sagner, M., Shook, R.P., Burgess, S., Blair, S.N., December, 2015. The prospective association between different types of exercise and body composition. Med. Sci. Sports Exerc. 47 (12), 2535–2541.

Duca, F.A., Sakar, Y., Lepage, P., Devime, F., Langellar, B., Dore, J., Covasa, M., 2014. Replication of obesity and associated signaling pathways through transfer of microbiota from obese – prone rats. Diabetes 63 (5), 1624–1636.

Egger, G., 2006. Are meal replacements effective as a clinical tool for weight loss? Med. J. Aust. 184 (2), 52–53.

Egger, G., Binns, A., 2001. The Experts Weight Loss Guide. Allen and Unwin, Sydney.

Egger, G., Swinburn, B., 1996. An ecological model for understanding the obesity pandemic. Brit. Med. J. 20, 227–231.

Egger, G., Thorburn, A., 2005. Environmental and policy approaches: alternative methods of dealing with obesity. In: Kopelman, P.G., Caterson, I.D., Dietz, W.H. (Eds.), Clinical Obesity in Adults and Children. Blackwell Publishing, London.

Egger, G., Bolton, A., O'Neill, M., et al., 1996. Effectiveness of an abdominal obesity reduction program in men: the GutBuster "waist loss" program. Int. J. Obes. 20, 227–235.

Egger, G., Dixon, J., Egger, G., Dixon, J., 2014. Beyond obesity and lifestyle: a review of 21st Century chronic disease determinants. BioMed Res. Int. Article ID: 731685. http://dx.doi.org/10.1155/2014/731685.

Elfhag, K., Rossner, S., 2005. Who succeeds in maintaining weight loss? A conceptual review of factors associated with weight loss maintenance and weight regain. Obes. Rev. 13 (6), 1070–1076.

Elfhag, K., Rossner, S., 2010. Initial weight loss is the best predictor for success in obesity treatment and sociodemographic liabilities increase risk for drop-out. Patient Educ. Couns. 79, 361–366.

Esteghamati, A., Mazaheri, T., Vahidi Rad, M., Noshad, S., April 20, 2015. Complementary and alternative medicine for the treatment of obesity: a critical review. Int. J. Endocrinol. Metab. 13 (2), e19678.

Foresight Report, 2007. Tackling Obesities: Future Choices. Government Office for Science, UK.

Furnham, A., 1994. Explaining health and illness: lay perceptions on current and future health, the causes of illness and the nature of recovery. Soc. Sci. Med. 39 (5), 715–725.

Geller, A.I., et al., 2015. Emergency department visits for adverse events related to dietary supplements. NEJM 373, 1531–1540.

Gudzune, K.A., Doshi, R.S., Mehta, A.K., 2015. Efficacy of commercial weight-loss programs. an updated systematic review. Ann. Intern. Med. 162, 501–512.

Hamed, T., 2009. Thinking in Circles about Obesity. Springer, NY.

Jakicic, J.M., Clark, K., Coleman, E., et al., 2001. Appropriate intervention strategies for weight loss and prevention of weight regain for adults. Med. Sci. Sports Exerc. 33 (12), 2145–2156.

Jones, B.J., Bloom, S.R., 2015. The new era of drug therapy for obesity: the evidence and the expectations. Drugs. http://dx.doi.org/10.1007/s40265-015-0410-1.

King, N., Caudwell, P., Hopkins, M., et al., August 2006. Plateaus in body weight during weight loss interventions. In: Proceedings, 10th International Congress on Obesity Sydney.

Kumar, R., Simpson, C.V., Froelich, C.A., Baughman, B.C., Gienapp, A.J., Sillay, K.A., July 2015. Obesity and deep brain stimulation: an overview. Ann. Neurosci. 22 (3), 181–188.

Kushner, R., Kuchner, N., 2008. Dr Kushner's Personality Type Diet. iUniverse, Bloomington, IN.

Latner, J., 2001. Self-help in the long-term treatment of obesity. Obes. Rev. 2, 87–97.

Mekary, R.A., Grøntved, A., Despres, J.P., De Moura, L.P., Asgarzadeh, M., Willett, W.C., Rimm, E.B., Giovannucci, E., Hu, F.B., February 2015. Weight training, aerobic physical activities, and long-term waist circumference change in men. Obesity (Silver Spring) 23 (2), 461–467.

Metz, J.A., Stern, J.S., Kris-Etherton, P., et al., 2000. A randomized trial of improved weight loss with a prepared meal plan in overweight and obese patients: impact on cardiovascular risk reduction. Arch. Intern. Med. 160 (14), 2150–2158.

Noffsinger, E.B., 2012. The ABCs of group visits: an implementation manual for your practice. Springer, New York.

NHMRC (National Health and Medical Research Council), 2013. Clinical Practice Guidelines for the Management of Overweight and Obesity in Adults, Adolescents and Children in Australia. https://www.nhmrc.gov.au/guidelines-publications/n57.

Park, S., Bae, J.H., July 2015. Probiotics for weight loss: a systematic review and meta-analysis. Nutr. Res. 35 (7), 566–575.

Phillips, C., Perry, I., 2013. Does inflammation determine metabolic status in obese and nonobese adults? J. Clin. Endocrin. Metab. 98. http://dx.doi.org/10.1210/jc.2013-2038.

Pittler, M.H., Ernst, E., 2004. Dietary supplements for body-weight reduction: a systematic review. Am. J. Clin. Nutr. 79 (4), 529–536.

Sharma, A., Kushner, R., 2009. A proposed clinical staging system for obesity. Int. J. Obes. 33, 289–295.

Sjöström, et al., 2012. Bariatric surgery and long-term cardio-vascular events. JAMA 307, 56–65.

Svensson, M., Hult, M., van der Mark, M., Grotta, A., Jonasson, J., von Hausswolff-Juhlin, Y., 2014. The change in eating behaviors in a web-based weight loss program: a longitudinal analysis of study completers. J. Med. Internet Res. 16 (11), e234.

Swinburn, B., Egger, G., 2004. The runaway weight gain train: too many accelerators, not enough brakes. Br. Med. J. 329 (7468), 736–739.

Thirlby, R.C., Randall, J., 2002. A genetic "obesity risk index" for patients with morbid obesity. Obes. Surg. 12 (1), 25–29.

Thomas, G.A., Bond, D.S., Phelan, S., Hill, J.O., Wing, R., 2014. Weight-loss maintenance for 10 years in the National Weight Control Registry. Am. J. Prev. Med. 46 (1), 17–23.

Vague, J., 1947. La différentiation sexuelle, facteur déterminant des formes de l'obesité. Presse Méd 30, 339.

Wing, R.R., 2007. Behavioural Approaches to the Treatment of Obesity. In: Bray, G.A., Bouchard, C. (Eds.), Handbook of Obesity: Clinical Applications, second ed. Marcel Dekker Inc., NY, pp. 147–168.

Wing, R., Hill, J.O., 2001. Successful weight loss maintenance. Annu. Rev. Nutr. 21, 323–341.

World Health Organization, 2016. www.who.int.

Wu, F.-Z., Wu, C.W., Kuo, P.-L., Wu, M.-T., 2016. Differential impacts of cardiac and abdominal ectopic fat deposits on cardiometabolic risk stratification. BMC Cardiovasc. Disord. 16 (20), 1–9.

Professional Resources

Measuring Body Fat and Fat Loss

Anthropometric Measures

Weight	Although a necessary measure, this does not always reflect body fat.
BMI (weight/height^2)	Not a highly specific individual measure alone. Can be combined with waist circumference and body fat by other methods. Can discriminate against some muscular individuals.
Waist circumference (WC)	A good measure in itself of metabolic risk. Can also be combined with other measures. Usually taken at midpoint between the bottom rib and the iliac crest but the umbilicus may be better for males.
Ideal weight	Requires two simple measures and a calculator. The measures are weight, as measured by ordinary scales, and percentage body fat, as measured by bioimpedance analysis scales (these can be purchased economically from www.lifestylemedicine.com.au; they also measure weight).

The formula for ideal weight is:
(lean body mass [kg])/(1- percentage ideal body fat in decimals)
where,
lean body mass = weight – fat weight in kg
fat weight = weight in kg – percentage body fat

Ideal body fat percentage is 12–24% (men) or 15–35% (women). Knowing the variation in ideal body fat makes it possible to set different ideal weight goals, depending on starting body fat.

Example: A 100 kg (220 lb) man who is 35% fat will have 100 kg (220 lb) × 0.35 = 35 kg (77 lb) of fat weight and therefore 100 kg (220 lb) – 35 = 65 kg (143 lb) lean body mass.

The ultimate goal should be to get to the upper limit of ideal fat % (i.e., 24% for men and 35% for women), but this can be done in three stages as shown for this 100 kg man with 35% fat:

Goal 1 may be to get to 30% fat; hence, goal weight will be 65/(1 – 0.3) = 92.8 kg (204 lb).
Goal 2 may be to get to 27% fat; hence goal weight will be 65/(1 – 0.27) = 89.0 kg (196 lb).
Goal 3 should be to get to 24% fat; hence goal weight will be 65/(1 – 0.24) = 85.5 kg (188 lb).
Goals 1, 2, and 3 might be short-term, medium-term, and long-term goals.
Using this formula, goal weight is not only dynamic but is based on fat rather than weight, which is a less important measure.

Measures of Genetic Involvement in Obesity

Genetics are often used as an excuse for overweight. The following test, developed by US surgeons Thirlby and Randall (2002), provides an indication of whether genetics may be a factor in obesity.

Continued

Professional Resources—cont'd

Patient Instructions

Answer the following four questions to see whether genetics can be blamed for your body weight. For measurement purposes, the term *overweight* means having a Body Mass Index or BMI over 25, *obese* means a BMI over 30, and *very obese* a BMI over 40, where BMI is calculated by taking your weight (measured in kg) and dividing by the square of your height (measured in meters).

1. As far as you know, were either or both of your parents overweight or very overweight for most of their lives?

	Points	
	Obese	**Very Obese**
Neither/don't know/no	0	0
Yes, one parent	7	14
Yes, both parents	14	28

2. Do you have any first-degree relatives who have been obese for most of their lives? Score two points for every obese first-degree relative up to a maximum of 10 points.
3. How would you describe the average BMI of your siblings?

	Points
No siblings obese (BMI,30)	0
Average sibling obese (BMI,30)	6
Average sibling very obese (BMI,40)	12

4. When did you first become overweight and/or obese?

	Overweight	**Obese**
Never	0	0
Before age 10	20	30
Before age 20	10	20
Before age 30	5	10

Interpreting your score:

<20: Your weight problem does not appear to be significantly genetically related. This means it is related to lifestyle and therefore should be quite easy to solve if you are committed to do so.

20–50: There appears to be a moderate hereditary component to your weight problem. This means you may find it a little harder to lose fat than some of your friends. You may need help from a dietitian but your problems should not be too difficult to overcome.

30–100: There appears to be a significant hereditary component to your weight problem. This means you may need special help and closer attention from a dietitian. With the proper approach and a long-range plan, you should be able to overcome your bad start.

Section II

LIFESTYLE AND ENVIRONMENTAL DETERMINANTS OF CHRONIC DISEASE

Chapter 8

Nutrition for the Nondietitian

Joanna McMillan Price, Garry Egger

Research about diet and nutrition seems to contradict itself with aggravating regularity.

Walter Willet, 2005

INTRODUCTION: THE CONFUSION OF NUTRITION

Peruse the pages of any mix of popular media and you will consistently find one thing: a lack of consistency on the supposed health benefits of different aspects of nutrition. One week we are told that coffee causes cancer, the next that it prevents it. All fat was once thought to be dangerous; now we are advised to eat more of some types to prevent heart disease. Dairy products were thought to be fattening; now they are actually promoted for weight loss.

So what is going on? Why is it so hard for scientists to reach a definitive conclusion on good nutrition? And how do we know that what we are promoting now is not likely to change in the future?

THE SCIENCE OF NUTRITION

Nutrition is a comparatively young science, with much to be discovered. Because of the wide range of ingredients in foods and the multidimensional physiological responses to this, research can be tedious and demanding. Progress is made in stages, with inevitable controversies in small issues but a gradual evolution of understanding of major health-related factors.

To discuss nutrition and health in a single chapter in the context of lifestyle-related chronic diseases is therefore a daunting task. Consequently, we have chosen to focus only on those aspects of nutrition that are relevant to metabolic health and, in particular, obesity and weight control (and the extent to which these influence other aspects of metabolic health). Even here, we cannot hope to cover the full range of information available, so we have chosen to concentrate on those new and practical findings that have implications for lifestyle medicine. Again, this is not meant as a substitute for competent professional advice but to complement such information.

Lifestyle Medicine. http://dx.doi.org/10.1016/B978-0-12-810401-9.00008-5

We look first at calorific volume in food and the factors contributing to this, then turn to new findings relating to specific aspects of the major nutrients and the implications of these findings for metabolic health. We have assumed a knowledge of nutritional components, and so do not elaborate on the basics. More detailed recommended nutritional reading is contained in the References and Professional resources. Energy intake from fluids and the influence of eating behaviors are considered in subsequent chapters.

THE CONCEPT OF (ENERGY) VOLUME

It has long been known, at least for animals, that longevity is increased with relatively low total energy intake over the course of a lifetime. Why this may be so is not clear. Because of ethical considerations in doing the research, it is also not clear whether the same applies to humans, although the indications are that it does. Several hypotheses have been proposed as to why this may be so (Civitarese et al., 2007; Morgan et al., 2007). It is also not known whether a restriction of all or just some types of calories may be important (Heilbronn and Ravussin, 2005). Nevertheless, while there are diet books on just about every component of nutrition, the truth is that the most significant component of nutrition for weight control (and metabolic health) is its (calorific) volume (Rolls, 2007).

Volume, as outlined in Chapter 7, in energy input is made up of three components:

energy density (how rich) × **portion size** (how much) × **frequency of eating** (how often)

This determines total energy input and, therefore, potential energy imbalance.

Energy Density

In the past, the main concern for weight control has been dietary fat in food. This was based on the notion that fat contains 9 kcal/g (38 kJ/g) whereas carbohydrate and protein are less than half of this (alcohol, which is 7 kcal/g or 30 kJ/g, has a more confusing story and is discussed in Chapter 9). Hence, while reducing fat in a diet can significantly reduce total energy input more than an equivalent weight reduction of other nutrients, this is not the only component that determines total energy and it can alter the metabolic response to food.

Food manufacturers in the 1990s capitalized on science's obsession with fat by promoting products as being a particular percentage "fat free," having replaced the fat with sugar and other carbohydrates. Paradoxically, the end result was often an increase in calories per gram or energy density (ED) of the food. ED, it has recently been realized, is the most significant component of energy intake, and attempts are being made to quantify levels that may be recommended for weight loss and good health.

Based on figures by Dr. Barbara Rolls and colleagues from Penn State University (Rolls et al., 2006), an upper limit cutoff of around 3 kcals/g (approximately 13 kJ/g) has been determined for those with a weight concern (Cameron-Smith and Egger, 2002). Other cutoffs are shown, with some food examples, in Table 8.1. ED can be easily calculated from food labels by dividing the energy per 100g by 100. ED in fluids is related to mL of fluid, with recommendations for cutoffs of this given in Table 9.1.

The principal factors decreasing ED in foods are fiber and water; the principal factors increasing it are fats and sugars. Hence, foods such as fruits, vegetables, pastas, and cereal usually have a low ED, whereas full cream dairy products, many processed foods, and those with added sugars, have a high ED.

TABLE 8.1 Recommended Energy Density (ED) Cutoffs for Food Intake

Energy Density	Measure	Recommendation	Examples
Low	<7.5 kJ (1.8 kcals)/g	Eat freely	Fruit, vegetables, cereals, baked beans, several fish, bread, pasta, cereal grains, porridge, lean meats (e.g., kangaroo, venison, wild meats)
Medium	7.6 kJ (1.81 kcals)/g–13 kJ (3 kcals)/g	Eat sparingly	Low fat ice creams, gelato, avocado, fruit pies, muffins, jelly sweets, coconut milk
High	>13 kL (3 kcals/g)	Eat only occasionally, if at all (with the exception of those foods with a high nutritional value, e.g., nuts)	Fatty meats, full cream dairy products, cheeses, spreads (butter/margarine), chocolate, pastries, pies, cakes biscuits, fried foods, potato crisps/thin chips, snack foods, coconut oil, ghee

Of course, the ED of a food does not tell one anything about its nutritional value. For example, foods containing healthy fats may have a high ED but should still be included in a healthy diet. There are also exceptions to the rule where more evidence is coming to light. Full fat dairy products are one such example. The most recent research does not show an association between these foods and overweight/obesity or cardiovascular disease (CVD). In fact, as part of a weight-reducing diet, all dairy foods have been shown to improve outcomes (Tremblay et al., 2015). Controversially, skim milk has been associated with overweight in epidemiological studies, however, the reasons for this are unclear. It may be that those consuming skim milk are also consuming more low fat, highly refined carbohydrate foods, basing food choices on fat content alone. Or, it may be that full fat dairy foods are more filling and therefore help to reduce later food intake. Information on ED must, therefore, be taught alongside other relevant nutrition information on aspects such as healthy fats and quality carbohydrates. Nevertheless, a reduction in the ED of foods in the diet is the first way to reduce total energy input.

However, without any other changes, successful weight loss is unlikely. The next step is to reduce portion size.

Portion Size

The size of a meal is obviously a relevant factor in total energy intake, and it is apparent that meal sizes have increased in recent times. The size of a standard McDonald's meal in the 1950s was 75 g of fries with 200 mL of Coke; by 2001, this had become 200 g of fries with 950 mL of Coke. Nieslen and Popkin (2003), in a survey of portion sizes in meals eaten at home, in restaurants or as takeaways from 1977 to 1998, found that most meals, including those cooked at home, increased in size over time, some by more than 50%. This is likely to have increased even more so since that time.

Larger meals can be eaten as part of a weight loss program if the energy density of the individual ingredients and the portion sizes of these ingredients are kept low. This is the secret of the Mediterranean diet, which is composed of low ED, high fiber foods eaten in reasonable amounts. However, even low ED foods, if eaten in large enough amounts, can make weight control difficult. This is illustrated in the differences in the lower average body weight of the French, who eat relatively small but high ED meals, compared to the Americans and Australians, who tend to eat lower fat but larger meals (sometimes referred to as the "French Paradox"). Portion sizes are therefore important and should ideally be reduced by choice, but they can be manipulated by behavioral factors such as the size of a meal plate (see Chapter 11).

Rolls et al. (2006) have found that food intake, at least in a laboratory situation, is clearly dependent on ED and portion size, as shown in Fig. 8.1. Larger portion sizes (and higher ED meals) tend to be eaten in company and where food is served at restaurants and external sources. Hence, advice as to how to manage portion size and choice of meals (particularly ED) in such settings needs to be provided.

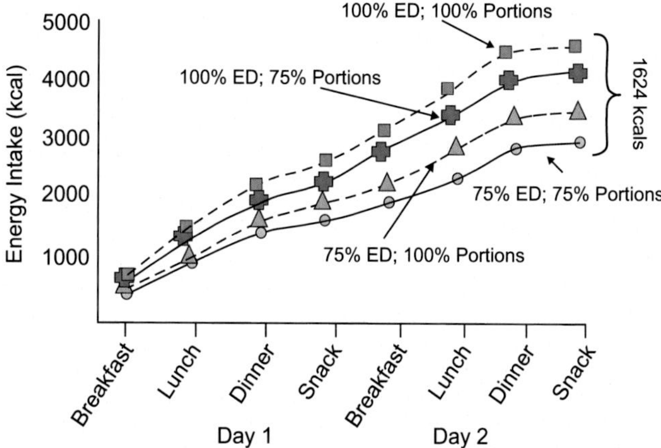

FIGURE 8.1 Influence of energy density and portion size on food intake. *Adapted from Rolls, B.J., Roe, L.S., Meengs, J.S., 2006. Reductions in portion size and energy density of foods are additive and lead to sustained decreases in energy intake. Am. J. Clin. Nutr. 83, 11–17 and Cameron-Smith, D., Egger, G., 2003. The Ultimate energy guide. Allen and Unwin, Sydney.*

Eating Frequency

The third factor in total energy input is the frequency of eating. Given that there is energy expenditure (EE) involved in digesting food, this can be relevant (depending on the total amount of energy consumed during the course of a day). Although controversial, some research suggests that it takes more energy for the body to digest and metabolize several small meals across the course of the day than one large meal containing the same amount of calories. Fig. 8.2 provides a theoretical display of this.

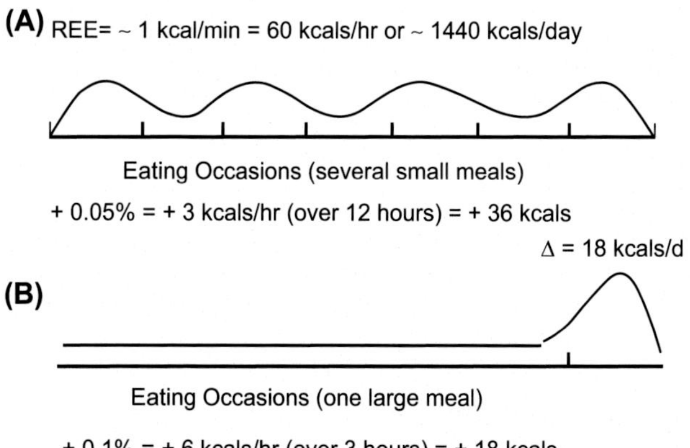

FIGURE 8.2 Energy expenditure associated with food frequency. *Adapted from Rolls, B.J., Roe, L.S., Meengs, J.S., 2006. Reductions in portion size and energy density of foods are additive and lead to sustained decreases in energy intake. Am. J. Clin. Nutr. 83, 11–17.*

If average EE is expressed as a resting metabolic rate of about 1 kcal (4.2 kJ) per minute in an average-sized (80 kg) person, the rate of increased EE associated with the thermogenic effect of an average meal intake has been calculated at 0.05%. If several small meals are eaten during the day, as shown in the top scale in Fig. 8.2, then this would account for an extra 36 kcals/day (150 kJ). Assuming that thermogenesis increases by twice as much after a single, larger meal (0.1% metabolic rate) and that this has an effect lasting for up to 3 h, the difference between graphs A and B accounts for about 18 kcals/day (76 kJ). It must be stressed that this is purely theoretical, but it does show the relevance of eating frequency, provided total energy intake is constant. The differences are also far less when comparing one meal a day to three meals a day, so the optimal frequency of eating is not clear and is likely to be different for everyone, based on several other factors.

The important point here, however, is that increasing eating frequency only works if the total (calorific) volume of energy intake remains the same. In practice, people are accustomed to eating a certain quantity of food at each meal and advice to eat more often can simply result in eating more. People struggling with their weight usually need structured advice about their eating, with information on what to eat and how much to eat at each meal or snack. A further advantage of eating regularly is that it prevents excessive hunger that can result in overeating. A useful clinical tip here is to advise patients not go for more than about 4 h without something (healthy) to eat, with the exception of the overnight fast.

FAST FOOD - THE NEXT STEP

AN ALTERNATIVE APPROACH

Recently a simple, but significant, advance has been made in public nutrition education by Brazilian nutritionist Dr. Carlos Monteriro as a way of determining healthy nutrition (Monteiro, 2009). Monteiro suggests that "the issue is not food nor nutrients, but processing." He considers three food groupings: Group

1 consists minimally processed whole and natural foods, such as plants, fruits, vegetables, roots, and tubers.

Group 2 is made up of substances that are extracted from whole foods such as oils, fats, flours, pastas, starches, and sugars, which aren't usually consumed by themselves but used in food preparation.

Group 3 consists of ultraprocessed foods, usually made up from the raw materials from group 2 to make low nutrient dense processed products such as breads, cookies (biscuits), ice creams, chocolates, confectionery (candies, sweets), breakfast cereals, cereal bars, chips (crisps), and savory and also sweet snack products in general, and sugared and other soft drinks.

Monteiro's simple approach is to eat more of group 1, less of group 2, and nil, or minimal of group 3 (Table 8.2).

TABLE 8.2 Food Categories and Nutrition Recommendations

Food Types	Description	Examples	Recommendations
1. Minimally processed foods	Whole foods submitted to processes that do not substantially alter nutritional properties (e.g., cleaning, freezing, drying, pasteurizing, packaging, etc.).	Fresh meat and milk, grains, pulses, legumes, nuts, fruits, vegetables, roots, tubers.	Eat often and freely
2. Substances extracted from whole foods	Mostly not consumed by themselves but used in the cooking or added to other foods. Used as raw material bases for group 3.	Oils, fats, flours, pasta, starches, sugars.	Eat sparingly
3. Ultra-processed foods	Made from group 2 substances to which either no or only small amounts of group 1 are added; plus salt, preservatives, flavors and colors. Unlike group 2, these are often consumed alone or in combination.	Breads, biscuits, ice creams, chocolates, confectionery, breakfast cereals, cereal bars, chips (crisps), sweet snack products, meat products such as nuggets, hot dogs, burgers, sausages, processed meats. These are branded, distributed, and heavily marketed.	Eat infrequently or avoid

This is confirmed through the approach based on an "antiinflammatory" eating plan as discussed in Chapter 3. Research (Egger and Dixon, 2014) has shown that there is a low grade, systemic, inflammatory reaction by the immune system to certain foods (generally Monteiro's group 2 and 3 foods) suggesting that humans have not evolved with these types of foods, but a neutral or antiinflammatory response to more natural, whole foods (Monteiro's group 1).

In general, the previous recommendations for food intake and in the "antiinflammatory" eating plan are summed up in Michael Pollan's simple statement, "Eat food. Not too much. Mainly plants" (Pollan, 2007).

RECENT DEVELOPMENTS IN NUTRIENT UNDERSTANDING

In simple terms, a "volume" approach to nutrition suggests that foods that are low in energy density (usually whole, natural foods) and eaten in small portion sizes, frequently, are likely to have weight control and metabolic health benefits. Within this, there is value in looking at the nutritional and metabolic components of certain nutrients.

Protein

Protein has perhaps been undervalued in nutrition in recent times because of the concentration on obesity and the influence of the obesogenic nutrients carbohydrate and fat. Protein and its component amino acids have an essential anabolic function in the building and replacement of muscle. Muscle, in turn, according to Wolfe (2006), has probably also been underappreciated for its key role in health and disease.

Studies of hunter-gatherer humans calculate protein intakes of 19–35% of energy (Cordain et al., 2002) in Paleolithic times. This has dropped to an average of 13–15% in modern societies, largely as a result of increases in fat and carbohydrate. There is now a shift back toward increased protein intake for metabolic and obesity benefits. Increasing protein is proving to be a successful strategy for weight loss, particularly maintaining weight loss over time. This is most likely due to several key factors: protein has more than three times the thermic effect of either fat or carbohydrate; it is the most satiating macronutrient; it has a relatively low energy content (4 kcal/g or 17 kJ); and it requires considerable energy loss for storage as fat (25%). Like other nutrients, however, protein has been found to come in a wide variety of forms.

Recently, this has been dichotomized into "fast" and "slow" protein, where the former is digested quickly and, through action on the portal circulation, appears to send satiety signals that result in early cessation of eating (Bendtsen et al., 2013; Layman and Baum, 2004). This is obviously advantageous for those looking for weight loss or maintenance, because of the reduced total energy intake likely to occur over the course of a day. The combination of "fast" protein and calcium also appears to increase fecal energy loss, adding to the weight loss benefits. While forms of fast protein are still being tested, it appears that dairy products such as whey, with high concentrations of the amino acid leucine, are some of the best. Other potential fast proteins are seafood, soy proteins, and perhaps lean wild game meats, although this is not currently known (Westerterp-Plantenga and Lejeune, 2005).

While the advice to eat more protein, like our hunter-gatherer ancestors, is valid, it is not easy to do. This is because of the major differences in the modern food supply. For example, the meat from domesticated animals is far higher in total and saturated fat and lower in essential omega-3 fats than the meat from wild animals eaten in the past. Many popular commercial high protein diets are far removed from our evolutionary diet and there are major health concerns from many such plans. An increased intake of protein need not be accompanied by a low carbohydrate intake or a high fat intake. For weight control and good health, a high protein diet should also be rich in plant foods (particularly vegetables and fruit) and provide a moderate amount of quality carbohydrates and healthy fats.

Nevertheless, an increase in protein content in the diet (preferably at the expense of decreased fat and high glycemic index (GI) carbohydrate, see following) is likely to have health and weight control benefits in the general population. An alternative way of doing this may be through meal replacements (such as those discussed in Chapter 7) where high protein, calcium, and other nutrient content can be contained in small energy packets.

Carbohydrate

Carbohydrate has also come in for its fair share of attention over the last decade or so, particularly in relation to weight control. Reducing carbohydrate intake is the basis of "low carb diets" such as the Atkins Diet, one of the most popular weight loss programs of recent years. Yet research in the late 20th century shows that carbohydrate is far from a homogeneous nutrient. Traditionally classified into "simple sugars" or "complex carbohydrate" based on saccharide chain length, the assumption was that it is better to consume complex carbohydrates and reduce the intake of sugars. More recent analysis shows that this classification is not useful in telling us how the carbohydrate affects us physiologically.

The GI has emerged as a better criterion because it measures directly the impact of foods on blood glucose. The GI is a ranking of foods on a scale of 100, using a standard measure of glucose for reference. While initially met with skepticism, particularly regarding its practical application, there is now growing evidence for a role of GI in preventing and treating diabetes, reducing the risk of cardiovascular disease and certain cancers, and weight control (McMillan-Price and Brand-Miller, 2006).

Two features of low GI foods (defined as those with a GI value <55) are potentially beneficial for weight control: their satiating qualities and their ability to promote fat oxidation at the expense of carbohydrate oxidation.

Both of these characteristics stem from slower rates of digestion and absorption—and correspondingly lower glycemic and insulinemic postprandial responses—of low GI compared to high GI foods. However, not all studies have agreed on the impact of the GI on weight loss. Part of the problem may be in knowing the GI of individual foods. While much research has gone into providing GI values, there are numerous foods yet to be tested. In general, low GI foods are those that have had less human interference or processing, although this is not always the case. Wholegrain products do not always have a low GI; for example, many brown breads and the majority of breakfast cereals have a high GI. Similarly, many processed, high fat, and sugar products have a low GI, but this does not mean they should be recommended. Rather, the GI is a tool to help reduce the glycemic and insulinemic impact of the diet. It should not be used in isolation but in accordance with other key nutrition messages.

Both the GI and the total amount of carbohydrate in a food affects the absolute blood glucose response. It has therefore been suggested that glycemic load (GL), defined as the GI/100 × grams of carbohydrate, is more useful. However, this has limited practical utility because it is clearly dependent on the serving size of the meal. In practice, reducing the GI of carbohydrate foods in the diet is effective in reducing the overall GL of the diet.

In essence, the GI is useful as a concept and can help improve the quality of carbohydrates in the diet, but obsession with the numbers should be avoided. GI values for individual foods can be obtained from www.glycemicindex.com.

However, it has recently been shown that there is a high interpersonal variability in postmeal glucose response (Zeevi et al., 2015). Using personal and microbiome features will enable a more accurate glucose response prediction in the future. Personalized nutrition interventions based on this new model already have been shown to successfully lower postmeal glucose (Zeevi et al., 2015).

What Can Be Done in a Brief Consultation?

- ☐ Ask patients where they see deficiencies in their current nutrition.
- ☐ Provide information for completing the DAB-Q (Diet, Activity and Behaviour - Questionnaire) (see Professional resources).
- ☐ Discuss the concept of calorific volume and its components for nutrition.
- ☐ Explain energy density and how to read labels for this.
- ☐ Advise on the importance of breakfast and regular meals (every 4 h) to prevent gorging.
- ☐ Provide specific nutritional information for health risks (e.g., reduced saturated fats for lipid lowering).
- ☐ Recommend a good nutrition Website, such as www.healthyeating.com.au.

Fat

The biochemistry and epidemiology of fat in human health is extremely complicated, and we will confine ourselves to the three main forms of dietary fatty acids: saturated (SFA), monounsaturated (MUFA), and polyunsaturated (PUFA), so named because of the relative degree of saturation of carbon atoms with hydrogen. All three are natural to food sources. Trans-fatty acids are those that are created by the hydrogenation of these fats, primarily through food processing. Trans fats have been shown to be dangerous for health, are now being restricted worldwide, and will not be discussed further here.

Fat has been the "bad guy" in nutrition for many years. Because all fats, irrespective of type, have an energy content of 9 kcals/g (38 kJ), a reduction in all forms from around 40% of dietary energy to 30% has been recommended for CVD prevention and weight control. As the average intake of fat in Western men is around 110 g a day and women around 90 g a day, this means a reduction to around 80 and 70 g, respectively. However, recently it has been found that not all fats have the same health or weight control properties (Khor, 2004). Saturated fats have for decades been thought of as the baddies. However, in recent years this has been questioned, particularly with respect to the association with heart disease. The confusion may have arisen at least in part from pooling all saturated fats together when in fact there are many different subtypes with different effects in the body (Zong et al., 2016).

We know from clinical trials that some of the saturated fats raise LDL cholesterol, while others don't. For example, stearic acid (18 carbon chain), dominant in dark chocolate, seems to have a neutral effect on blood cholesterol, while lauric, myristic, and palmitic acids (12, 14, and 16 carbon chains, respectively) all raise both LDL cholesterol and HDL cholesterol.

The other problem is we eat foods, not individual nutrients like saturated fat, alone. Different saturated fat–containing foods have different effects within the body. Consuming saturated fat from a piece of cheese is not the same as consuming it from a commercial party pie or a meat-lovers pizza. Eating a steak is not the same as eating a processed hot dog.

We have long been advised not to eat too much full fat dairy on account of its saturated fat content. However, research studies suggest that full fat cheese and milk do not raise cholesterol as once thought. Butter on the other hand does. It seems that the combination of saturated fats alongside the high protein and calcium of cheese, milk, and yogurt alters the way it is being metabolized in the body. Plus dairy foods, including the full fat varieties, have been shown to help with weight loss when part of a kilojoule-reduced diet.

So what conclusions should we draw? The strongest evidence is for dietary patterns, and the Mediterranean diet comes out as a shining example. Characterized by a high intake of vegetables, extra virgin olive oil, nuts, legumes, whole grains, moderate alcohol, fish and seafood, with more moderate amounts of meat and dairy. These are clear dietary winners in keeping the cardiovascular system healthy.

As for saturated fats, our ancestral diets (true "Paleo diets") did not contain high levels of saturated fats and we probably should not be doing so today. In fact, none of the healthiest diets in the world contain high levels of saturated fat. But that doesn't mean total avoidance. The best advice is to put a focus on foods rather than nutrients. Advising patients to cut down on pastries, biscuits, cakes, crisps/chips, and processed meats is more effective than a directive to eat less saturated fat. These foods not only contain saturated fats but the undoubtedly harmful trans fats, not to mention preservatives, salt, sugar, and whatever other undesirable additives are in such processed foods.

Because of this, recommendations have moved away from the rather simplistic "eat less fat" to more specific advice on the type of fat that is best to eat. MUFA intake should be maintained or recommended in place of SFA. MUFAs are found in olive and canola oils, many nuts and their oils including peanuts, almonds, and macadamias, and in meat from wild game such as kangaroo, crocodile, and in venison. PUFAs can be split into the omega-6 and omega-3 fractions and the ratio of these two types of fat seems to be important. Our hunter-gatherer ancestors ate far more omega-3 and far fewer omega-6 fats than we do today, and this is likely to have played an important role in preventing CVD (Cordain et al., 2005). The long-chain omega-3 PUFAs are particularly important and are primarily found in oily fish (e.g., salmon, herring, mackerel, trout, tuna) or seafoods and dietary supplements made from oily or fish liver (e.g., cod liver oil). There are smaller amounts of long-chain omega-3s in lean red meats, especially grass-fed and game meats, and eggs. Short-chain omega-3s do not have the same beneficial effects but can be elongated in the body to some degree and are also recommended. Linseed (sometimes called flaxseed) and green leafy plant food (especially seaweeds) are the best sources. Since a high intake of the omega-6 PUFAs prevents maximum absorption and use of the omega-3s, it is important to reduce these fats at the same time to achieve the desired balance. This means fewer PUFA margarines and most seed oils, including sunflower, safflower, and corn oil.

While these recommendations apply for CVD health, work particularly by Dr. Kerin O'Dea et al. at Melbourne University showed paradoxically that isocaloric diets high in monounsaturated fats compared to saturated fats can also help reduce body fat, particularly abdominal fat (Piers et al., 2003). As this goes against a law of thermodynamics, one explanation may be that energy expenditure is increased somehow on a high MUFA diet (such as the Mediterranean diet), possibly because of increased feelings of well-being. Alternatively, there may be differences in the absorption and utilization of MUFA and SAFA that may explain such results. Regardless, it appears that the best recommendations on dietary fats now are:

- Reduction in total fats to 25–30% of energy in the total diet;
- Reduction in saturated fats to <10% of energy in the total diet;
- Maintenance of monounsaturated fats in proportion;
- Reduction of some polyunsaturated fats, particularly seed oils;
- Increase in omega-3 PUFAs, particularly from fish and seafood.

To support these recommendations, recent reviews have shown that metaflammatory responses are more common following ingestion of saturated and trans fats (Egger and Dixon, 2010). This might be expected given that such substances have traditionally been extremely low or nonexistent in historically eaten food sources.

Salt

Modern Western diets are relatively high in sodium even without the addition of extra salt. However, adding salt is often part of a meal and this brings total salt intake to well above the 154 mmol per day that has been indicated for good health. In the past, salt has been implicated as a cause of high blood pressure, but findings relating to this have been variable. More recently, it has been found that salt sensitivity tends to have a genetic component and that salt intake in some families is likely to have more of an impact on blood pressure than in others.

Only recently has salt been considered as a factor in obesity. This does not relate to its energy content but to other appetitive factors in salt. Because it increases thirst, a high salt intake in children, in particular, is likely to lead to a high fluid intake. And because a large proportion of the fluid intake of children in modern Western societies is in calorific form, this is likely to increase energy intake over the course of a day. Salt, whether it be sodium chloride, potassium chloride, or MSG, is also known to decrease satiation of food and therefore lead to greater energy intake and increase hunger. But salt may also have an effect on the learned aspects of nutrition discussed in Chapter 9. The craving for salty foods can be learned, leading to a desire for an increase in those foods; for example, salted peanuts are craved more and are able to be eaten in much larger quantities than unsalted nuts (consuming them with alcohol adds to the total energy intake because alcohol can have a decreased inhibitory effect).

SPECIFIC DIETARY REQUIREMENTS

It is beyond the scope of this chapter to discuss dietary requirements and personalized nutrition for specific health issues; referral to a qualified dietitian is recommended. Dietary advice based on estimates of energy intake and calorie reduction is also not considered here because it is not generally necessary, except in specific cases where more skilled dietetic input may be involved.

The basic dietary advice considered herein is applicable to weight control and good metabolic health in general. A generic dietary approach for weight loss and diabetes management is shown in Table 8.3, although it should be stressed that individual dietary advice may be required for the management of specific diabetes cases. Note also that this table is useful only for the practitioner and is not intended to be used with the patient. Patient resources should always be specific about foods to consume as percentage energy levels are meaningless.

TABLE 8.3 General Dietary Recommendations for Weight Loss and Diabetes Management

	Healthy Normals	Diabetes	Weight Loss
Total Energy	Reduce to balance EE	—Reduce to cause energy deficit (e.g., 250-500 kcals/day) —	
Fat			
Saturated	Reduce to <10%	— Reduce to <10% —	
Polyunsat.	Reduce to <10%	— Reduce to 10% or less —	
Monunsat.	Maintain or increase to >10%	10–16% if not overweight	Reduce to 10%
Carbohydrate	~50%	45–50% but reduce total amount if weight still a problem after fat reduction	
Sweeteners	Not necessary	Use if weight is still a problem after fat reduction	
Alcohol	2–4 Drinks/day	— Reduce if weight is stil a problem after fat reduction —	

ASSESSING FOOD INTAKE

Food intake can be assessed through a number of means: skilled clinical questioning, dietary diaries, or food frequency questionnaires. None are perfect because they require the recall or honesty of the patient in an area where psychological barriers can play a role.

In the first instance, if an energy imbalance has occurred through food intake, it is useful to consider *why* and *how* this may have happened. Some typical reasons why are:

- An obesogenic environment encouraging overconsumption (e.g., a stressful environment with constant availability of highly palatable, high energy foods);
- Genetic influences, such as hunger and substrate utilization;
- Medications leading to hyperphagic behavior (see Chapter 26);
- Life stages, such as adolescence, pregnancy, or menopause;
- Life events, such as quitting smoking or early abuse;
- Age and its influence on food consumption.

How excess energy intake occurs can be through a range of means such as:

- too much total food
- too much fat
- too much high energy density food
- too much high GI food
- binge eating/night eating/excessive dieting
- holiday eating/social eating/feasting
- too much alcohol/food with alcohol
- too much soft drink
- too little fiber
- lack of awareness of intake
- the "eye–mouth gap"
- the "exception rule"

The "eye–mouth gap" refers to the finding that overweight individuals tend to underestimate their food intake (Lissner, 2002). The "exception rule" describes a tendency in some patients to discount days of excess as exceptions, although they are quite common. Determining the factors common to an individual patient is tantamount to recommendations to correct this.

The Dietary, Activity, and Behavior Questionnaire

The advent of computers and the Internet have aided dietary measurement in recent times. Several computerized dietary analyses are available, but these generally require skilled interpretation and prescription by a dietitian. One self-help assisted program for patients was developed by Egger et al. (2006): the Dietary, Activity, and Behavior Questionnaire (DAB-Q) enables patients to complete a number of questionnaires at a public Website (www.lifestylemedicine.com.au/alma-resources/dab-q/) and bring printouts to the clinician for discussion and negotiation about change. These rank items and behaviors that are frequently consumed or carried out and that are potent in causing weight gain but can reasonably be modified.

SUMMARY

Nutrition is a relatively young science that is constantly changing in response to ongoing research. The key aspects of nutrition for metabolic health are the total volume of energy consumed and the nutrient composition mix. New findings suggest that protein, carbohydrate, and fat need to be examined in terms of specific properties, such as speed of digestion, substrate utilization, and rate of metabolism. Finally, salt is a significant component of modern foods and may have an effect on weight control through its effect on overconsumption of other energy-dense fluids and foods. New research in the field of nutrigenomics will drive a more personalized approach to nutrition that will help to make interventions more effective.

Practice Tips

Providing Nutritional Advice

	Medical Practitioner	Practice Nurse/Allied Health Professional
Assess	• Whether diet is a significant contributor to ill-health • Potential dietary influences on risk • Why dietary imbalance may have occurred • How dietary imbalance may have occurred	• Food intake through such processes as the DAB-Q (Diet Activity and Behaviour Questionnaire) (see Professional resources) or a diet diary • Impact of energy density (ED), portion size or frequency
Assist	• By treating dietary-influenced comorbidities • By identifying general nutritional goals • By reviewing progress of drug therapy for weight loss/metabolic treatment • By reviewing ongoing special dietetic programs such as meal replacements • By using templates as appropriate	• General education about eating • By providing good nutritional literature • Monitoring food intake if possible • By using appropriate tactics to move through stages of readiness to change
Arrange	• Referral for specific dietary advice where required • List of other health professionals for a care plan	• Discussion of motivational tactics with other involved health professionals • Consultation with a dietitian/nutritionist • Contact with self-help groups

Key Points

Clinical Management

- Use tests such as the DAB-Q (see Professional resources) to quantify incorrect eating and activity patterns.
- Ascertain which aspect of food volume may be important for a particular patient.
- Teach cutoffs for energy density (at least "high" cutoff).
- Discuss ways of reducing portion size in the patient's environment.
- Stress importance of ED and portion size together.
- Advise on frequent, low ED food intake (i.e., every 4 h).
- Change nutrient proportions for weight loss.
- Suggest low GI carbohydrates for diabetes management and possibly weight control.
- Reduce saturated and some polyunsaturated fat but maintain the proportion of monounsaturated fats.
- Check on salt intake and reduce if necessary.

REFERENCES

Bendtsen, L.Q., Lorenzen, J.K., Bendsen, N.T., Rasmussen, C., Astrup, A., 2013. Effect of dairy proteins on appetite, energy expenditure, body weight, and composition: a review of the evidence from controlled clinical trials. Adv. Nutr. 4 (4), 418–438.

Cameron-Smith, D., Egger, G., 2002. The Ultimate Energy Guide. Allen and Unwin, Sydney.

Civitarese, A.E., Carling, S., Heilbron, L.K., CALERIE Pennington Team, et al., 2007. PLoS Med. 4 (3), e76. http://dx.doi.org/10.1371/journal.pmed.0040076.

Cordain, L., Boyd Eaton, S., Brand-Miller, J., et al., 2002. The paradoxical nature of hunter-gatherer diets: meat-based, yet non-atherogenic. Eur. J. Clin. Nutr. 56 (Suppl. 1), S42–S52.

Cordain, L., Boyd Eaton, S., Sebastian, A., et al., 2005. Origins and evolution of the Western diet: health implications for the 21st century. Am. J. Clin. Nutr. 81, 341–354.

Egger, G., Dixon, J., 2010. Inflammatory effects of nutritional stimuli: further support for the need for a big picture approach to tackling obesity and chronic disease. Obes. Rev. 11 (2), 137–145.

Egger, G., Dixon, J., 2014. Beyond obesity and lifestyle: a review of 21st Century chronic disease determinants. BioMed Res. Int. 2014:731685. http://dx.doi.org/10.1155/2014/731685.

Egger, G., Pearson, S., Pal, S., 2006. Individualising weight loss prescription: a management tool for clinicians. Aust. Fam. Phys. 35 (8), 591–594.

Heilbronn, L.K., Ravussin, E., 2005. Calorie restriction extends lifes span – but which calories. PLoS Med. 2 (8), 231.

Khor, G.L., 2004. Dietary fat quality: a nutritional epidemiologist's view. Asia Pac. J. Clin. Nutr. 13 (Suppl), S22.

Layman, D.K., Baum, J.I., 2004. Dietary protein impact on glycaemic control during weight loss. J. Nutr. 134, 73S–968S.

Lissner, L., 2002. Measuring food intake in studies of obesity. Public Health Nutr. 5 (6A), 889–892.

McMillan-Price, J., Brand-Miller, J., 2006. Low-glycaemic index diets and body weight regulation. Int. J. Obes. 30, S40–S46.

Monteiro, C., 2009. Nutrition and health. The issue is not food, nor nutrients, so much as processing. Public Health Nutr. 12 (5), 729–731.

Morgan, T.E., Wong, A.M., Finch, C.E., 2007. Anti-inflammatory mechanisms of dietary restriction in slowing aging processes. Interdiscp. Top. Gerontol. 35, 83–97.

Nieslen, S.J., Popkin, B.M., 2003. Patterns and trends in food portion sizes. JAMA 289 (4), 450–453.

Piers, L.S., Walker, K.Z., Stoney, R.M., et al., 2003. Substitution of saturated with monounsaturated fat in a 4-week diet affects body weight and composition of overweight and obese men. Br. J. Nutr. 90 (3), 717–727.

Pollan, M., 2007. The Omnivore's Dilemma: A Natural History of Four Meals. Penguin, NY.

Rolls, B., 2007. The Volumetrics Eating Plan: Techniques and Recipes for Feeling Full on Fewer Calories. Harper, NY.

Rolls, B.J., Roe, L.S., Meengs, J.S., 2006. Reductions in portion size and energy density of foods are additive and lead to sustained decreases in energy intake. Am. J. Clin. Nutr. 83, 11–17.

Tremblay, A., Doyon, C., Sanchez, M., 2015. Impact of yoghurt on appetite control, energy balance, and body composition. Nutr. Rev. 73 (S1), 23–27.

Westerterp-Plantenga, M.S., Lejeune, P.G.M., 2005. Protein intake and body weight regulation. Appetite 45 (2), 187–190.

Wolfe, R.R., 2006. The under appreciated role of muscle in health and disease. Am. J. Clin. Nutr. 84, 475–482.

Zeevi, et al., November 19, 2015. Personalized nutrition by prediction of glycemic responses. Cell 163 (5), 1079–1094.

Zong, G., Li, Y., Wanders, A.J., et al., 2016. Intake of individual saturated fatty acids and risk of coronary heart disease in US men and women: two prospective longitudinal cohort studies. BMJ 355, i5796.

Professional Resources

Measuring Nutrition

Diet, Activity, and Behavior Questionnaire (DAB-Q)

The Diet, Activity, and Behavior Questionnaire (DAB-Q) is a simple, free online questionnaire to help patients determine the best way to lose weight. The DAB-Q asks about eating and activity patterns. There are five tests, covering:

- foods eaten in excess
- foods that could be increased
- passive activities carried out in excess
- activities that could be increased
- eating behavior

Results are scored on frequency, weight gaining potential, and changeability (i.e., the ease with which this food or activity could be changed by this patient), giving a maximum score out of 100. High-scoring items are then targeted for change.

The DAB-Q is available from http://lifestylemedicine.org.au/about/lifestyle-medicine/learning-and-tools/.

More information is available from Egger et al. (2006).

Chapter 9

Fluids, Fitness, and Fatness

Garry Egger, Suzanne Pearson

Let them drink Coke.

Variation on a theme from a modern Marie Antoinette

INTRODUCTION: FLUIDS AS ENERGY

During the early stages of the obesity epidemic in the 1980s and 1990s, having excess fat was attributed to eating excess fat. This made sense, because it was well known that dietary fat is twice as energy dense as carbohydrate or protein in physical terms and possibly even more in biological terms (i.e., less energy is required to store fat as fat than to store carbohydrate or protein as fat). Hence, reducing fat, for most people, would result in a reduction of total energy intake and a decrease in body weight. However, this diverted our attention away from high energy containing beverages. As a result, soft drink manufacturers were emboldened when marketing their products.

As indicated in this Chapter, it has become clear that fat, alone, is not solely responsible for the obesity epidemic. It is the energy density of foods and beverages, together with portion size and frequency of consumption, that tends to increase total energy intake and, in the absence of a high energy output, contribute to the growing waistlines of the population. This applies to the kilojoules in each milliliter of drink as much as to the kilojoules per gram of food. In fact, it could be more so with beverages because of the likelihood of "passive overconsumption" of fluid energy due to its nonsatiating effect. This is compounded by the fact that soft drink varieties and consumption have doubled over the last 20 years and constitute more than 50% of the increase in energy intake since the 1970s (Popkin et al., 2006). The spotlight has thus turned to the obesogenic and other health effects of beverages, as well as foods.

In this chapter, we look at some of the different types of beverages now available and the effects of these, mainly on energy intake but also on other aspects of health. As with other chapters, we focus only on fluids for general health. We do not look at fluid needs for specific diseases, such as heart failure or renal insufficiency, or specific purposes, such as endurance sports.

Lifestyle Medicine. http://dx.doi.org/10.1016/B978-0-12-810401-9.00009-7

HOW MUCH FLUID IS REQUIRED?

A common misconception is that humans require eight glasses of water, each with 227.2 mL (8 oz) of fluid, every day for survival and good health. In a considered review of this, Valtin (2002) found no scientific basis for this suggestion. Fluid requirements have a wide number of influences, including age, size, activity level, and temperature, so a generic prescription is of little value (Stevenson et al., 2015). While eight glasses of water may be adequate for an individual of middle age and average size in a temperate climate, the same may not be true for an active older adult in a hot climate or an inactive adolescent in a cool but moist climate. In general, it is believed that a range of fluid intake of about 1–4 L a day is required (Popkin et al., 2006). The average intake in the United States in 2000 was 2.18 L (0.57 gal). This represents an increase of approximately 30% from the mid-1970s, largely as a result of increases in the consumption of bottled water, soft drinks (doubled), alcoholic beverages, and juices (2.5-fold) (Valtin, 2002).

It is important to remember that fluid is obtained from many different sources. In addition to water, tea, coffee, and other beverages, fluid is a component of most foods, especially fruits and vegetables. Hence, the volume of fluid consumed as beverages does not necessarily reflect the total daily fluid intake. Furthermore, water is produced in the body as a "waste product" during the breakdown of nutrients.

In general, fluid requirements are well controlled by thirst in a natural environment (Millard-Stafford et al., 2012). Though low-level dehydration is not uncommon in many older people (defined as a >2% decrease in body fluid), overhydration is difficult to achieve in healthy people. Consequently, a recommendation to maintain fluid intake is unlikely to be deleterious. It is therefore the type of fluid that is most relevant for health and energy intake.

HOW MUCH ENERGY IN FLUID IS RECOMMENDED?

A healthy diet does not require fluids to contain energy or nutrients—potable water is adequate to provide all the fluid needs of humans. As explained in this Chapter, energy density is a prime determinant influencing daily energy balance. As the energy in fluids can be passively overconsumed, excessive intake is of great concern for preventing weight gain and associated problems. The energy density of beverages and their intake recommendations are shown in Table 9.1.

RECOMMENDED FLUID INTAKES

Currently, about 20% of the total dietary energy intake in countries such as the United States and Australia is derived from beverages. As a general guideline, it is recommended that this figure be reduced to around 10%. Current and recommended levels of six different categories of fluids designed to get to this level

TABLE 9.1 Recommended Energy Density Cutoffs for Fluids (Beverages Are Categorized Here By Energy Density Alone, Not on Nutrient Content or Nutritional Value)

Energy Density	Measure	Recommendation	Examples
Low	<1 kJ (0.2 kcal)/mL	Drink freely	Water tea/coffee Low calorie soft drinks/cordials Some iced teas Skim milk Vegetable juice
Medium	1 kJ (0.2 kcal)/mL – 1.49 kJ (0.3 kcal)/mL	Drink sparingly	Low fat milk (unflavored) Some sports drinks Alcohol (with provisos) Some iced teas
High	>1.5 kJ (0.4 kcal)/mL	Drink occasionally (if at all)	Sweetened soft drinks Fruit juice Full cream milk Kids' "energy" drinks Sports drinks Flavored milk

From Cameron-Smith, D., Egger, G., 2002. The Ultimate Energy Guide. Allen and Unwin, Sydney.

are shown in Fig. 9.1. These are based on the findings of a Beverage Guidance Panel (Popkin et al., 2006) formed in the United States in 2006 (Popkin et al., 2006). While some criticism of these guidelines has been made (e.g., Weaver et al., 2006), the suggestions given next provide a broad outline of recommended fluid requirements.

SUGGESTED CHANGES IN BEVERAGE INTAKES

Beverages That Can be Increased or Maintained

Level 1—Water should be increased from about 40% to at least 50% of total fluid intake (or to 2–16 glasses a day depending on factors such as age and climate).

Water is and always has been the staple fluid. Historically, it has been consumed from the source of catchment, with variations in mineral quality depending on runoff surfaces and other geographical conditions. To cater to larger populations in more recent times, water has been stored and maintained over longer periods with additives (such as fluoride in some areas as a public health measure to prevent dental decay). Other minerals are also often found in water, including calcium, magnesium, and numerous trace elements.

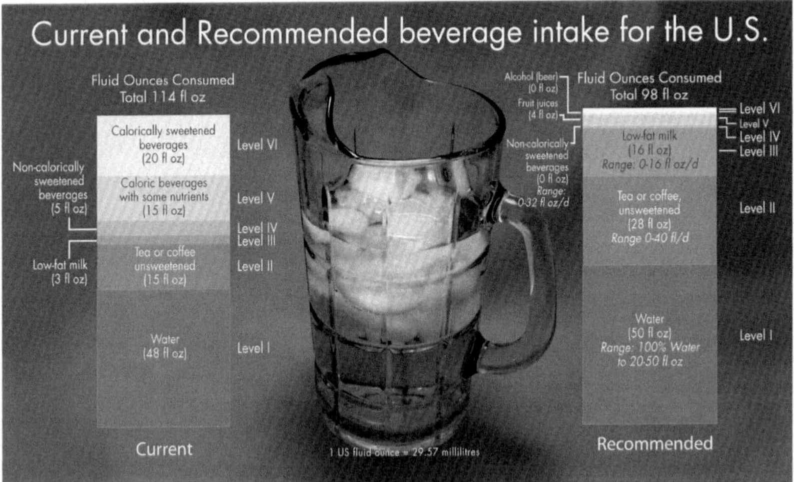

FIGURE 9.1 Current and recommended fluid intakes. *Reprinted with permission from Popkin, B., Armstrong, L.E., Bray, G.M., et al., 2006. A new proposed guidance system for beverage consumption in the United States. Am. J. Clin. Nutr. 83, 529–542.*

Water is becoming a scarce commodity in some parts of the world, and this has led to an increased mineralization of the water supply in these areas. The increased calcification (hardness) of water, which is natural in these areas, has been shown to have a protective effect for cardiovascular disease. However, increased salination, which results from a rising of the water table in some areas, can have adverse effects on blood pressure in some people.

Commercially sold bottled water has become popular in recent times, satisfying consumer anxieties about increasing contamination of water supplies. The lack of fluoride in bottled water helps alleviate worries about added fluoride. This, however, now means that bottled water may lack the protection against dental decay that comes from fluoride. In most areas of the developed world, transport and production of bottled water is a waste of natural resources, and this water does not provide any additional health benefits.

Level 2—Tea/coffee can be increased from 13% to about 30% of total fluid intake, or up to six to eight cups/day.

There is an inverse link between tea consumption and heart disease, with both green and black tea having health benefits, particularly in the prevention of cardiovascular disease. Although the evidence on coffee is more confusing, moderate consumption of coffee is thought to have some health benefits, particularly in cardiovascular health (Chrysant, 2015), although it is not clear whether this is due to components in the coffee or other factors such as weight loss, and reduction of type 2 diabetes, which may also be a minor benefit of coffee consumption. Whether the benefits are related to caffeine in coffee or other components is also not clear, given that decaffeinated coffee has also been shown to have health and weight loss benefits. Caffeinated drinks without added sugar (i.e., artificially sweetened cola drinks) or fat (i.e., low fat milk added to coffee/tea) have a zero or very low energy content and therefore add little or nothing to total energy balance. Teas or coffees made with whole milk or cream, however, lose the low energy advantage.

The limitation of tea/coffee consumption seems to be in the level of caffeine consumed, which may adversely affect some people. A limit of 400 mg of caffeine a day for most people (about three to four cups of strong coffee or six to eight of tea) and 300 mg in pregnant women has been suggested. As an indication of caffeine quantity, a standard cup of brewed coffee contains around 150 mg caffeine, a cup of tea around 80 mg, and a glass of cola about 40 mg. "Caffeinism," which is characterized by anxiety, nervousness, sleep disturbance, irritability, agitation, and gastrointestinal disturbances, and which can cause withdrawal symptoms such as headache and irritability, can result from high doses of caffeine, the level of which may be individually determined. A daily intake level of about 1000 mg (1 g) of caffeine is thought to be a cutoff for this. Withdrawal from caffeine, where this may be contraindicated (such as in anxiety), can be achieved gradually to reduce any withdrawal effects, which disappear within a few days.

Recent studies show a genetic interaction with caffeine metabolism. A gene identified as CYP1A2 has been implicated in adverse effects of caffeine consumption for heart disease (Cornelis et al., 2006). While not necessarily related, about 50% of the population respond to caffeine intake with increased sympathetic nervous system activity (e.g., an inability to sleep after drinking coffee) that others do not appear to be affected by and can sleep soundly after caffeine. The implications of this are not clear, but it may be that the thermogenic, lipolytic, and metabolic effects of caffeine may be greater in some people than in others, explaining why caffeine intake can facilitate moderate weight loss in some people. Regardless, low energy caffeinated drinks can be recommended in moderation in most people without a health risk.

Level 3—Low fat/skim milk and soy beverages can be increased from 3% to approximately 15% of total fluid intake, or 1–1.5 full glasses a day.

Dairy products have had a mixed press over recent years, largely because of the high levels of saturated fat in dairy products from domesticated cattle. The dairy industry has been quick to correct this through technological advances that allow the production of most dairy products at different levels of fat content. This has reduced the energy content while maintaining the nutritional value of dairy foods.

Low fat dairy is a particularly good source of calcium and vitamin D. Because whey, the main protein source in milk, is regarded as a "fast" protein (see Chapter 8), it may have value in increasing satiety and thus assisting weight loss. Another mechanism for weight loss through dairy products may be the combination of whey protein and calcium, which has been shown to increase fecal energy loss (Lorenzen et al., 2007). The weight loss benefits of low fat dairy products have been verified in several, but not all, recent studies on dairy products and weight loss. Dairy product intake has also been found to be inversely related to the metabolic syndrome. Soy beverages have health benefits but contain less calcium than dairy.

Level 4—Noncalorically sweetened beverages (diet drinks) currently form 4% of total intake but can be lowered to 0%, or raised to 30%, or up to 0–3 glasses a day, if replacing levels 5 and 6.

Noncalorically sweetened beverages are those that are artificially sweetened with aspartame, sucralose, or more recently stevia. "Diet" beverages have traditionally had a slightly different taste to versions sweetened with sugars. However, new developments in food technology have allowed the molecular restructuring of artificial sweetener molecules to provide a much more similar taste.

While there are no real health benefits from consuming such drinks, their use as a "treat" or as an alternative to high energy and sweetened drinks is likely to continue, so these would be recommended over other caloric beverages. Widely published scare campaigns regarding the use of artificial sweeteners are not supported in the scientific literature, although there are known effects on dental caries if they are consumed regularly and in large quantities (see Chapter 27). The artificial sweeteners used may also increase a desire for sweet foods, although this is currently unproven.

Beverages That Should be Reduced and/or Drunk Sparingly

Level 5—Caloric beverages with some nutrients should be decreased to less than 3% of total fluid intake, or one glass or less, a day.

Caloric beverages with some nutrients include fruit juices, milk, and alcohol. Fruit juices do provide some of the nutrients of their natural source, but their high concentration of fruit sugars contributes to their having an energy density equivalent to that of sweetened soft drinks. According to Popkin et al. (2006), "There is no specific need to consume fruit juice and consumption of whole fruits should be encouraged for satiety and energy balance." Fruit "smoothies" and drinks from the new type of "juice bars" are usually high calorie versions of fruit drinks and are therefore also not recommended.

Vegetable juices are generally lower in energy than fruit juices and so are better than fruit juices in terms of energy balance. However, processed vegetable juices can have high levels of added sodium, making consumption of whole vegetables preferred. Whole milk is both energy dense and high in saturated fat. Sports drinks usually contain about 50–90% of the energy of sweetened drinks but (sometimes) contain some minerals that are advantageous for endurance events. The high energy content of most sports drinks, however, suggests they should be used sparingly.

Alcohol is also energy dense, but the biological value of this energy is not as clear as with other drinks. Alcohol may have some other health benefits, which will be discussed in greater detail later.

Level 6—Calorically sweetened beverages should be decreased to less than 3% of total energy, or 0–1 glass/day.

Sweetened soft drinks are those least recommended for beverage intake because they add a high energy burden, have a low nutrient value, can be passively overconsumed, and may contribute to other deleterious health concerns (e.g., dental caries; see Chapter 27). While consumption of these fluids is likely to remain high in some communities, soft drink manufacturers including Coca Cola and PepsiCo are aware that this is a diminishing overall market. At least 50% of their total markets now come from low energy drinks or water, and market growth in these products far exceeds that of sweetened beverages.

Unlike water, where intake is finely regulated by thirst, sweetened beverages can be craved and consumed in the absence of thirst. Conditioned learning can be the cause of this (see Chapter 10), such that people who drink soft drinks when thirsty tend to then crave their sweet taste rather than water in subsequent thirsty states. Encouraging the drinking only of water when thirsty may help decondition this learned process; if sweetened drinks have to be taken, they can be used as a treat when people are not thirsty.

Although not the case in Australia, fructose from corn syrup is used to sweeten soft drinks in some parts of the world, including the United States. This not only has the effect of increasing energy density (as with sucrose) but could have an even greater role in obesity through the special effects of fructose on the metabolism of fat in the liver. Spanish scientists have shown that when fructose affects a specific nuclear receptor (PPAR-alpha), the liver's ability to degrade the sweetener is decreased, leading to even greater fat deposition (Roglans et al., 2007).

Soft drinks can also have a corrosive effect on teeth. However, in a novel study examining to what extent this occurs, researchers measured the weight of enamel lost with various types of soft drinks (Jain et al., 2007). This showed that noncola soft drinks are more corrosive than cola drinks but that, within cola drinks, the full sugared versions are more corrosive than the diet versions, supporting even further the use of nonsweetened soft drinks if these have to be consumed. There appears to be no health justification in recommending any level of calorically sweetened beverages as part of a normal diet.

What Can be Done in a Brief Consultation?

- ☐ Check on soft drink, fruit juice, and cordial consumption.
- ☐ Administer a fluids questionnaire (see Professional resources), if appropriate.
- ☐ Explain the importance of fluid intake in energy consumption and weight gain.
- ☐ Discuss when and how much alcohol is consumed.
- ☐ Discuss the high energy value of fruit juice, full cream milk, and sports drinks.
- ☐ Provide a handout on fluids and health, such as the one shown in Fig. 9.1.
- ☐ Recommend a good nutrition website, such as www.healthyeating.com.au.

ALCOHOL

Alcohol is obtained from the fermentation of carbohydrate from a variety of plant sources. In energetic terms, alcohol is measured in a bomb calorimeter as containing 7 kcals (31 kJ) per gram. However, due to the complex metabolism of alcohol, it is controversial as to whether, when, and how this "physical" energy is utilized biologically.

There are three biochemical cycles for the metabolism of alcohol, none of which result in its conversion to fat or other nutrients but all of which involve conversion to energy in some form. As a toxin, alcohol has priority in the energy cycle and will displace other nutrient sources. Hence, foods high in energy density will be "spared" if consumed with alcohol, meaning that the "beer gut" in men is probably more of a beer-and-peanuts or beer-and-chips gut than a beer gut per se. The methylated oxygenated system of alcohol metabolism in the liver results in a greater futile cycle of energy wastage from alcohol in big, regular drinkers, but losses of alcohol from heat, urine, breath, and sweat may also reduce the biological value of alcohol energy, even in casual drinkers. Epidemiological studies consistently show an inverse association between alcohol consumption and weight gain, which would tend to support this. A further possible effect of alcohol on weight gain is through stimulating appetite. However, a review of studies carried out in this area in 2006 indicated that this is not the case (Gee, 2006). Hence, the inconsistent biochemical and clinical evidence and the possibility of intraindividual differences mean a firm conclusion on alcohol and weight gain cannot yet be drawn (Sayon-Orea et al., 2011).

While the metabolic value of alcohol is disputed, there is no dispute about the energetic value of alcoholic mixes. Mixes made with sweetened soft drinks, fruit juice, or full cream dairy products provide the energy of the mix as well as some of the energy from the alcohol itself. Because alcohol is metabolized differently, however, the level of fermentation of a carbohydrate source (and thus the alcohol content) may influence the energetic value of a drink. Consequently, a low carbohydrate beer may contain less metabolizable energy, although greater alcohol, than a high carbohydrate version.

Alcohol also has been shown to have health benefits for atherogenesis and the prevention of cardiovascular disease and type 2 diabetes when consumed in moderation. It has been proposed that this is due in part to the antioxidants contained in alcohol (Tolstrup et al., 2006). Red wine, in particular, has been suggested as a high source of a polyphenol antioxidant called resveratrol, which has been found to have health benefits in humans, particularly in raising HDL cholesterol levels, as well as delaying the aging process in mice sustained on a "junk food" diet (Sinclair and Guarente, 2006). However, comparisons between French drinkers of red wine and Irish drinkers of anything (Marques-Vidal et al., 1996) suggest that all alcohol may have a protective effect.

All this means that alcohol may need to be considered differently from other beverages. There are apparent health advantages from its consumption in moderation but also potential risks (see Table 9.2). There are also quite obvious social

TABLE 9.2 Health Advantages and Disadvantages of Alcohol Consumption

Advantages (Moderate Consumption): Decreases	Disadvantages (High Consumption): Increases
Mortality	Birth defects
Heart disease	Breast cancer
Stroke	Cirrhosis
Kidney disease	Hypertension
Type 2 diabetes	Atrial fibrillation

disadvantages from alcohol abuse. Overall, this suggests caution in the recommended use of alcohol. The World Health Organization guidelines of a maximum of four standard alcoholic drinks a day for males and two for females, with one to two alcohol-free days, are a good basis for recommendations until further evidence is forthcoming.

SUMMARY

Fluids provide a significant contribution to total energy intake and need to be carefully considered and managed in any lifestyle-based program. The general prescription of eight glasses of fluids a day is not scientifically supported and individual needs are likely to vary widely. However, a rough guide for fluid intake is around 1–4 L a day, mainly from water, tea, and coffee, and low fat dairy and soy. Alcohol has some health benefits in moderation but also considerable social and other disadvantages when overconsumed.

Practice Tips

Fluid Intake

	Medical Practitioner	Practice Nurse/Allied Health Professional
Assess	• Fluid intake, not just alcohol. Whether fluids are used compulsively. • The role of fluids in weight control for this patient. • Dental health if using bottled water.	• Fluid intake (see Professional resources). • Habitual use of energy-rich fluids (i.e., if drunk when thirsty). • Need for changes in fluid intake for this patient. • Alcohol intake. • Drink mixes with alcohol. • Body fluid with a BIA measure if available.

Continued

Practice Tips—cont'd

Fluid Intake

	Medical Practitioner	Practice Nurse/Allied Health Professional
Assist	• Cutting back on high-energy fluids. • Appropriate alcohol use. • Eliminating fluids such as fruit juice and choosing appropriate alternatives with artificial sweeteners if needed for tea/coffee.	• Modification of fluid intake for weight control. • Deconditioning of habitual drinking of sweetened fluids. • By developing a strategy for putting this into action. • Education about alternatives to fruit juice. • Access to artificial sweetener samples. • Meeting fluid needs (1–4 L/day) after determining fluid needs by patient parameters (e.g., age, sex) to change.
Arrange	• Referral to a dietitian if necessary. • Referral to a practice nurse for deconditioning and education if appropriate.	• Consultation with a dietitian with knowledge of fluids. • Care plan if necessary. • Contact with and coordination of other health professionals.

Key Points

Clinical Management

- Do not ignore beverage energy intake when discussing weight control and/or health.
- Aim to reduce total energy intake from beverages to less than 10% of total energy.
- Discuss fluid energy requirements and the specific needs of specific patients.
- Encourage use of low energy drinks (preferably water).
- Get patients to recognize thirst as a healthy control mechanism.
- With young patients, in particular, recommend drinking water when thirsty and drinking other drinks as a treat, if necessary, but not to quench thirst.
- Teach patients how to read drink labels to recognize (and avoid) high energy-dense cutoffs.

REFERENCES

Cameron-Smith, D., Egger, G., 2002. The Ultimate Energy Guide. Allen and Unwin, Sydney.

Chrysant, S.G., 2015. Coffee consumption and cardiovascular health. Am. J. Cardiol. 116 (5), 818–821.

Cornelis, M.C., El-Sohemy, A., Kabagambe, E.K., et al., 2006. Coffee, CYP1A2 genotype, and risk of myocardial infarction. JAMA 295, 11135–11141.

Gee, C., 2006. Does alcohol stimulate appetite and energy intake. Br. J. Community Nurs. 11 (7), 298–302.

Jain, P., Nihill, P., Sobkowski, J., Agustin, M.Z., 2007. Commercial soft drinks: pH and in vitro dissolution of enamel. Gen. Dent. 55 (2), 150–154.

Lorenzen, J.K., Nielsen, S., Holst, J.J., et al., 2007. Effect of dairy calcium or supplementary calcium intake on postprandial fat metabolism, appetite and subsequent energy intake. Am. J. Clin. Nutr. 85, 678–687.

Marques-Vidal, P., Ducimetiere, P., Evans, A., et al., 1996. Alcohol consumption and myocardial infarction: a case-control study in France and Northern Ireland. Am. J. Epidemiol. 143, 93–1089.

Millard-Stafford, M., Wendland, D.M., O'Dea, N.K., Norman, T.L., 2012. Thirst and hydration status in everyday life. Nutr. Rev. 70 (Suppl. 2), S147–S151.

Popkin, B., Armstrong, L.E., Bray, G.M., et al., 2006. A new proposed guidance system for beverage consumption in the United States. Am. J. Clin. Nutr. 83, 529–542.

Roglans, N., Vila, L., Farre, M., et al., 2007. Impairment of hepatic Stat-3 activation and reduction of PPAR-alpha activity in fructose-fed rats. Hepatology 45 (3), 778–788.

Sayon-Orea, C., Martinez-Gonzalez, M.A., Bes-Rastrollo, M., 2011. Alcohol consumption and body weight: a systematic review. Nutr. Rev. 69 (8), 419–431.

Sinclair, D.A., Guarente, L., 2006. Unlocking the secret of longevity genes. Sci. Am. 294 (3), 48–51.

Stevenson, R.J., Mahmut, M., Rooney, K., 2015. Individual differences in the interoceptive states of hunger, fullness and thirst. Appetite 95, 44–57.

Tolstrup, J., Jensen, M.K., Tjønneland, A., Overvad, K., Mukamal, K.J., Grønbæk, M., 2006. Prospective study of alcohol drinking patterns and coronary heart disease in women and men. BMJ 332, 1244–1248.

Valtin, H., 2002. "Drink at least eight glasses of water a day." Really? Is there scientific evidence for "8 x 8"? Am. J. Physiol. Regul. Integr. Comp. Physiol. 283, R993–R1004.

Weaver, C., Lupton, J., King, J., et al., 2006. Dietary guidelines versus beverage guidance system. Am. J. Clin. Nutr. 84 (5), 1245–1246, 1246–1248.

Professional Resources

Measuring Fluid Intake

	Points				
Questions	**4**	**3**	**2**	**1**	**0**
How often do you drink alcohol?	Every day	4–5 days/weeks	2–3 days/weeks	1–2 days/weeks	Never
How much do you usually drink (one standard drink is one middie of beer, one nip of spirits, or one glass of wine	More than 10 drinks	5–10 drinks	2–4 drinks	1–2 drinks	Don't drink alcohol
How often do you drink soft drinks? (don't count diet drinks)	Every day	4–5 days/weeks	2–3 days/weeks	1–2 days/weeks	Never

Continued

Professional Resources—cont'd

Measuring Fluid Intake

	Points				
Questions	**4**	**3**	**2**	**1**	**0**
When you do drink soft drinks (not counting diet drinks), how much do you usually drink?	More than 10 drinks	5–10 drinks	2–4 drinks	1–2 drinks	Don't drink soft drinks
How often do you drink fruit juices?	Every day	4–5 days/ weeks	2–3 days/ weeks	1–2 days/ weeks	Never
When you do drink fruit juices, how much do you usually drink?	More than 10 drinks	5–10 drinks	2–4 drinks	1–2 drinks	Don't drink fruit juices
How often do you drink full cream milk?	Every day	4–5 days/ weeks	2–3 days/ weeks	1–2 days/ weeks	Never
When you do drink full cream milk, how much do you usually drink?	More than 10 drinks	5–10 drinks	2–4 drinks	1–2 drinks	Don't drink full cream milk

Scoring: Within each drink category (alcohol, soft drink, fruit juice, milk), add the score for questions 1 and 2. Interpreting your score:

- 1 or less on any combination—drinks are not an influence on your weight
- 1–5 on any combination—drinks could be a problem
- >5 on any combination—drinks are definitely a problem

All the drinks mentioned here can add to energy intake and hence body weight. Consequently, reducing the drink or amount drunk will help with weight loss.

Chapter 10

Behavioral Aspects of Nutrition

Neil King, Garry Egger

No matter how clearly one states the principles of self-help, people often misunderstand or distort them.

Ellis and Harper (1975)

INTRODUCTION: EATING AS A BEHAVIOR

In a laboratory at the University College London in 1999, health researchers fed free chocolate twice a day for 2 weeks to a handful of student volunteers (Gibson and Desmond, 1999), not necessarily because they liked the students, but to show that eating is not just related to biological hunger but is also influenced by psychology and learning.

The researchers took a group of people identified as chocolate "cravers" and a second group who could either "take it or leave it." They split the groups again and trained them by feeding them chocolate twice a day either when they were hungry (i.e., 2 h after their last meal) or after they had just eaten. After 2 weeks, everyone was asked to rate their cravings when they were hungry or, on another day, when they were full. Cravers and noncravers who ate chocolate exclusively when hungry increased their chocolate craving posttraining but, at least for the cravers, only when ratings were made while they were hungry. For those trained when they were full, chocolate craving decreased posttraining, but this decrease did not depend on whether the subjects were currently hungry or full.

As complicated as this may sound, it simply suggests that cravings for chocolate or other foods may be an expression of a strong appetite or a learned desire for that food, elicited by hunger and acquired by repeatedly eating the craved food when hungry. In other words, eat a highly palatable and desirable food (e.g., chocolate) when you are hungry, and hunger will come to be associated (i.e., conditioned) with that food. Similarly, drinking a sweetened soft drink when you are thirsty could lead to thirst being associated with a desire for sweetened soft drink, rather than water.

The implications of this are that eating and exercise behaviors are often learned or acquired. In most individuals, eating behavior (e.g., food preference, meal pattern, eating frequency) has been established over many years

Lifestyle Medicine. http://dx.doi.org/10.1016/B978-0-12-810401-9.00010-3

and is therefore resistant to short-term changes. Of course, it is possible that they can be unlearned through such techniques as self-monitoring, stimulus control, problem solving, contingency management, cognitive restructuring, social support, and stress management (Wadden and Clark, 2005). However, it is important to recognize that eating behavior is embedded in one's behavioral lifestyle—it will be difficult to change overnight. In this chapter, we look at the influence of learning on eating behavior, as well as the influence of other factors such as mood states and eating context, and how these can all be managed in the clinical situation.

HUNGER AND APPETITE: ARE THEY THE SAME?

In the scientific literature and general conversation, the terms "hunger" and "appetite" are often used synonymously. This is also the case in a typical medical dictionary. However, intuitively there seems to be a difference. Appetite is a generic term, which encompasses a range of factors associated with eating. Hunger is a more specific term, usually associated with the biological drive or motivation to eat; hence, it is representative of a state. As defined in the *Macquarie Dictionary*, hunger is "the painful sensation or state of exhaustion caused by need of food" whereas appetite is "an innate or acquired demand or propensity to satisfy a 'want'."

The distinction between *need* and *want* is important. Hunger is genuinely physiological and is based on a biological need that cannot be willed away, which increases with the passage of time during food deprivation, and which alternative distractions cannot reduce. Only in extreme cases (e.g., anorexia nervosa, hunger strikers) is it possible to override the intense, nagging pain of hunger. Therefore hunger is associated with homeostasis and reflects the state of energy needs (Blundell and Finlayson, 2004).

"Want" could be described more as a reflection of desire and may be uncoupled with the need to eat. The terms "liking" and "wanting" have been used to differentiate between the hedonic and homeostatic association with food intake (Berridge, 2004; Mela, 2006). Liking and wanting have been defined as affective and motivational factors of food intake respectively (Finlayson et al., 2007a,b).

Hunger is of genuine concern in an environment where overconsumption can lead to energy imbalance and obesity, especially when we know that individuals eat in the absence of hunger or satiety cues (e.g., buying a pastry when passing a pleasant-smelling bakery or overconsuming at buffets). Therefore hunger may be a sufficient but not a necessary cue for eating. The availability of food in the current environment encourages eating, particularly in the absence of hunger. Some ways of manipulating nutrient content to reduce (or delay) genuine hunger are discussed in Chapter 8. A list of these and other approaches is contained in Table 10.1.

TABLE 10.1 Possible Ways of Reducing/Delaying Biological Hunger

Tactic	Process
Use a recognized meal replacement	Replace meals or take before a main meal or before going out for a social meal
Increase high-fiber, low-energy dense foods	Increase food bulk while reducing calorific volume (e.g., by eating more fruit, vegetables, pasta, cereal)
Increase "fast" protein	Use, for example, whey from dairy, soy, and possibly seafood to replace fat and high glycemic index carbohydrate
Distinguish between hunger and appetite	Rate hunger on a virtual scale (see later); only eat when biologically hungry
Select low-fat venues when eating out	Select generally low-fat foods for social eating
Eat more food earlier in the day	Higher intake of low-energy dense, high-fiber foods in the morning reduces later intake and is "burned off" more readily
Decondition "cravings"	Separate stimulus and response through standard behavioral principles (see later)
Eat breakfast	A low-energy dense, high-fiber breakfast reduces later food intake and helps "kick-start" the metabolism
Snack regularly	Regular healthy (low-energy dense, high-fiber) snacks every 4 h or so reduce chances of overeating through becoming too hungry
Avoid salty foods and foods high in monosodium glutamate	Foods with added salt (e.g., peanuts, chips) tend to have less effect on satiation, meaning more can be eaten
Increase spicy foods	Taken as an entrée or part of a main meal, these can increase metabolism and decrease total food intake
Drink caffeine (for some people)	Caffeine may reduce hunger and food intake in some people, depending on their genetic-based reaction to caffeine
Become "friends" with a mild feeling of hunger	Mild levels of hunger are innervating and stimulating. This should not be seen as something to always be avoided
Use hunger-suppressing medication	Some medications can have a genuine hunger-suppressing effect. Phentermine can have short-term effects

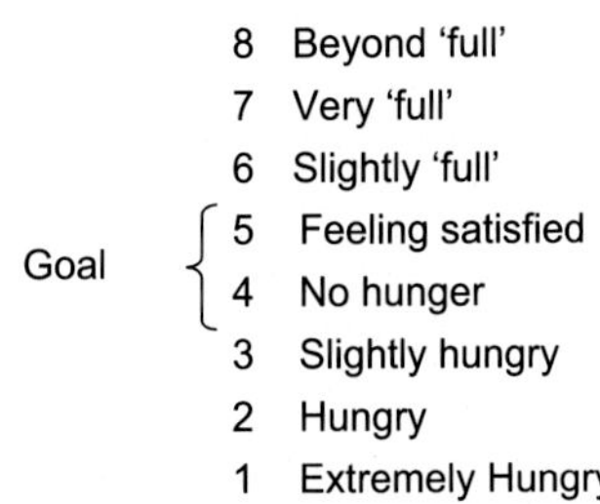

FIGURE 10.1 Suggested hunger scale.

Genuine hunger is best determined by the use of a Likert scale (Fig. 10.1).

Patients should be encouraged to maintain hunger levels between 3 and 5 on the hunger scale, becoming neither too hungry nor too full. If they become too hungry, they could potentially select the wrong kinds of food to fill up on and subsequently consume more energy over the day.

Another type of drive, whether it be labeled appetite or learned hunger, is more psychological, and is likely to diminish in intensity if one is distracted and does not necessarily increase over time. It is the identification of these two drives that enables a differentiation between the different types of drives for food and drink.

DETERMINANTS OF ENERGY INTAKE

In animals, food seeking is primarily initiated by feelings of hunger, mediated by hormones, such as ghrelin and neuropeptide Y, originating from the gastrointestinal tract and the hypothalamus. Satiation and satiety signals also arise from the intestinal tract and, in between meals, from adipose tissue and the liver. Several years ago, Blundell (1990) made a clear distinction between satiation and satiety. Satiation refers to the processes involved in the termination of a meal (within meal) and satiety refers to the processes after a meal that influence the next eating episode. Satiety signals arrest in the processing of food in the intestine and leads to the termination of eating. In a natural environment, hunger thus becomes a major determinant of energy intake (EI). However, psychological factors (e.g., dietary restraint, dietary disinhibition) could also influence the onset and termination of eating episodes. The processes involved in satiation and satiety were clearly described using the satiety cascade (Blundell, 1990). Fig. 10.2 displays the various psychological and biological processes involved in satiation and satiety.

In a modern environment, hunger is just one of several factors known to be associated with EI. Some of the other factors are shown in Fig. 10.3.

Some of the physical factors influencing intake shown in Fig. 10.3 are discussed in Chapter 8. However, a problem with regulating food intake in the modern environment is the readily available, highly palatable, and energy-dense nutrients in foods, such as sucrose and fat. This upregulates the expression of

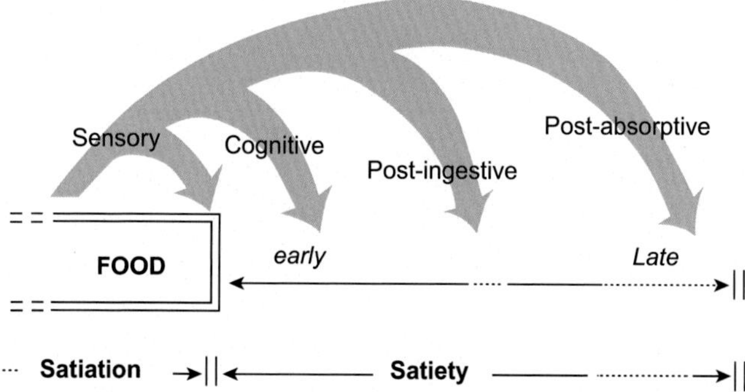

FIGURE 10.2 Satiety cascade: the temporal profile of processes involved in bringing a meal to an end (satiation) and influencing the onset of the next eating episode (satiety). *From Blundell, J.E., 1990. How culture undermines the biopsychological system of appetite control. Appetite 14 (2), 113–115.*

hunger, at the same time blunting the response to satiety signals and activating the reward system, making eating as much a learned as a biological drive in the modern world. This can create habits relating to eating and drinking, some of which are caused by simple conditioning (referred to here as behavioral habits) but others may be more complex emotionally (described here as cognitive habits).

"... but this was an exception. I only ever eat this much on birthdays, public holidays, weekends and weekdays ending in a 'Y.' "

HABITS

Habits are learned actions that become automatic—they are ways of responding that enable humans (and animals) to get through the day without being too distracted by repetitive actions. In the context of eating behaviors, habits can be

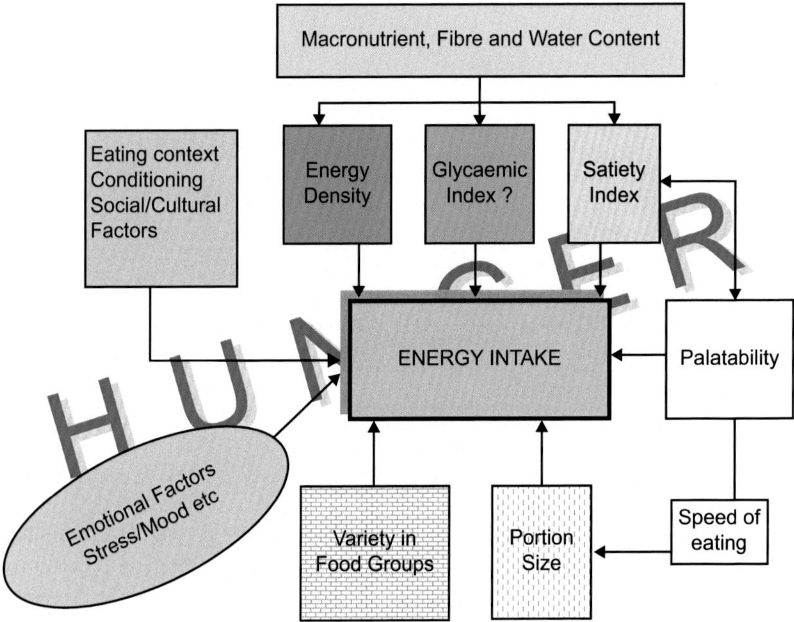

FIGURE 10.3 Determinants of energy intake in humans.

either behavioral or cognitive. However, unlike walking, eating food (the act of putting food in the mouth) is a conscious behavior of which individuals are usually aware. Of course, an individual may have awareness of the act of eating per se while remaining unaware of the mechanisms (e.g., ghrelin) that determine the expression of appetite and the pattern of eating. However, eating is a deliberate and volitional behavior, and individuals should be made aware of this and thus control their own eating behaviors.

Behavioral Habits

In their simplest form, habits develop as an association between a stimulus and a response. Initially, a stimulus that initiates a response is known as an unconditioned stimulus (US), resulting in an unconditioned response (UR). The stimulus is then paired with a conditioned stimulus (CS) to give a conditioned response (CR). In psychological terms this is known as classical conditioning, perhaps the best known example of which is the salivatory experiences of Pavlov's dogs, who became conditioned to the ringing of a bell (CS) such that it soon signaled food (US) and caused a salivatory response (CR) even without presentation and digestion of the meal (UR).

Similar conditioning processes can occur in human eating behavior. An example of a typical learned response to eat in modern society, for example, would be getting up from a chair to get something to eat every time an advertisement interrupts a program on television. The association between the advertisement and eating is a surprisingly simple connection but is the kind of thing that people do every day without conscious thought. Other examples of eating behaviors that are conditioned include:

- Eating at the same time every day whether hungry or not;
- Reading while eating (and thus getting hungry while reading);
- Always finishing everything on the plate;
- Eating energy-dense, savory snacks (e.g., peanuts, chips) with alcohol; and
- Eating energy-dense, sweet snacks (e.g., cakes, biscuits) with tea or coffee.

Behavioral habits are relatively easily dealt with through behavior modification principles. These include deconditioning, stimulus control, self-monitoring, goal setting, cognitive restructuring, and relapse prevention.

Deconditioning

This is based on the notion that eating behaviors are often conditioned by a stimulus–response connection. The first stage of recognizing this is by monitoring behavior or "stalking the behavior, like a hunter stalks his prey" (Tupling, 1995) through self-monitoring (see later). This can be done by keeping a diary that not only lists the foods eaten but the emotions at the time of eating. Once a conditioned behavior (and accompanying mood states) has been identified, it can then be modified by:

- Limiting exposure to the stimulus: If having a beer leads to eating peanuts, the beer would be avoided, or limited, hence leading to "extinction" of the response.
- Changing the response: In the preceding case, an alternative response, such as eating less energy-dense foods (e.g., Japanese peas) or not eating with drinks, could be encouraged.
- Changing the stimulus–response connection: In this case, a glass of wine might be paired with fruit to develop a new connection.

Stimulus Control

Based on the notion that environmental antecedents control behaviors, stimulus control involves changing the environments that encourage unhealthy eating. Participants in weight control programs, for example, are encouraged to increase their purchase of fruit and vegetables, wash and prepare these for easy eating, and place them prominently in the refrigerator. In contrast, unhealthy foods would be kept out of the house or in places with difficult access.

What Can Be Done in a Brief Consultation?

- ☐ Provide a food diary to check on consequent behavior with food/drink intake.
- ☐ Check the health habits checklist provided in Professional Resources.
- ☐ Determine the importance of eating and drinking habits for particular patients.
- ☐ Ask about the conditions under which special high-energy treats are consumed.
- ☐ Advise not to drink sweet drinks when thirsty or eat treats when hungry.
- ☐ Encourage patients to self-monitor their eating and exercise habits.
- ☐ Prescribe the DAB-Q questionnaire (see Chapter 8 and www.lifestylemedicine.com.au).

Self-Monitoring

It is well documented that most people underestimate the amount they eat. This is even more so with the overweight (and underreporting is also nutrient specific, that is, individuals selectively underreport the fatty foods, misrepresenting total food intake) (Goris et al., 2000). Evidence shows that even dietitians, who are trained to accurately report food intake, misreport their own food intake (Goris and Westerterp, 1999). Self-monitoring is one way of overcoming this and also providing feedback that reinforces behavior change. This can be in the form of a diet diary, eating record, or diet recall. Self-monitoring of any behavior increases self-awareness and, in the case of unhealthy behaviors, is likely to improve that particular behavior. This is the principle on which most commercial weight management groups work. A reduction in body weight is a likely short-term outcome of self-reporting food intake.

Because of what can be labeled the "eye–mouth gap," there are limitations in accepting records on which to base prescription from a diet diary, given that not all foods are necessarily included. However, with the correct training and stressing the importance of compliance, self-reported food records can be a valid and reliable indicator of food intake. Indeed, evidence has shown that confronting people with their own misreported food intakes leads to improvements in the accuracy of their self-reporting (Goris et al., 2000). Furthermore, the process of recording food intake can lead to a change in behavior such that dietary records are not indicative of usual dietary intake. On the other hand, dietary diaries can be used to help patients ensure they are on track with their dietary intakes.

By including provision for recording mood status at the time of eating, information can also be gained about possible emotional causes of extra EI. An example of a self-report food diary is shown in the Professional Resources at the end of this chapter. Generally, patients complete the date, time, and detailed information about the food or drink consumed. Other methods of assessing food intake include the food frequency questionnaire, which assesses the frequency of food consumption (again, see Professional Resources).

Goal Setting

Short-term, medium-term, and long-term goals help provide an incentive for behavior change. These can be process goals (such as reducing chocolate intake by one-half each week or increasing fruit to three different pieces a day) or outcome goals (such as changes in body weight or blood fats). Goals should be realistic and achievable.

Cognitive Restructuring

This is similar to the cognitive behavior therapy approaches considered in Chapter 14 and includes modifying cognitive as well as physical cues for eating, such as the sight and smell of food. Internal thoughts such as "because I have had a bad day I deserve a treat" need to be reframed to lead to more appropriate eating behavior. Dividing foods into good or bad, developing excuses for irrational behavior (see later), and making comparisons with other people can all serve as negative thoughts that need to be restructured, often with the help of an experienced psychologist or counselor.

Relapse Prevention

This involves teaching patients to anticipate the types of situations that may cause them to lapse back into poor eating habits and to plan strategies for coping with these situations. Based on principles of behavior change, relapse should be regarded as a normal process of relearning.

Some simple approaches to modifying eating behavior are shown in Table 10.2.

Cognitive Habits

Cognitive habits are more complex and are learned patterns of thinking, often involving emotion and including thoughts of depression, failure, worthlessness, frustration, and unrealistic ideals. These can lead to cyclical ways of thinking that feed back on themselves, such as through a diet cycle as shown in Fig. 10.4.

Dealing with cognitive habits often involves detailed psychological techniques, ranging from psychoanalysis to cognitive behavior therapy, that enable the cycle of thinking shown in Fig. 10.4 to be broken. Many modern psychology practitioners, however, recognize that it is thinking about what may happen as much as what actually has happened, or is about to happen,

TABLE 10.2 Alternative Approaches to Modifying Eating Behavior

Cue Elimination and Physical Environment	Manner of Eating	Food Choice	Alternative Activities
Eat only in designated place	Slow rate of eating	Cut snacks in half	Exercise
Eat only when sitting	Swallow each bite before taking a second	Measure portions	Relaxation
Set regular eating times	Put utensils down between bites	Serve only amounts preplanned	Meditation
Plan snacks and meals ahead	Pause in eating	Share dessert	Drink water
Rate hunger before eating	Relax before eating	Include favorite foods	Imagery
Dissociate eating from other activities	Savor foods—enjoy each bite	Eat a variety of foods	Tasks
Plan restaurant meals ahead	Eat only until satisfied	Have snacks ready	Call someone
Store foods in inaccessible places	Allow time for eating	Serve dressings "on-the-side"	Write a letter
Store treats in opaque containers	Leave some meal uneaten	Use spices instead of condiments	Reevaluate goals
Use small plates and bowls	Push food aside ahead of time	Use garnishes	Practice assertiveness
Record food intake	Cover food when finished	Use low-calorie substitutes	Chart progress
Shop when not hungry			
Use a list when shopping			
Avoid "problem" places and people			
Remove plates after meal			
Clean plates immediately			
Write notes as reminders			

that causes many pathologies, phobias, and simple unconstructive behaviors. This is not new. Indeed, there are a range of famous quotations dating back over the years to suggest this is the case (Table 10.3). The basics of one approach, called rational emotive therapy (RET), which forms the basis of many modern therapeutic techniques, are considered later.

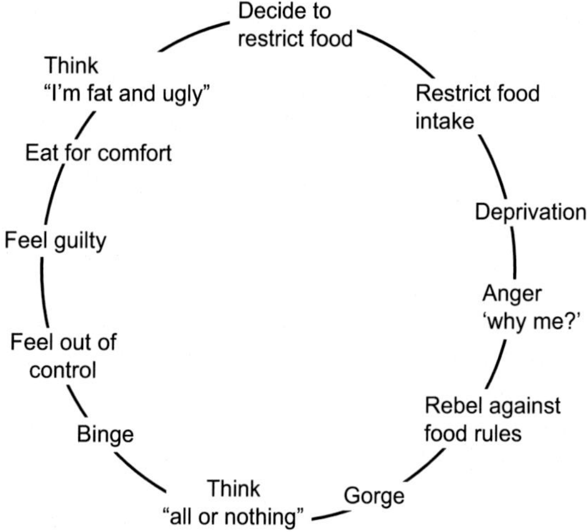

FIGURE 10.4 A vicious cycle of thinking associated with obesity.

TABLE 10.3 The Significance of Thought Patterns in Behavior
We are a result of everything we have thought. Guatama Siddhartha (the Buddha, 500 BC)
People are disturbed not by things, but by the views they take of them. Epictetus, AD 50
There's nothing either good or bad, but thinking makes it so. Shakespeare (*Hamlet*)
My life has been full of catastrophes, most of which have never happened. Mark Twain

RATIONAL EMOTIVE THERAPY

RET was developed by US psychologist Albert Ellis (Ellis and Harper, 1979) to help counter the irrational beliefs that cause much minor psychopathology. RET is based around a simple format using the letters ABCD:

A Adversity
B Beliefs about that adversity (both rational and irrational)
C The Consequences of beliefs about that adversity
D Disputing of irrational beliefs about the adversity

Ellis claimed that, "We consciously and unconsciously choose to think and hence feel in certain self helping and self harming ways" and "once you understand the basic irrational beliefs you create to upset yourself, you can use this understanding to explore, attack and surrender your other present and future emotional problems." This has been picked up by others working in the area, such as Carlson (2003), who suggested that, "being upset by your own thoughts is similar to writing yourself a nasty letter—and then being offended by that letter!"

In relation to eating behavior, adversity may be a situation that breaks the pattern of healthy eating, such as a binge or overeating session. Adversity then, according to Ellis, usually takes a "jump," leading directly to C or the consequences of this adversity. After bingeing, an overweight person might think, "I'm useless. I can never get anything right. It could only happen to me." However, it is the jump here from A to C that results in the form of thinking that may lead to resignation and relapse.

Weight cycling, resulting from binge eating and then dieting, is a common phenomenon with potential influences from this kind of thought process. While some degree of restrained eating to prevent weight gain may be required, there is often an irrationality of cognitive processes, instigated by unrealistic social pressures. It is the beliefs about this that are the real cause of much of the consequences. According to Ellis, a person's irrational beliefs are generally associated with thoughts of must, should, and have to:

- I *must* always be perfect.
- You *should* treat me well or you are an awful person.
- The world *has to* always be good to me.

Ellis also discusses 10 common forms of irrational beliefs that are self-limiting in this sense and which make up the basis of RET. These, and some examples related to weight control, are shown in Table 10.4.

RET is a relatively simple approach to a complex problem. This, or the principles coming from it, can be applied to several aspects of behavior, including eating and good nutrition. Psychologists skilled in this approach should be part of the referral cycle for detailed counseling on changing cognitive habits; however, other clinicians can often recognize the irrationality of beliefs relating to food and eating and their adverse effects on good nutrition Table 10.5.

TABLE 10.4 Ten Common Forms of Irrational Thinking Relating to Body Fat Maintenance

- All or nothing. ("If I don't starve, I'll get fat.")
- Overgeneralization. ("Things always go wrong.")
- Mental filter. ("It's not the program, it's me that's wrong.")
- Disqualifying the positive. ("That was just luck.")
- Jumping to conclusions:
  - Mind reading. ("She thinks I'm useless.")
  - Fortune telling. ("If I did lose fat, he wouldn't like me.")
- Catastrophizing. ("The world will end tomorrow if I fail.")
- Emotional reasoning. ("I feel so upset because I'm a failure.")
- "Musturbating." ("I must never eat anything fattening.")
- Labeling. ("I'm a loser.")The 3Ps: personal, permanent, and pervasive. "It only happens to me; I'll always be like this; It's the same with everything I do."

From Ellis, A., Harper, R.A., 1979. A New Guide to Rational Living. Wilshire Book Company, California.

TABLE 10.5 An Example of a Food Frequency Questionnaire

How Often Do You Eat the Following Foods	2 or More Times a Day	Every Day	3–5 Times a Week	1–2 Times a Week	1–3 Times a Month	Rarely	Never
White bread	□	□	□	□	□	□	□
Brown/grainy bread	□	□	□	□	□	□	□
Wholemeal bread	□	□	□	□	□	□	□
Sweet biscuits	□	□	□	□	□	□	□
Crackers/crispbread	□	□	□	□	□	□	□
Cakes, buns, pastries, etc.	□	□	□	□	□	□	□

SUMMARY

Food and drink are not the only aspects of healthy nutrition. Much is dependent on eating behavior. Without knowledge of this, little long-term benefit is likely to be achieved. Two significant components of eating behavior are behavioral habits (wrong ways of doing) and cognitive habits (wrong ways of thinking). Strategies for managing eating behavior, particularly in the presence of a plentiful food environment, are a significant component of Lifestyle Medicine.

Practice Tips

Changing Eating Behaviors

	Medical Practitioner	Practice Nurse/Allied Health Professional
Assess	• Whether eating behavior is an issue • Emotional/cognitive issues involved in eating patterns • Diet record/diary • Comorbidities and treat them • Potential eating disorders (e.g., using DAB-Q (see www.lifestylemedicine.com.au)	• Type of eating behavior that may be an issue • Emotional/cognitive issues involved in eating patterns • Conditioned stimuli causing wrong eating • Mood factors influencing eating (from a diet diary)
Assist	• Understanding of eating patterns and their effect on health • In differentiating between simple habits and complex cognitive/emotional issues	• Education about eating behavior • Development of a plan for stimulus–response control if appropriate • Patient understanding of the need for more detailed psychological help where required
Arrange	• Referral for psychological help where appropriate • Consultations with other health professionals for a care plan	• Discussion of motivational tactics with other involved health professionals • Referral to a behavioral specialist skilled in eating behaviors and/or weight control issues • Involvement with self-help groups if wanted

Key Points

Clinical Management

- Recognize the importance of eating patterns in nutritional advice.
- Try to find and deal with the cause of emotions leading to emotional eating.
- Distinguish between wrong behavioral and wrong cognitive habits (wrong ways of acting vs. wrong ways of thinking).
- Refer to a psychologist with an understanding of eating behaviors for detailed care.
- Break stimulus–response pairing in simple conditioned eating behavior.

REFERENCES

Berridge, K., 2004. Motivation concepts in behavioural neuroscience. Physiol. Behav. 81, 179–209.

Blundell, J.E., 1990. How culture undermines the biopsychological system of appetite control. Appetite 14 (2), 113–115.

Blundell, J.E., Finlayson, G.S., 2004. Is susceptibility to weight gain characterised by homeostatic or hedonic risk factors for over consumption? Physiol. Behav. 82, 21–25.

Carlson, R., 2003. Stop Thinking. Start Living, second ed. Harper Collins, London.

Ellis, A., Harper, R.A., 1975. A New Guide to Rational Living. Wilshire, North Hollywood, CA.

Ellis, A., Harper, R.A., 1979. A New Guide to Rational Living. Wilshire Book Company, California.

Finlayson, G.S., King, N.A., Blundell, J.E., 2007a. Is it possible to dissociate "liking" and "wanting" for foods in humans? A novel experimental procedure. Physiol. Behav. 90, 36–42.

Finlayson, G.S., King, N.A., Blundell, J.E., 2007b. The role of implicit wanting in relation to explicit liking and wanting for food: implications for appetite control. Neurosci. Behav. 31, 987–1002.

Gibson, E.L., Desmond, E., 1999. Chocolate craving and hunger state: implications for the acquisition and expression of appetite and food choice. Appetite 32 (2), 219–240.

Goris, A.H., Westerterp, K.R., 1999. Underreporting of habitual food intake is explained by undereating in highly motivated lean women. J. Nutr. 129 (4), 878–882.

Goris, A.H., Westerterp, K.R., 2000. Improved reporting of habitual food intake after confrontation with earlier results on food reporting. Br. J. Nutr. 83 (4), 363–369.

Goris, A.H., Westerterp-Plantenga, M.S., Westerterp, K.R., 2000. Undereating and underrecording of habitual food intake in obese men: selective underreporting of fat intake. Am. J. Clin. Nutr. 71 (1), 130–134.

Mela, D.J., 2006. Eating for pleasure or just wanting to eat? Reconsidering sensory hedonic responses as a driver of obesity. Appetite 47, 10–17.

Tupling, H., 1995. A Weight off Your Mind. Allen and Unwin, Sydney.

Wadden, T., Clark, V.L., 2005. Behavioral treatment of obesity: achievements and challenges. In: Kopleman, P.G., Caterson, I.D., Dietz, W.H. (Eds.), Clinical Obesity in Adults and Children. Blackwell Publishing, Oxford.

FURTHER READING

Egger, G., Swinburn, B., 1996. The Fat Loss Leader's Handbook. Allen and Unwin, Sydney.

Professional Resources

Measuring Eating Behavior

DAB-Q for measuring food/drink intake and eating habits

See Chapter 8 Professional Resources and www.lifestylemedcine.com.au/DAB-Q.

Example of a food diary

Time	Food/Drink	Portion/Serve Size	Description of Cooking Method
______	__________	______________	______________
______	__________	______________	______________
______	__________	______________	______________

Chapter 11

Physical Activity: Generic Prescription for Health

Mike Climstein, Garry Egger

If we could give every individual the right amount of nourishment and exercise, not too little and not too much, we would have found the safest way to health.

Hippocrates

INTRODUCTION: EXERCISE AND EVOLUTION

According to the great 20th century Swedish exercise physiologist Professor Per-Olof Åstrand, if the whole of the history of *Homo sapiens* was a 400 m (1312 ft) race, a 100-year-old person living today would have participated only in the last 10 mm (0.4 in.). For all but the last 2 mm (0.07 in.), anyone participating in this race would have been active enough not to need gyms, swimming pools, or fancy running shoes. Indeed, except for the very rich or elite (who had servants to do things for them), physical activity was part of the daily grind of staying alive.

This has led scientists to propose that a genome with genes selected for physical activity has "fed forward" to healthy metabolism in humans over hundreds of thousands of years (Booth et al., 2002; Rowland, 1988) but this has been disrupted by the modern inactive lifestyle, which began in the second half of the 20th century and accelerated with the advent of advanced technologies (e.g., the Internet) in its final years. Fig. 11.1 illustrates the theoretical trend and possible reasons for changes in ambient activity levels in Western populations over the last century. Estimates from this and other paleoanthropological studies suggest a decrease in activity levels in recent times of up to 60% compared with about 100 years ago. This is the equivalent of around 4200 kJ (1000 kcal) per day less or walking about 16 km (9.94 mL) less each day.

The disruption to healthy gene expression from this drastic change in physical activity and consequent energy expenditure has contributed to the rise of obesity throughout the world and its accompanying metabolic and mechanical pathologies. According to Peters et al. (2002), this has resulted from a change in the environment that historically has "pushed" humans to be active to one that

Lifestyle Medicine. http://dx.doi.org/10.1016/B978-0-12-810401-9.00011-5

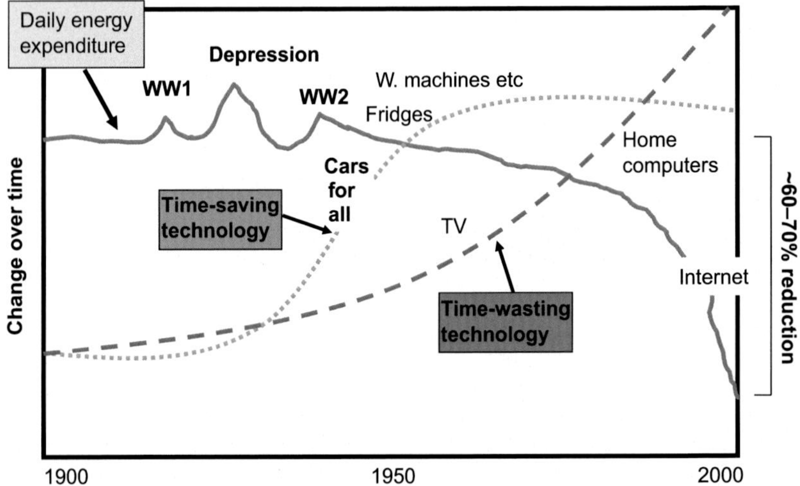

FIGURE 11.1 Changes in activity levels over time. *Adapted from Vogels, N., Egger, G., Plasqi, G., et al., 2004. Estimating changes in daily physical activity levels over time: implications for health interventions from a novel approach. Int. J. Sports Med. 25, 607–610.*

requires "pull," and hence a shift from "instinct" to "intellect" to attain a healthy gene expression. In other words, most humans now have to consciously plan extracurricular physical activity in their daily lives because this does not happen naturally. Seen in this light, it is not the prescription of a (unusual) high level of exercise that is required to reduce disease but the reduction of inactivity caused by industrialization and modern technology. Similarly, it is not so much the benefits of exercise or physical activity (see box for the distinction between the two) that we need to promote as part of a disease prevention, but the disbenefits of being sedentary or inactive.

DEFINING TERMS

Movement	Any musculoskeletal activity of a person.
Physical activity	Any musculoskeletal activity that involves significant movement of body or limbs.
Exercise	A type of physical activity defined as a planned, structured, and repetitive bodily movement done to improve or maintain physical fitness.
Fitness	The capacity of the heart and lungs to pump blood to the working muscles and of the muscles to use this oxygen supply to carry out work.
Health	Metabolic well-being together with the absence of physical impairment and enjoyment of physical and psychological well-being.

ACTIVITY LEVELS, MORTALITY, MORBIDITY, AND WELL-BEING

Although physical activity has been seen as healthy since the days of Hippocrates, the first modern evidence of its value came from a study of 31,000 London bus drivers by English epidemiologist Dr. Jerry Morris in the 1950s (Morris et al., 1953). Morris established that bus drivers wore larger uniforms than conductors, who were much more active on the job. The drivers also had a higher incidence of heart disease, suggesting a link between heart disease, early death, and inactivity.

Since then, the link between inactivity and early death, as well as a range of diseases and measures of well-being, has been clearly established in a number of major early publications (e.g., see Bouchard et al., 1994; Warburton et al., 2006). Much of this work has come from the great epidemiological researcher Prof. Steven Blair from the Cooper Aerobics Institute in Dallas, Texas (e.g., see Blair et al., 2001; Lee et al., 2005). Blair's team measured the baseline fitness of large cohorts of men and women who attended the institute for health screening. They followed them for up to 30 years, measuring mortality and morbidity and relating these to measures of fitness and activity. Fig. 11.2 illustrates mortality from cardiovascular disease in men and women categorized by tertiles of activity and fitness levels at baseline and follow-up. In general, the gradients are steeper for fitness than activity, but both are significant. This suggests that, while there are health benefits in being active

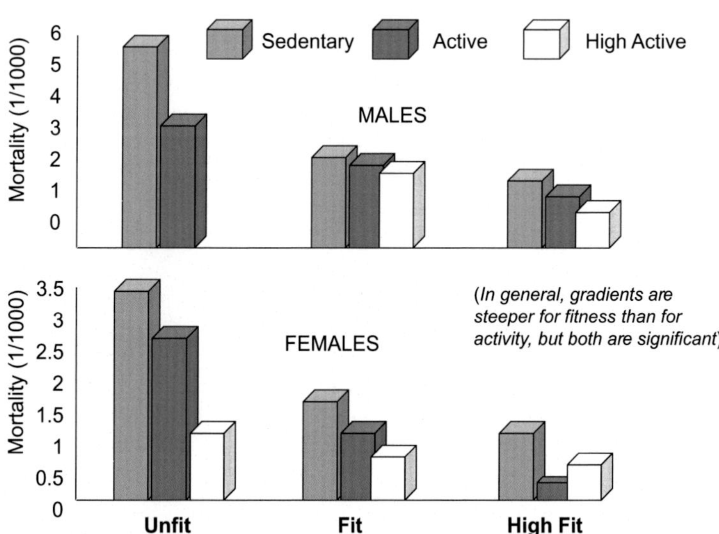

FIGURE 11.2 Mortality by fitness and physical activity categories. *Adapted from Blair, S.N., Cheng, V., Holder, J.S., 2001. Is physical activity or physical fitness more important in defining health benefits? Med. Sci. Sports Exerc. 33 (6 Suppl. 2), S379–S399.*

over being inactive, there are even greater benefits from being aerobically fit, which comes from being vigorously active on a regular basis.

This information has been used in developing National Physical Activity Guidelines (ABS, 2013), which have been adapted to provide a prescription that can be given to patients (as shown in Fig. 11.3).

The Guidelines provide a broad generic prescription for movement, physical activity, and exercise. This is summed up in a simple formula relating the health benefits of physical activity to the "volume" of energy expended (the formula balances the one used in this chapter for nutrition):

> **Volume of Physical Activity** (energy expended) = frequency (how often) × intensity (how hard) × duration (how long)

In today's environment, there are two prescriptions for accomplishing this. The first is to do less passive activity (e.g., watching TV, using computers, searching the Internet, playing video games). The second is to do more *incidental activity* or *planned activity*. "Incidental" activity refers to physical effort that could otherwise be carried out by machines or other people (e.g., walking instead of driving, using stairs instead of the lift) and relates well to the first of the Guidelines shown in Fig. 11.3. Clearly, incidental activity can contribute to the required energy volume for attaining health benefit(s). However, in the modern environment it is often quite difficult to achieve the required volume without some form of planned physical activity, such as going for a walk, riding a bike, or going to the gym.

Name: .... D.O.B........................ Male/Female

1. "***Think of movement as an opportunity, not an inconvenience***": Instead of using machines or other people, try adding the following to your day:

a... b..

c... d..

2. "***Be active every day in as many ways as possible***" Try more of the following:
Gardening ☐ Moving more around the house ☐ Standing instead of sitting ☐ Doing things by hand ☐ Playing with kids ☐ Playing games ☐ Active recreation ☐
Other ....

3. "***Put together at least 30 minutes of mild-moderate intensity activity on most, preferably all, days.***"
Walking ☐ No. of steps/day .... Other activity(s) ....

4. "***If you can, also enjoy some regular, vigorous exercise for extra health and fitness.***" For this patient Yes ☐ No ☐
If yes, list activity(s)..
Times/week 1-2 ☐ 3-4 ☐ 5-7 ☐
Special Requirements:
....
Review: Next visit ______

FIGURE 11.3 Physical activity guidelines. *Adapted from NH&MRC (National Health and Medical Research Council), 2005. National Physical Activity Guidelines. Aust Govt Press, Canberra ACT.*

Organizations such as the American College of Sports Medicine have also made use of widespread research findings to develop more specific prescriptions for a range of chronic diseases and disabilities (Durstine and Moore, 2003), from primary prevention to rehabilitation. A broad outline of these is given in Chapter 12, but before addressing specific exercise prescriptions for chronic diseases and disorders we will look at a system for analyzing the different types of activities and their impacts on different areas of health.

TYPES OF FITNESS: THE FIVE S'S

Fitness is a broad, general term referring to the ability to carry out a physical task. In exercise science, this can refer to a number of different types of physical abilities ranging from agility to strength, many of which are related to skill or athletic ability. While important for sport scientists for improving athletic performance, these are of less concern for the health of the population. Health-related fitness can be loosely categorized into five S's: stamina, strength, suppleness, size, and stability. Although not strictly a type of "fitness," size has been added here to indicate the importance of physical activity to body fatness in the modern environment and because it is a critical parameter in reducing falls in the elderly and improving glycemic control in people with diabetes, as will be seen in the next chapter.

Stamina

Stamina is another term for aerobic or cardiovascular fitness. This refers to the capacity of the heart and lungs to pump oxygenated blood to the working muscles. Although there is a mountain of reference material describing how to increase this parameter of fitness (e.g., see Egger at al., 1999), our chief concern here is how to bring patients to a level of physical activity that is likely to be of benefit to their health. As described by Blair et al. (2001) and the Australian Bureau of Statistics (2013), physical activity alone confers some benefits; however, fitness (from more intense activity or moderate to intense exercise) adds further benefit. In terms of the formula shown previously, this suggests that volume—determined by frequent activity over long (preferably, continuous) periods at low intensity—will confer benefits for weight loss and health (and provide some level of fitness in the totally inactive). However, activity at a higher intensity, carried out less often and for less time, will provide added fitness and cardiovascular benefits.

At a population level, the greatest health gains will come from the most sedentary and inactive people in the population becoming even slightly more active. This is shown graphically in Fig. 11.4. As can be seen, the benefits for the already active becoming even more active occur at a diminishing rate of returns. In other words, for a totally sedentary person, a daily walk around the block will significantly benefit his or her health, whereas an athlete exercising

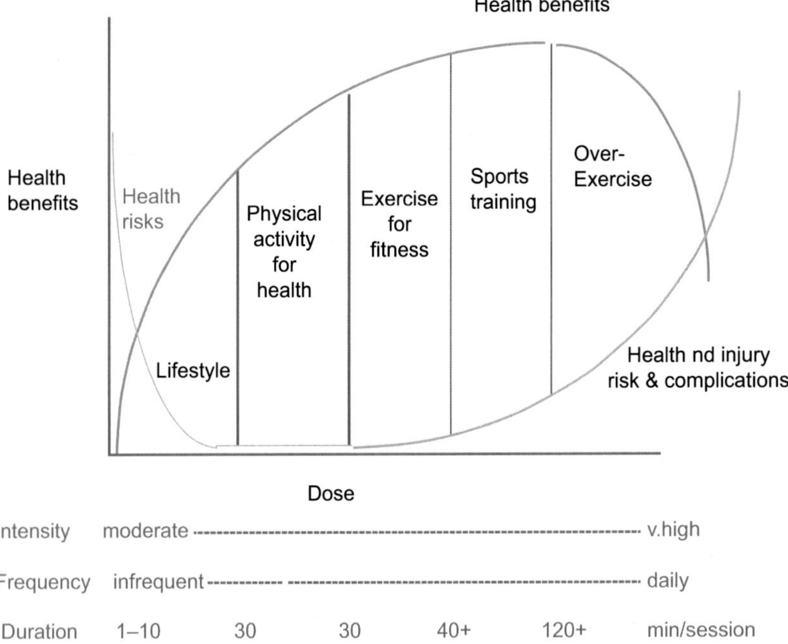

FIGURE 11.4 Benefits of increasing physical activity.

3 h a day may need an extra hour to attain even a minor increase in benefit. There is also undoubtedly a point beyond which injury risk and complications rise and tend to make added exercise less beneficial (O'Connor and Puetz, 2005).

As a generic prescription for stamina, aerobic activity (e.g., walking, jogging, cycling) should be carried out on most days of the week for 30–60 min (depending on intensity and desired outcome), or a minimum of around 2500 calories or 11,000 kJ of energy expended per week. At the lower end of intensity (see prescription for "size," following), this can be accumulated over the course of a day and is determined by time spent or the distance covered in an activity that uses the large muscles of the body, such as walking, cycling, rowing, or swimming. At this level, a measure of intensity, such as heart rate, is not necessary. For improvements in fitness, higher levels of intensity are required, and a heart rate training intensity of 60–80% of maximum heart rate (determined by the formula 220—age) is recommended. An alternative to estimating exercise intensity via heart rate is to aim for a rating of 12–13 on a 20-point scale of rating of perceived exertion (Borg, 1998), compared to around 8–10 for improved health.

Vigorous activity should not be recommended for those initiating exercise or for patients with cardiovascular risks or other problems and should be reached only after gradual progressive increases in intensity. While low intensity activity increases for health in the general population do not usually require specific precautions (people are normally carrying out this level of intensity in everyday activity), a qualified exercise physiologist should be involved in any exercise prescription for increased fitness.

High Intensity Interval Training

Recent studies have given prominence to the possibility of a form of high intensity training over short intervals (e.g., 20–30 s bursts of maximum activity followed by 30–40 s of rest) as an economical, and possibly more effective, way of increasing aerobic fitness.

Controlled trials have not shown convincing effects of high intensity interval training over moderate intensity training (Fisher et al., 2015), particularly in those who might need exercise most (e.g., the elderly, the obese, etc.), although it may be an effective and efficient method for increasing activity levels in the young and reasonably active (Logan et al., 2015). In relation to weight loss, it appears that the volume of total exercise (i.e., when measured isocalorically) is the important factor (Martins et al., 2015). Potential injury effects also decrease the value of such training in those with potential metabolic problems.

Strength

Strength refers to the ability of the muscles of the body to contract against resistance. In the past, adequate strength was gained from daily living—lifting, carrying, working—however, as with aerobic exercise, this has decreased significantly as part of daily life because of technological developments. For this reason, and to increase muscle mass and reduce functional decline and fall-related injuries in older adults, the American College of Sports Medicine has set as a national goal of increasing to 30% the proportion of adults who perform physical activities that enhance and maintain muscular strength and endurance on at least 2 days/week (up from 21.5% in men and 17.5% in women in 2004) (Peters, 2006). Restoring the appropriate amount of muscular activity to prevent disease generally requires some form of "planned" resistance training, where resistance can be in the form of:

- calisthenics (using the body as resistance, e.g., push-ups, sit-ups)
- water (aquarobics)
- rubber straps or bands (as commonly used in physiotherapy)
- tins or bags of household goods (e.g., beans, rice)
- guided weight machines (pin loaded, hydraulic, pneumatic)
- free weights (barbells and dumbbells)

Although planning a resistance training program can be complicated, it should be individualized to ensure it is safe and clinically effective and is thus generally best left to an exercise specialist. The general principles outlined following will apply to most regimens.

Resistance Training Aims and Terms

Resistance training can be used to increase muscular strength and power, muscular endurance, cardiovascular fitness (and produce weight loss) and body bulk (i.e., lean muscle mass), or a combination of these. Training involves the following:

Repetitions (reps)	The number of times an exercise is repeated.
Sets	The number of groups of repetitions.
Speed	Speed of contraction (generally slower on lowering, or "eccentric," exercise).
Rest	Rest time between sets.
Load	The amount of weight or resistance used.
Repetition maximum (RM)	The amount of weight able to be lifted a set number of times (i.e., 80% of 10 RM is 80% of the weight that can just be lifted 10 times).

Resistance Regimens

Generic regimens for improving strength/power and endurance are outlined next. Muscle bulking will not be considered here because this is generally not associated with health outcomes. Regimens need to be individualized for specific purposes and conditions.

Strength/Power Regimens	Generally involve low reps (4–6) and sets (2–3) with relatively heavy weights (i.e., 60–80% RM) and 1–2 min rest between exercises. This requires a gradual progression from the endurance regimen discussed following. Power outcomes involve greater speed of contractions and are generally used by athletes to improve performance.
Endurance regimens	These are based around higher reps (10–20) and fewer sets (1–2) with lighter weights (40–60% RM) and minimal rest between exercises. This can include "circuit training," which involves moving quickly through a number of different exercises involving different muscle groups.

Most resistance training exercises are completed in a bilateral manner, meaning both arms or legs are completing the repetition at the same time. However, for patients who are on prescribed antihypertensive medications, have a history of stroke, or have low ejection fractions, it is recommended that they complete (if appropriate) all resistance training exercises unilaterally (i.e., one arm or leg at a time). The rationale for this recommendation is that the increase in blood pressure (systolic and diastolic) is directly related to both the intensity of the weight lifted and the amount of muscle mass involved. The intensity of the weight lifted can be reduced to minimize the effect on blood pressure by making the movement unilateral, essentially reducing the active muscle mass by 50%.

EXERCISES

There are a large number of resistance exercises available for different muscle groups, and these are found in various publications (e.g., Egger et al., 1999). In general, they are split into "compound" exercises, involving large muscle groups or several major muscles, and "isolation" exercises, which isolate single muscle group(s). For health and fat loss purposes, compound exercises are more effective because they involve more muscle mass and hence greater total energy use. Isolation exercises are used more often for rehabilitation or hypertrophy of specific muscles.

While resistance training has typically been used for strength and power sports and for body bulking, there is recent empirical evidence for its application in middle-aged and older people to enhance functional living (Henwood and Taaffe, 2006). Falls, for example, make up the single biggest cause of injury in older adults. Resistance training has been shown to reduce fall injuries and hence health costs (Liu-Ambrose et al., 2004). Resistance training increases lean body mass in normal-fat participants and reduced fat mass in overweight and obese adults (Drenowatz et al., 2015). In a prospective study examining prospectively, weight training has the strongest association with less waist circumference increase (Mekary et al., 2015). Resistance training, in some form, for all elderly people is likely to become a trend of the future.

Suppleness (Flexibility)

Flexibility is important to maintain musculoskeletal integrity and to reduce the likelihood of injury to muscles in athletes. Its main value in health is in maintaining muscle length in muscle groups that may be strengthened but not lengthened from aerobic and resistance exercise, thus preventing injury due to sudden tearing of these muscles. Flexibility is improved by stretching and holding muscles (without pain) for up to 15 s at a time. Three main forms of stretching useful for lifestyle medicine are:

- Static stretching: A muscle is stretched and held for 10–15 s, and this is repeated a number of times.
- Ballistic stretching: This is a dynamic form of stretching that involves a "bouncing" movement of the muscle in preparation for sport. This is not recommended for general health.
- Proprioceptive neuromuscular facilitation (PNF) stretching: This is a two-phase process where a muscle is stretched and held for 8–10 s, then relaxed. Next, the opposing muscle is isometrically contracted for 8–10 s, then relaxed, before the original muscle is stretched further. This "tricks" the stretched muscle to extend further through the complicated recruitment of neural mechanisms designed to prevent overstretching and overcontraction. PNF stretching leads to the greatest gains in flexibility (Sharman et al., 2006).

MUSCLE GROUPS FOR STRETCHING

Although it is functional to stretch all major muscle groups, this is often difficult for overweight people. The goal of a lifestyle medicine program is to stretch, at a minimum, those muscles likely to be shortened and/or strengthened through aerobic or strength work. The main functional muscles here are the hamstrings, peroneal muscles at the front of the leg, and gastrocnemius (calf) muscles at the back of the leg. Simple stretches to recommend and demonstrate to patients as a minimum to accompanying an aerobic exercise program are shown in Fig. 11.5.

For best effect, muscles should be stretched after a brief period of "warming up." This is analogous to stretching chewing gum while it is still moist compared with stretching it after it has dried and is likely to tear. Patients should complete a 5–10 min brisk walk prior to completing stretching exercises.

Size

Size, in the context in which it is used here, refers to excess body fatness. Reducing this can be accomplished by either stamina (aerobic) or strength (resistance) exercise, using the formula for exercise "volume" described earlier. The development of stamina and strength, however, and the added cardiovascular benefits, requires systematically increasing the intensity of an exercise. Reductions in size (body fat) are influenced more by the total volume of physical activity.

Weight loss (at least theoretically) is a linear function of physical activity volume (although this does not necessarily happen in practice, as described in

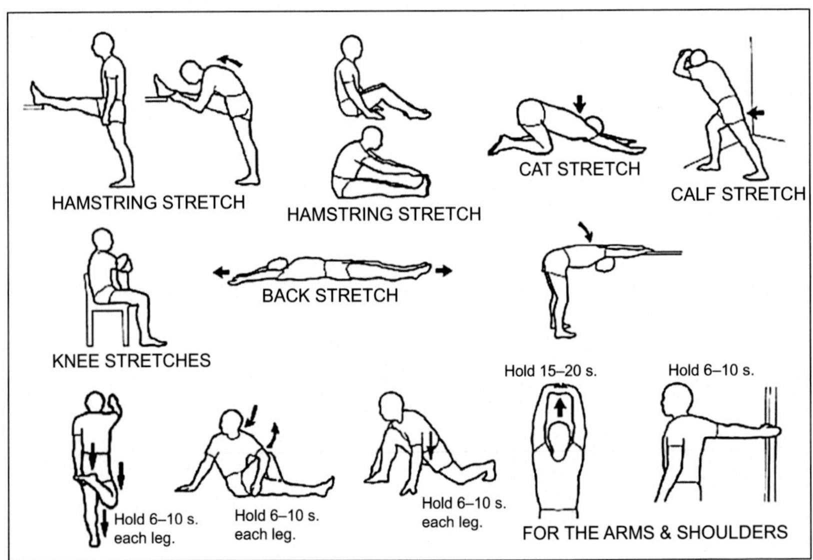

FIGURE 11.5 Some basic stretches for increasing suppleness.

Chapter 7, because of physiological adjustments that occur with changing energy balance). The best types of activities are the aerobic, weight-bearing activities (e.g., walking, jogging, skiing), in contrast to those where the body weight is supported (e.g., cycling, rowing, canoeing). For the majority of the inactive population, the greatest benefit in fat reduction will come from increased daily walking. Because this is influenced by volume and not intensity, it does not need to be continuous but can be accumulated over the course of a day.

ERIC THE IDLE TESTS HIS LATEST ENERGY SAVING INVENTION.

Pedometers are ideal tools for measuring this in steps, irrespective of length or speed of stride. As an average 80 kg man will burn roughly 1 kcal (4.4 kJ) for every 25 steps, relatively precise calculations can be made for total daily or weekly step counts by body weight. A recommended table of steps used in the GutBuster's Weight Loss Program for Men is shown in Table 11.1.

TABLE 11.1 Recommended Weekly Steps by Body Weight

Body Weight	Minimum Steps/Day	Optimum Steps/Week
70–80	8500	>100,000
81–90	7500	>100,000
91–100	7000	>90,000
101–110	6500	>80,000
111–120	6000	>70,000
120–130	5500	>60,000
130	5000	>50,000

TABLE 11.2 Suggested Recommendations for Physical Activity and Size Reduction

General	Combine exercise with a hypocaloric diet. Increase "incidental" as well as "planned" exercise.
For preventing weight gain	Accumulate 30 min of added moderate intensity activity per day (approximately 50,000 steps/week).
For weight loss	Accumulate 30–60 min of added moderate intensity activity per day (approximately 70,000 steps/week).
For weight loss without	Accumulate 3500 kcals (15,400 kJ) of moderate changing diet intensity activity per week (100,000 steps/week).
For weight loss maintenance (postobese)	Accumulate 60–90 min of moderate intensity activity per day (>100,000 steps/week). Alternatively, measure baseline activity and increase progressively by 30%.

A weekly stepping goal is recommended, in contrast to the daily goal often suggested for health benefits, because this allows for "good" days and "bad" days when time may be limited. Broad activity recommendations for different outcomes are discussed in Table 11.2.

Resistance training can add to the benefits of aerobic exercise for fat loss. Because work against resistance results in the maintenance or growth of muscle tissue, and because muscle is more metabolically active than fat, resistance training can result in a raised metabolic rate with a consequent increase in daily energy expenditure. In the early stages of a fat loss program, however, this may not be reflected in weight loss because muscle is over two times more dense (and therefore heavier) than fat. It is important to communicate to the patient that fat loss can actually occur in the presence of weight gain (from increased muscle bulk).

Stability

Stability refers to the ability to be stable, particularly with aging, and with changing orientation of the limbs. With decreasing muscle and joint strength, stability becomes important for the prevention of falls and bone breakages that influence the aging population. With young, fit individuals, stability can be tested by raising arms and legs individually while in a "planking" position (i.e., lying prone while taking the weight on elbows and toes) and holding this position for 15 s each. In older individuals, this can be tested by the Balance Error Scoring System (BESS) where the subject is asked to stand on one leg for up to 30 s (for further details, see NCSA (2016)). Increased stability can be gained by practicing the poses used to measure this (i.e., regular standing on one leg or prone 3-point position).

PROGRESSION OF EXERCISE TRAINING

For best effect, a physical activity program should be dynamic, changing to suit the needs of an individual as she or he becomes more fit. This requires ongoing supervision and attention in most cases where specific prescription is required. Even in the case of generic exercise prescription, ongoing supervision and attention is recommended to guarantee compliance and motivation. This can be done with the enlistment of an accredited exercise physiologist.

PURPOSE, MEANING, AND PHYSICAL ACTIVITY

In an important paper published in 2001, Dr. William Morgan suggested that the missing factor in exercise prescription—what he calls Factor P—is a purpose for being active (Morgan, 2001). Morgan claimed that in the past, activity was tied to survival and so had meaning and purpose. The reason adherence to exercise prescription today is so low is because of a lack of meaning and purpose in exercise for exercise sake. Morgan admitted that exercise can develop its own meaning and that a goal of exercise prescription may be to get to the point of exercise "addiction."

Other writers in this field have suggested that there are phases people go through in reaching a level of fitness that is intrinsically rewarding and thus provides purpose in an otherwise nonpurposeful environment. In the early stages of fitness development, however, this is unlikely, and other purposes (with extrinsic forms of reward) should be sought. Morgan refers to a quote by Norwegian psychiatrist Dr. Egil Martinsen that "if the patient is considering the purchase of an exercise machine, a dog would make for a good selection." Lifestyle-related activities, instead of those developed as part of a nonpurposeful regimen, have been shown to work best in this respect.

What Can be Done in a Brief Consultation

- ☐ Ask, "Do you ever engage in physical activity for the sole purpose of improving your health and fitness?" to get an indication of fitness level.
- ☐ Measure waist circumference, weight and body fat (if possible), and relate these to physical activity involvement.
- ☐ Explain the difference between exercise for fitness and movement for fat loss.
- ☐ Suggest an immediate program of accumulated (preferably weight-bearing) activity for health benefits.
- ☐ Lend, rent, or suggest the purchase of a pedometer to measure activity level baseline and to provide step goals.
- ☐ Discuss the possibility of getting a dog—if appropriate.
- ☐ Work through the four Physical Activity Guidelines prescription (see Fig. 11.3) and give the prescription to the patient.
- ☐ Provide written material on lifestyle areas identified.

THE CLINICIAN'S ROLE

Physical activity has typically been regarded in a "motherhood" light, with a clinician's exhortation to "do some exercise" thought to require no further elaboration. With the rapidly changing ambient exercise environment of the 21st century, however, this is no longer sufficient. Doctors and allied health professionals all now require at least a partial knowledge of the fundamentals of generic exercise prescription and an understanding of the requirements for referral for specific prescription. More importantly, because there is no pharmaceutical substitute for exercise, there is a need for a detailed understanding of what will best motivate a patient to become more active, not only for the short term but for the rest of his or her life in an environment that demands some form of intellectually driven activity plan. While doctors may recognize the need for this, and practice nurses or other health professionals may help assess and measure this need, the best support in the long run is going to come from involving qualified exercise physiologists or personal trainers, at least until the point where intrinsic motivation can stimulate the patient to continue.

SUMMARY

Because the human body has evolved in an environment of regular activity, this has "fed forward" to a healthy metabolism. However, the decrease in activity in the modern environment is now reflected in metabolic abnormalities. Reestablishing a regular pattern of physical activity is thus imperative for metabolic health. The five S's of fitness—stamina, strength, suppleness, size, and stability—summarize exercise needs. Together, these provide a generic prescription for activity for most people. Individual prescription may then be needed for dealing with specific problems, and this is the basis of the next chapter.

Practice Tips

Prescribing Physical Activity and Exercise for Health

	Medical Practitioner	Practice Nurse/Allied Health Professional
Assess	Need for increased activity (e.g., with an active question; see Professional Resources). Need for more detailed fitness testing (i.e., graded exercise test). Need for more specific prescription. Factors limiting activity (contraindications and/or limitations).	Need for activity/exercise. Factors limiting activity. The five S's; do a basic step test if needed. Need for advanced testing. Need for referral.

Practice Tips—cont'd

Prescribing Physical Activity and Exercise for Health

	Medical Practitioner	Practice Nurse/Allied Health Professional
Assist	By identifying activity goals with the patient (short and long term). By recommending a generic program. By suggesting a type of program. By suggesting types of alternatives (e.g., gym, exercise physiologist, personal trainer).	By providing a generic exercise prescription (see Professional Resources). Removal of impediments to movement. Areas for improvement in the five S's. By developing motivational goals. By planning of an exercise regimen. By providing ongoing support/ encouragement.
Arrange	An exercise physiologist referral to a specialist if required for injury (e.g., podiatrist, orthopedic specialist, physiotherapist).	Consultation with: • An exercise physiologist to develop an exercise plan. • A personal trainer if detailed care and motivation required. • A local fitness center membership if desired.

Key Points

Clinical Management

- Generic prescription for physical activity is vital for all patients; specific prescription may require more skilled and ongoing input by referral to and working with an exercise physiologist (see next chapter).
- Consider motivational levels and ongoing commitment to being physically active, not just immediate commitment.
- Suggest small changes that may stimulate compliance.
- If possible walk (or exercise) before breakfast for optimal weight loss (unless diabetic) because this recruits fat into the energy cycle earlier than if food is used for energy.
- Exercise lightly clad in cooler months because the body is required to used more energy to maintain core body temperature.
- Vary the type/intensity/duration of physical activity for fat loss to reduce the effects of "efficiency" and hence reduce energy requirement.
- Exercise before a main meal to reduce hunger and therefore overeating.
- For weight loss, use weight-bearing exercise such as walking where possible.
- If joint or other problems do not allow weight-bearing exercise, carry out weight-supportive exercise such as walking in water, swimming, cycling, or rowing over longer periods.

REFERENCES

Australian Bureau of Statistics (ABS), 2013. Australian Health Survey: Physical Activity, 2011–12. ABS Cat. No. 4364.0.55.004. ABS, Canberra.

Blair, S.N., Cheng, V., Holder, J.S., 2001. Is physical activity or physical fitness more important in defining health benefits? Med. Sci. Sports Exerc. 33 (6 Suppl. 2), S379–S399.

Booth, F.W., Chakravarthy, M.V., Gordon, S.E., et al., 2002. Waging war on physical inactivity: using modern molecular ammunition against an ancient enemy. J. Appl. Physiol. 93, 3–30.

Borg, G., 1998. Borg's Perceived Exertion and Pain Scales. Human Kinetics, Champaign, IL.

Bouchard, C., Shephard, R., Stephens, T. (Eds.), 1994. Physical Activity and Health: International Proceedings and a Consensus Statement. Human Kinetics, Champaign, IL.

Durstine, J.L., Moore, G.E., 2003. ACSM's Exercise Management for Persons with Chronic Disease and Disabilities. Human Kinetics, Champaign, IL.

Drenowatz, C., Hand, G.A., Sagner, M., Shook, R.P., Burgess, S., Blair, S.N., December 2015. The prospective association between different types of exercise and body composition. Med. Sci. Sports Exerc. 47 (12), 2535–2541.

Egger, G., Champion, N., Bolton, A., 1999. The Fitness Leader's Handbook, fourth ed. Simon and Schuster, London.

Fisher, G., Brown, A.W., Bohan Brown, M.M., Alcorn, C., Winwood, L., Resiehr, H., Geaorge, B., Jeansomme, M.M., Allison, D.B., 2015. High intensity interval- vs. moderate intensity-training for improving cardiometabolic health in overweight or obese males: a randomized controlled trial. PLoS One 10 (10), e038853. http://dx.doi.org/10.1371/journal.pone.0138853 eCollection 2015.

Henwood, T.R., Taaffe, D.R., 2006. Short-term resistance training and the older adult: the effect of varied programmes for the enhancement of muscle strength and functional performance. Clin. Physiol. Funct. Imaging 26 (5), 305–313.

Lee, S., Kuk, J.L., Katzmarzyk, P.T., Church, T.S., Ross, R., 2005. Cardiorespiratory fitness attenuates metabolic risk independent of abdominal subcutaneous and visceral fat in men. Diabetes Care 28 (4), 895–901.

Liu-Ambrose, T., Khan, K.M., Eng, J.J., et al., 2004. Resistance and agility training reduce fall risk in women aged 75 to 85 with low bone mass: a 6-month randomized, controlled trial. J. Am. Geriatr. Soc. 52 (5), 657–665.

Logan, G.R., Harris, N., Duncan, S., Plank, L.D., Merien, F., Schofield, G., October 17, 2015. Low-active male adolescents: a dose response to high-intensity interval training. Med. Sci. Sports Exerc. 48 (3), 481–490.

Martins, C., Kazakova, I., Ludviksen, M., Mehus, I., Wisloff, U., Kulseng, B., Morgan, L., King, N., October 19, 2015. High-intensity interval training and isocaloric moderate-intensity continuous training result in similar improvements in body composition and fitness in obese individuals. Int. J. Sport Nutr. Exerc. Metab. 26 (3), 197–204.

Mekary, R.A., Willett, W.C., et al., February 2015. Weight training, aerobic physical activities, and long-term waist circumference change in men. Obesity (Silver Spring) 23 (2), 461–467.

Morgan, W.P., 2001. Prescription of physical activity: a paradigm shift. Quest 53, 366–382.

Morris, J.N., Heady, J.A., Raffle, P.A., 1953. Coronary disease and physical activity of work (part 2). Lancet 265, 1111–1120.

NCSA, National Stength and Conditioning Association, 2016. Guide to Tests and Measurements. Human Kinetics, IL.

NH&MRC (National Health and Medical Research Council), 2005. National Physical Activity Guidelines. Aust Govt Press, Canberra ACT.

O'Connor, P.J., Puetz, T.W., 2005. Chronic physical activity and feelings of energy and fatigue. Med. Sci. Sports Exerc. 37 (2), 299–305.

Peters, J.C., 2006. Obesity prevention and social change: what will it take? Perspectives for progress. Exerc. Sport Sci. Rev. 34 (1), 4–9.

Peters, J.C., Wyatt, H.R., Donahoo, W.T., et al., 2002. From instinct to intellect: the challenge of maintaining healthy weight in the modern world. Obes. Rev. 3, 69–74.

Rowland, T.W., 1998. The biological basis of physical activity. Med. Sci. Sports Exerc. 30 (3), 392–399.

Sharman, M.J., Cresswell, A.G., Riek, S., 2006. Proprioceptive neuromuscular facilitation stretching: mechanisms and clinical implications. Sports Med. 36 (11), 929–939.

Vogels, N., Egger, G., Plasqi, G., et al., 2004. Estimating changes in daily physical activity levels over time: implications for health interventions from a novel approach. Int. J. Sports Med. 25, 607–610.

Warburton, D.E., Nicol, C.W., Bredin, S.S., 2006. Health benefits of physical activity: the evidence. Can. Med. Assoc. J. 174 (6), 801–809.

Wizen, A., Farazdaghi, R., Wohlfart, B., 2002. ICO Conference, Brazil, 2002.

Professional Resources

Measuring Fitness

A Simple Screening Question

Ask, "Do you ever engage in physical activity for the sole purpose of improving your health and fitness?" to get an indication of fitness level.

GutBuster's Clinical Test Battery

This is a brief test battery for measuring the five S's and involving a 10-min fitness test. Individual scores are given for each of the six tests. There is also a total score for comparison with later results. With the exception of tests 1 and 2, all should be added to give a total score; tests 1 and 2 scores should be multiplied before being added to the other scores.

Size: Tests 1 and 2

Perform a bioimpedance analysis using body fat scales.

Measure BMI calculated as weight (kg) divided by height squared (M^2).

Measure waist circumference, measured at the midpoint of lowest rib and iliac crest or umbilicus for men and midpoint for women. Measures should be consistent.

Measurement	Score
Test 1. BIA (%)	Score out of 5
Test 2. Waist circumference (cm/in)	Score out of 6 see table below
Combined size score	Score out of 30
(BIA 3 waist circumference)	

Explanation: BIA (bioimpedance analysis) measures body fat through a process of electrical conductance. Normal ranges for men are 12–24% and for women 15–35%. Waist circumference is a measure of fat distribution. For best health, Caucasian men should be <100 cm and women <90 cm around the waist. Cutoffs for Indians and Asians are 10 cm less and for Pacific Islanders 10 cm more. The size score (BIA × WC) gives a measure of medically dangerous fat distribution.

Continued

Professional Resources—cont'd

How to calculate a score for these tests:

	Men	Women	Contribution to Total Score
BIA (%)			
Good	<24%	<32%	5
OK	24–30%	32–35%	4
Poor	>30%	>35%	3
Waist Circumference (cm)			
Good	<90 cm (35.4 in.	<80 cm (31.5 in.)	6
OK	90–100 (35.4–39.4)	80–90 (31.5–35.4)	3
Poor	>100 (39.4 in.)	>90 cm (35.4 in.)	1

Stamina: Tests 3 and 4

Take a resting pulse over a minimum of 15 s to get a score per minute.

Do the Scale of Physical Activity Level (SPAL) Questionnaire (Wizen et al., 2002), i.e.,

Ask, "Which is the highest level activity on the scale that you are able to carry out for 30 min or more?"

1. Sit
2.
3. Walk slowly
4.
5. Walk at normal pace/cycle slowly
6.
7.
8. Jog/cycle
9.
10. Run
11.
12. Run fast/cycle fast
13.
14.
15. Run very fast (0.15 km/h)
16.
17.
18. Perform aerobic activity at a professional level (women)
19.
20. Perform aerobic activity at a professional level (men)

The measures here are in METs (1 MET = 1 metabolic unit, or the amount of energy burned at rest). For greater precision the following formula (METpred) is available for corrections with age:

$$\text{METpred} = 7.37 + (0.595 \times \text{MET}) - (0.057 \times \text{age})$$

Measurement	Score
Test 3. Pulse (beats per minute)	Score out of 8
Test 4. SPAL (METS)	Score out of 34

Explanation: Pulse is a rough measure of physical fitness and hence receives a low overall score. SPAL is a pencil and paper measure of aerobic fitness. Measures are in METS, or multiples of metabolism at rest. Average scores for men are 6–9 METS and for women 5.5–8.5. High scores indicate better fitness.

How to score for these tests:

	Men	Women	Contribution to Total Score
Resting Pulse (bpm)			
Good	<65	<75	8
OK	66–95	76–105	5
Poor	>95	>105	0
SPAL			
Very good	>12	>11.5	34
Good	9.0–11.9	8.5–11.4	28
OK	6.0–8.9	5.5–8.4	17
Poor	3.0–5.9	2.5–5.4	12
Very poor	<3.0	<2.5	6

Suppleness (sit-and-reach or sit-and-stand test): Test 5

Lower back flexibility can be measured with a sliding flexibility rule or a ruler placed between the center of the feet. The procedure is then as follows:

1. Warm up the body before measuring.
2. Sit on the floor with legs straight and feet in the foot holes provided on the flexibility rule (or with ruler between the feet).
3. Slowly bend forward at the trunk with legs straight and measure the distance that can be reached with fingertips either before or beyond the feet (maximum 115 cm; minimum 215 cm).
4. If the fingertips do not reach the feet, the measure is regarded as a minus score. If the fingertips pass the feet, the measure is regarded as a plus score.
5. Check the score in the table provided and get a total contribution score.

Measurement	Score
Test 5. Sit and reach (cm)	Score out of 16

Explanation: This test measures the flexibility of hamstrings and lower back, which is a good indication of overall flexibility. Low flexibility suggests possible musculoskeletal problems. Higher scores are better than lower and lower limits for people under 40 years old are +4 cm (1.6 in.) and for people over 40 year olds, −1 cm (0.4 in.).

How to score for this test:

	Age <35	Age >35	Contribution to Total Score
Very good	>15	>7	16
Good	+15 to +5	+7 to 0	12
OK	+4 to −1	0 to −10	8
Poor	−1 to −10	−11 to −15	4
Very poor	less than −10	less than −15	2

Continued

Professional Resources—cont'd

Strength (sit-ups or stand-ups): Test 6

In relatively active people, sit-ups to measure abdominal strength over 20s is the preferred test (although grip strength using a dynamometer is becoming increasingly more popular). This is done by counting the number of completed sit-ups that can be done in a set period (in this case 20s). The procedure is:

1. Lie on your back on the floor with feet on a standard height chair so that the knees are bent at right angles.
2. Fold the arms in front of the chest with elbows pointing forward.
3. Raise the shoulders off the ground in a "crunch" position until the elbows touch the thighs. Return then to the position of shoulders flat on the ground.
4. Carry out as many complete sit-ups as possible in a 20-s period. If the elbows do not touch the thighs, or the shoulders are not returned to the flat-on-the-ground position, the sit-up is not counted.
5. Check the number of sit-ups completed in the table provided and get a total contribution score.

For the more frail and the extremely overweight or injured, a sit-and-stand test, rising from a chair to full extension of the legs is an alternative. Scores are similar.

Measurement	Score
Test 6. Stand-ups (numbered)	Score out of 12

Explanation: Abdominal strength as measured by the sit-up test and leg strength as measured by stand-up test are good measures of overall body strength. Acceptable levels for people under 30 years old are 12 sit-ups and stand-ups in 20s; for people aged 30–39, 11; and for people over 40, 10.

How to score for this test:

	Age <29	Age 30–39	Age >40	Contribution to Total Score
Good	>17	>15	>13	12
OK	12–17	11–15	10–13	8
Poor	<12	<11	<10	4

Total Fitness Score

The total fitness score (out of 100) 5 (test 1 3 test 2) 1 test 3 1 test 4 1 test 5 1 test 6. The total score indicates fitness level:

0–20 Very poor: needs to be considerably improved
21–40 Poor: needs work in all areas
41–60 Average: can be improved for better results
61–80 Good: improvements possible in some areas
81–100 Very good: fitness is generally not a concern

Weaknesses That Could Be Improved

body fat
aerobic fitness
flexibility
strength
total fitness

Professional Resources—cont'd

Patient Handout

Reducing body fat	There is no best single way to lose fat. Some people benefit from one approach while others benefit from another. The one thing we can say is that if you make changes that cannot be maintained for life, they are not likely to help you reduce weight long term—and that's what it's all about. Check with your doctor to see what is likely to be the best approach for you to lose weight out of about a dozen techniques available.
Increasing aerobic fitness	In the first instance this is not hard. It just requires increasing amount of movement you do over the course of a day. You can use a pedometer to measure the number of steps you do each week, and then try to increase this over the next few weeks by 30% or more. If you want to lose weight, you should try to accumulate 70,000 steps a week or more. But "accumulate" means this doesn't have to be all in one go. Once you have started to increase the amount of movement you do, you may then want to increase your fitness a little more by increasing the intensity of the movement. Having a personal trainer or using an accredited gym can be other ways of increasing fitness.
Increasing flexibility	Flexibility is important to prevent stiffness and muscle and joint pain. Flexibility is increased by stretching muscles as shown in the diagram (see Fig. 11.5). Do not stretch to a point of pain, and hold the stretch for 8–10 s before relaxing and repeating. You can also increase flexibility through yoga, pilates, and a range of similar programs.
Increasing strength	Strength is increased by carrying out exercises against resistance. This can be resistance against your own body weight (called calisthenics), resistance against water (such aquarobics), using large rubber bands or other heavy objects, or becoming involved in a weight training program at a gym or with a personal trainer. Resistance training helps to maintain muscle mass and reduce body fat. This type of exercise is vital for older people to prevent falls and muscular weakness.

Continued

Professional Resources—cont'd

Physical Activity Guidelines Prescription

To be discussed with the patient.

1. Think of movement as an opportunity, not an inconvenience	List things you can do for yourself, instead of using machines or other people.
2. Be active everyday in as many ways as possible	Try more of the following (tick as appropriate): gardening moving more around the house standing instead of sitting doing things by hand playing with kids playing games active recreation other
3. Put together at least 60 min of moderate intensity activity	Walking (number of steps/day) Other activity or activities
4. If you can, also enjoy some regular, vigorous activity for extra health and fitness	Yes/no (tick as appropriate) If yes, list the activity or activities 1–2 3–4 5–7

Special requirements (list)

Chapter 12

Physical Activity: Specific Prescription for Disease Management and Rehabilitation

Mike Climstein, Garry Egger

It seems to me that we need to begin thinking about exercise prescription and the problem of adherence to physical activity in a very different sort of way.

Morgan (2001)

INTRODUCTION: SPECIFIC EXERCISE PRESCRIPTION

In the summer of 2005, while exercise physiologists around the world were "bunkered down" in their expensively equipped research laboratories, Dr. Heinz Drexel and his colleagues from the Vorarlberg Institute for Vascular Investigation in Feldkirch, Austria, were out in the Alps, handing out lift passes and measuring a group of volunteers' reactions to walking up or down steep ski slopes (Drexel, 2006). After a month of training half the group walked up and half walked down a ski run daily for a month, catching the lift the other way, the researchers measured changes in their subjects' blood chemistry. They then switched the groups for a month and measured their reactions to this.

To the surprise of many, the responses were different. The concentric contraction of muscles walking up hill (shortening under load) caused a significant decrease in triglycerides in the blood (a risk factor for heart disease) whereas the eccentric contraction of muscles required in walking downhill (lengthening under load) caused a greater decrease in blood sugars (a risk factor for diabetes). Low-density lipoprotein (LDL) cholesterol was reduced by approximately 10% in both cases.

The significance of these findings is in the specificity of the metabolic responses to various types of physical activity. For a long time, it was thought that a generic prescription for exercise for health was adequate. In fact a consideration of historical landmarks in physical activity research (Fig. 12.1) illustrates that while the ambient environment in earlier times necessitated activity for survival, humans required little extra exercise. Exercise prescription in the

Lifestyle Medicine. http://dx.doi.org/10.1016/B978-0-12-810401-9.00012-7

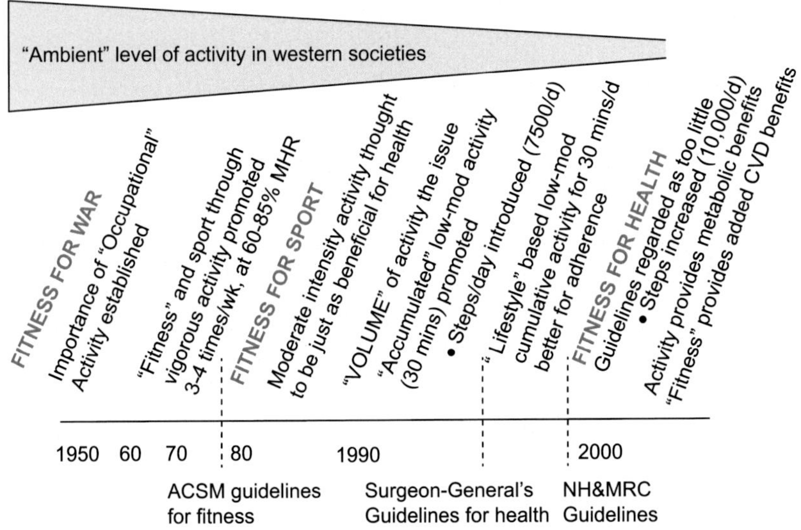

FIGURE 12.1 Historical landmarks in physical activity research.

mid-20th century and before was based mainly in the interests of getting young men "fit for war"; during the 1970s and, 1980s, the concentration was on training for athletes and sportspeople; and only since the publication of the US Surgeon General's Guidelines in 1994 has the orientation changed to physical activity for health (US Surgeon General, 1996).

These shifts in direction resulted in the development of new national physical activity guidelines in several countries, including Australia (NHMRC, 2005; shown in prescription form in Fig. 11.3). There has been an even greater decrease in ambient activity levels since the first Australian guidelines were published in 1999, a result of the Internet. The American College of Sports Medicine's (ACSM) manual *Exercise Management for Persons With Chronic Diseases and Disabilities* (ACSM and Dirstine, 2009) suggests the need for a modified series of guidelines (shown in Fig. 12.2), with some additions to the standard guidelines as well as a different dimension for specific prescription.

While the generic prescription shown in Fig. 12.2 is absolute, more specific prescription of exercise for metabolic health is now also indicated. This has been illustrated in a number of reviews (e.g., Pedersen and Saltin, 2006), and meta-analyses (e.g., Snowling and Hopkins, 2006). The ACSM report (2009) lists more than 40 medical conditions for which there is evidence of benefit from regular exercise for prevention or therapy (Table 12.1).

This points to the importance of involving skilled exercise personnel as an integral part of any allied health team presenting lifestyle medicine, while still acknowledging the importance of generic prescription for general

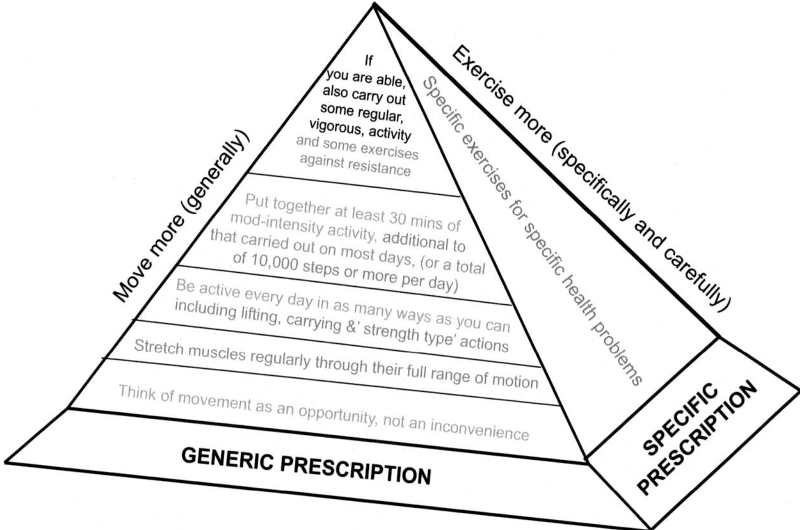

FIGURE 12.2 Potential additions to the 1999 *National Guidelines for Physical Activity* including specific prescriptions for disease management. *Adapted from NHMRC (National Health and Medical Research Council), 2005. National Physical Activity Guidelines. Canberra, DoHA. Available from:* www.health.gov.au/internet/wcms/Publishing.nsf/Content.

metabolic health outlined in Chapter 11. This chapter is designed to illustrate some of the major areas where specific prescription may be indicated and how and when to involve an accredited exercise physiologist in the allied health team.

EXERCISE AND DISEASE MANAGEMENT

Table 12.2 outlines a small subsample of the ailments listed in Table 12.1, with asterisks indicating the level of evidence for potential benefit from each of the first four forms of fitness (from the five S's) discussed in Chapter 11 for prevention and/or treatment.

No attempt is made here to cover all of the ailments that benefit from exercise or the fine detail of prescription: discussion is confined to a small number of commonly occurring problems with predominantly lifestyle-based etiologies that are discussed in detail elsewhere throughout this book. Each condition is considered in the context of the first four of the five S structure discussed in Chapter 11 and the "volume" equation also outlined in Chapter 11. For more details on specific programming, readers are referred to the ACSM manual *Exercise Prescription for Disease Management* (ACSM, 2011), *Clinical Exercise Physiology* (Ehrman et al., 2003), and other publications in the references at the end of this chapter.

TABLE 12.1 Medical Conditions With Evidence Regarding the Benefits of Regular Exercise for Prevention of Therapy

Cardiovascular conditions	congestive heart failure, coronary heart disease, cardiac transplantation, high serum lipids (total cholesterol, HDL cholesterol, VLDL cholesterol, triacylglycerol), hypertension, peripheral vascular disease, stroke
Rheumatoid conditions	ankylosing spondylitis, rheumatoid arthritis, osteoarthritis, osteoporosis
Infectious diseases	AIDS, postpolio syndrome
Geriatric conditions	sarcopenia
Cancers	breast, colon, prostate
Respiratory conditions	chronic obstructive pulmonary disease
Mental health	anxiety, depression
Neurological conditions	migraine, multiple sclerosis, myasthenia gravis, Parkinson disease
Metabolic conditions	obesity, diabetes
Hematological conditions	sickle cell anemia, hemophilia
Miscellaneous	chronic fatigue, low back pain, chronic pain, physical disability, renal transplant, sleep disorders
Urogenital conditions	premenstrual syndrome, urinary incontinence

Based on ACSM (American College of Sports Medicine), Durstine, J.L., 2009. ACSM's Exercise Management for Persons with Chronic Diseases and Disabilities, third ed. Human Kinetics, Champ Ill.

GENETICS AND RESPONSE TO EXERCISE

Human variation in responsiveness to regular exercise was first recognized more than 30 years ago (Bouchard, 1983). Some people respond better to exercise interventions than others (nonresponders). Even though there has been significant progress in acquiring technologies that are necessary to probe genomic features and metabolic pathways and systems responsible for the individuality of adaptation to exercise, and despite the progress made to date in defining these genomic markers and pathways, future research will help to better define the molecular mechanisms involved and to translate them into useful clinical and public health applications. In the realm of physical activity research and therapies, whole exome and genome sequencing and omics profiling will allow deep profiling of an individual's genetic makeup that may impact the tolerance, effects, and performance expected from pursuing physical activity, in addition to tracking of the biological changes that occur with activity and intentional

TABLE 12.2 Evidence for Specific Benefits of Different Types of Exercise Prescription in Management of 10 Major Diseases

Disease	Stamina		Strength		Suppleness		Size	
	Prevent	Treat	Prevent	Treat	Prevent	Treat	Prevent	Treat
Alzheimer	–	*	–	*	–	–	–	–
Anemia	–	*	–	*	–	–	–	–
Arthritis	*	*	*	**	*	***	**	**
Asthma	–	**	–	**	–	–	*	*
CHD	**	***	*	**	–	*	**	**
COPD	*	**	–	*	–	*	*	*
Chronic fatigue	–	**	–	–	–	–	–	–
Diabetes	***	***	*	**	–	–	**	***
Frailty	*	*	*	*	*	*	–	–
Fibromyalgia	–	*	–	–	–	*	*	*
Hyperlipidemia	*	**	*	*	–	–	**	*
Hypertension	**	**	–	*	–	–	**	**
Lower back pain	–	–	*	*	*	*	*	*
Obesity	***	***	*	**	–	–	***	***
Parkinson	–	*	–	*	–	*	–	–
Stroke	**	**	–	**	–	*	*	*

–, no evidence; *, limited evidence; **, reasonable evidence; ***, strong evidence. *CHD,* coronary heart disease; *COPD,* chronic obstructive pulmonary disease.
Adapted from Durstine, J.L., Moore, G.E., 2000. ACSM's Guidelines for Exercise Testing and Prescription, Human Kinetics, Champaigne Ill.

exercise. This personalization will help to optimize preventive and therapeutic exercise interventions (Bouchard et al., 2015).

EXERCISE AND DIABETES

Exercise has now emerged as one of the key treatment options for type 2 diabetes mellitus (T2DM), even in the absence of weight loss (Thomas et al., 2006), with effective and progressive exercise protocols being evolved for dealing with different stages of the disease. According to Nieman (1998), "with optimum diet and exercise therapy, and achievement of ideal body weight, only about one in 10 type 2 diabetes patients would require any medication." Paffenbarger et al. (1997) showed that diabetes develops progressively and that for every 500 kcal (2200 kJ) increase in energy expenditure per week, the risk of T2DM is reduced by 6%.

Exercise prevents diabetes through a number of different mechanisms, many of which duplicate the effects of insulin in glucose disposal (Signal et al., 2004). Aerobic exercise on a daily basis (including increased incidental exercise) should be encouraged as the basis of management. Increased frequency, duration, and intensity results in greater glycemic control (Boule et al., 2003). However, resistance training has also been shown to be highly effective (Dunstan et al., 2003), and a progression from light aerobic exercise through to more intensive resistance training in a motivated patient is likely to have optimum benefits. Resistance training exercises should be completed on two to three days per week. Potential types and kinds of training for general T2DM management are shown in Table 12.3. Prescription needs to take account of the specific requirements of the patient and any limitations or contraindications the patient may have pertaining to exercise (Eves and Plotnikoff, 2006).

A position statement released by the American Diabetes Association (ADA, 2006), recommends the following for physical activity:

- 150 minutes/per week of moderate-intensity aerobic physical activity (50–70% of maximum heart rate) is recommended and/or at least 90 min per week of vigorous aerobic exercise (>70% of maximum heart rate). The physical activity should be at least three days/week, with no more than two consecutive days without physical activity.
- In the absence of contraindications, people with T2DM should perform resistance training exercise three times a week, targeting all major muscle groups, completing three sets of 8–10 reps at a weight that cannot be lifted more than 8–10 times.

According to the ADA position statement, vigorous exercise may be contraindicated for individuals with diabetic retinopathy and nonweight-bearing exercise is recommended for people with diabetic peripheral neuropathy. The ACSM position statement for diabetes states that increased intensity, additional

TABLE 12.3 Typical Exercise Needs for Type 2 Diabetes Management

Exercise Type	Examples	Frequency	Intensity	Duration	Cautions
Size and stamina	Walking Cycling Aquarobics Walking in water Incidental exercise	Preferably daily. Don't miss two consecutive days.	Mild–moderate: 40–65% HR_{max} or <13 on an RPE scale, increasing with fitness	30–60 min accumulated. More is better. 8–10,000 steps/d or increase 30% from baseline	Effort should be predictable. Monitor medication. Check extremities with neuropathy
Strength	Resistance Weights Circuit training (large muscle groups)	2–3 days/week	1 ste, 10–15 reps, 60% of 1 RM, increasing weight with strength	8–10 exercise until complete	Eye pressure, gradual progression
Suppleness	Static stretching, Yoga, proprioceptive neuromuscular facilitation	Before and after aerobic exercise, extra if needed.	Don't stretch to the point of pain	Until complete	Over-stretching

sets, or a combination of volume and intensity may produce greater benefits and may be appropriate for certain individuals (Albright et al., 2000).

Exercise has also been found to be effective for both the prevention and treatment of gestational diabetes mellitus (GDM). While the precautions of normal exercising during pregnancy apply, resistance exercise (using elastic or rubber bands) has been found to be effective in reducing GDM into the second trimester of pregnancy (Brankston et al., 2004). There is also an inverse association between pregravid activity (amount and duration) with later GDM, suggesting a role for exercise in management of GDM before and throughout pregnancy (Dempsey et al., 2004).

Precautions

Generally, the danger associated with not exercising in diabetes is greater than that from exercising, although special and individual cautions do apply. Because exercise reduces blood sugars, consistent self-monitoring of blood glucose levels prior to, during, and after exercise is required to avoid hypoglycemia. For additional recommendations for exercise and diabetic patients, refer to the

ACSM's Guidelines for exercise testing and prescription (ACSM and Durstine, 2009). Special care also needs to be taken with peripheral neuropathy (weight-bearing exercise is generally contraindicated) and blood and eye pressure during resistance training. Increases in blood pressure may be minimized by having the patient complete all resistance training exercises unilaterally and using lighter intensity (i.e., 40–50% of 1 RM).

EXERCISE AND HEART DISEASE

There is strong scientific evidence to support exercise in the prevention (Giannuzi et al., 2003) and rehabilitation of coronary heart disease (Scrutinio et al., 2005). The identified benefits include improving coronary risk factors, reducing symptoms, and reducing the risk of new coronary events. Furthermore, exercise has been shown to reduce overall and cardiovascular mortality by approximately 25% (Giannuzi et al., 2003).

Cardiac Rehabilitation

Comprehensive cardiac rehabilitation is appropriate for many patients with heart disease, including postrevascularization (CABG, coronary stents, angioplasty), post-AMI, angina, heart failure, heart transplantation, valve replacement/repair, cardiomyopathy, and pacemaker insertion. The concept of comprehensive cardiac rehabilitation was developed in the 1950s and includes a medical evaluation, prescribed exercise, education, and counseling. Cardiac rehabilitation traditionally consists of three phases:

- **Phase 1** is conducted in hospital and initiated within the first 24–48 h. Exercise is generally limited to range of motion (ROM) exercises and self-care activities multiple times per day. With progression, walking is included, generally limited to an exercise intensity of resting heart rate 120 bpm.
- **Phase 2** is also generally provided in hospital and initiated two to six weeks postevent. Exercise is generally completed three days per week with constant monitoring (ECG and BP). Educational lectures pertaining to medications, heart disease, risk factor reduction (smoking cessation, stress management), and exercise are also important components of the program. This phase of rehabilitation generally lasts from four weeks to six months.
- **Phase 3** is generally conducted in a community setting in health and fitness facilities. Exercise that is deemed clinically effective and safe may include treadmill walking, cycling, rowing, swimming, and resistance training. A clinical exercise physiologist generally prescribes exercise routines, with progress reports provided to both the patient and referring doctor on a regular basis.

Doctors are advised to contact their local hospitals for phase 1 and 2 cardiac rehabilitation programs available in their area; contact local health and fitness facilities for structured phase 3 programs, ensuring that they meet recognized guidelines and are staffed by suitably qualified exercise physiologists. Emergency equipment including a defibrillator should also be available.

The beneficial effect of rehabilitative exercise for patients with diagnosed coronary heart disease is well established (Jolliffe et al., 2000). Pederson and Saltin (2006) reported that exercise-based cardiac rehabilitation reduced all-cause mortality by 20% and cardiac mortality by 26%.

Resistance training has also been found to be beneficial in acute myocardial infarction patients. Elhke (2006) summarizes the recommendations for resistance training in this population:

- frequency: 2–3 days/week
- intensity: initially, 50% of 1 RM, progressing to 60–70% 1 RM
- volume: 8–12 reps in younger adults, 10–15 in the elderly
- progression: upper body, 1–2 kg; lower body, 2–4 kg
- recovery: 30–90 s between sets and >48 h between sessions

Resistance training may be contraindicated in patients with unstable angina, exertional angina, or low ejection fractions. Consult with the patient's cardiologist for the appropriateness of resistance training in the rehabilitation program. Rehabilitative exercise recommendations for heart disease and related disorders are quite diverse, but Adams et al. (2006) have shown that most resistance exercises are safe for most cardiac patients and that the American Cardiology Association suggestion that patients should not "lift anything greater than 5 lbs" is totally unrealistic and based on a lack of scientific evidence. Medical professionals are advised to seek the advice of their local exercise physiologist or cardiac rehabilitation programs.

EXERCISE AND ARTHRITIS

Arthritis is one of the most common forms of musculoskeletal problems in modern society. Figures from the United States suggest that 22% of the adult population have medically diagnosed arthritis and around 8% have arthritis-attributable activity limitations (Hootman et al., 2006). The two main forms of the disease are osteoarthritis and rheumatoid arthritis, but there are more than 100 other forms, making a universal prescription difficult. Nevertheless, in most cases—and often contrary to popular opinion—exercise in some form, with supervision, can improve the progression of the disease.

Basic exercise parameters for arthritis are shown in Table 12.4. While vigorous exercise is contraindicated in the presence of acute joint inflammation, the more common presentation is of subacute or chronic joint symptoms. Specialist advice should first be sought about the value of an exercise intervention, noting that the most immediate benefit may be to diminish the ill-health effects of inactivity, as much as improving the problem. Low impact activities

TABLE 12.4 Exercise Recommendations for Arthritis

Exercise Type	Examples	Frequency	Intensity	Duration	Cautions
Suppleness	Range of motion and proprioceptive neuromuscular facilitation stretching of affected joints Warm water exercises Mild yoga and tai chi if appropriate	1–2 sessions/week	Very mild. Moderated by pain	As long as it takes	Avoid overstretching. Warm up slowly
Strength	Circuit training Machines Rubber bands Isometric exercise	2–3 days/week with at least one day's rest	Start mild with gradual increases	2–3 reps to begin, building to 10–12 reps	Avoid media/lateral forces, high reps and high resistance
Size and stamina	Large muscle aerobic activities. Combine full, partial and nonweight-bearing aquarobics	3–5 days/week	40–60% HR_{max}	5 min session building to 30 min	Increase duration, not intensity, with progression

should be selected, with the emphasis on suppleness and joint range of motion. Aquarobics or warm water exercise can be used for stamina, flexibility, and weight loss. Resistance training should aim to stabilize the musculature around affected joints and, while the load may need to be reduced if pain or swelling occurs, this should not mean eliminating the exercise per se. Speed of movement is not an issue and exercise dose can be accumulated during several sessions throughout the day.

EXERCISE AND HYPERTENSION

A diagnosis of hypertension often leads to a recommendation not to exercise because it can raise blood pressure. However, while this is true in the acute phase, chronic exercise has been shown to reduce both systolic and diastolic blood pressure by up to 10 mm Hg in Stage 1 or 2 hypertension, as defined by

TABLE 12.5 Exercise Recommendations for Hypertension

Exercise Type	Examples	Frequency	Intensity	Duration	Cautions
Size and stamina	Large muscle endurance activities (e.g., walking, cycling)	3–7 days/ week	Mild–moderate (40–60–80% HR_{max})	Can be accumulated for weight loss 30–60 min session for BP	Don't use HR if on beta-blockers. May cause post exertional hypertension. Not for Stage 3 hypertension
Strength	Possibly circuit training if desired	3–4 days/ week with at least one day's rest	High reps, low resistance	30–60 min	Avoid isometrics. No high intensities
Suppleness					Only as related to aerobic and strength exercises

the NHMRC. This can halve the risk of cardiovascular death and, together with the reduced risk from lowered body weight, can have significant benefits for the one in three adults known to have higher-than-recommended blood pressures. The benefits shown most commonly come from aerobic exercise, with or without weight loss (Pedersen and Saltin, 2006). However, recent evidence has also identified improved blood pressure from resistance training exercise (Fagard, 2006). With the exception of circuit training, which involves an aerobic component, chronic strength or resistance training has not been shown consistently to lower BP. As stated previously, all hypertensive patients carrying out resistance training exercises should complete all of the exercises in a unilateral manner and at a lower intensity.

As shown in Table 12.5, some precautions need to be taken. The benefits for those with Stage 3 or essential hypertension may not be direct but may come from reduction of other risk factors. However, this is also likely to require ongoing medication.

EXERCISE AND DYSLIPIDEMIA

Because exercise enhances the ability of muscles to oxidize fat, it is a primary recommendation for the correction of dyslipidemia, including hypertriglyceridemia and raised subfractions of cholesterol. Because many patients with

TABLE 12.6 Exercise Recommendations for Dyslipidemia

Exercise Type	Examples	Frequency	Intensity	Duration	Cautions
Size and stamina	Weight loss with low saturated fat diet. Large muscle aerobic statin-related exercises for raising HDL and lowering Tg.	Daily	Mild–moderate (40–70% HR_{max})	40–60 min. Frequency and duration are more important than intensity	Exercise may potentiate the propensity for statin-related muscle damage.
Strength	Resistance training to lower LDL.	2–3 days/week with at least one day's rest	Start mild with gradual increase	2–3 reps building to 10–12 reps	Changes may require several months.
Suppleness					Only as related to aerobic and strength exercises.

this problem also have hypertension or symptomatic heart disease, the recommendations for exercise prescription need to be quite specific. This prescription follows recommendations for the general public, but the amount of physical activity needs to be increased and there may be some benefits in different forms of exercise for different dyslipidemic states (Table 12.6). The benefits in hypercholesterolemia are primarily attained from aerobic exercise, however, it is recommended that both aerobic and resistance training exercises be completed (Kim and Lee, 2006).

Aerobic exercise has been shown to have less effect in lowering LDL cholesterol, for example, than in lowering triglycerides or raising HDL (although there is some suggestion that raising HDL may be better done through resistance training). Inclusion of a low-fat diet and weight loss can enhance the effects, and inactive patients are likely to benefit most from increases in activity levels.

As with other ailments where medication is necessary, exercise should be regarded as adjunct to drug therapy, except in mild cases where it may be sufficient in itself. One particular disadvantage with the use of statins—which are used effectively for lipid correction—is an increased risk of myalgia and this appears to be even greater in heavy exercisers (Thompson et al., 2003). The implications of this are discussed further in Chapter 24.

What Can be Done in a Brief Consultation

- ☐ Check on the different types and levels of exercise currently carried out.
- ☐ Advise that there are different exercise requirements for specific health problems (the five S's; see Chapter 6).
- ☐ Ask about factors limiting potential exercise involvement (e.g., sore joints, muscular pain, time).
- ☐ Where risk factors may be mild, prescribe aerobic exercise (e.g., walking) for a short period and check effects before using medication.
- ☐ Consider simple measures such as flexibility or strength exercises. Refer to an exercise physiologist for specific exercise prescription.

EXERCISE, STRESS ANXIETY, AND DEPRESSION

Stress imposes a threat on the body and elicits a set of "flight or fight" neuroendocrine responses, mobilizing energy and inducing a state of insulin resistance in the body. Although this was essential in ancient times for survival, such threats have largely been overtaken in modern society by psychosocial threats. Identical stress responses are initiated by emotional or psychological stressors, but the energy mobilized is not used up and can be stored as visceral fat in the body. Tsatsoulis and Fountaoulakis (2006) suggest that physical activity should be the natural means for preventing or ameliorating the metabolic and psychological comorbidities induced by chronic stress. Physical alleviation of the body's reaction to stress can come through either aerobic (flight) or resistance (fight) activity; the level of specificity has not been detailed.

Current recommendations for dealing with stress are to follow standard exercise prescription protocols for the general population. Because anxiety (a possible outcome of stress) can create a heightened level of arousal, "dampening down" of the sympathetic nervous system through more intensive activity—either stamina or strength based, and carried out earlier in the day to maintain reduced sympathetic tone—may be the most effective approach. Given that a redistribution of energy substrates occurs to visceral stores, endurance activities such as walking or jogging may have the greatest effects on metabolic homeostasis, as well as anxiety moderation. Flexibility activities like yoga or tai chi can add a relaxation component to this.

Similarly, there is no clearly defined exercise prescription for depression, although there is clear evidence that this may be as effective as modern antidepressant medications (Motl et al., 2005), with the added benefit of improving physical functioning (Brenes et al., 2007). Recent research suggests a close and possibly even causal link between inactivity and depression (see Chapter 15), and a consistent effect of exercise that is as effective as medication over the long term (e.g., Pedersen and Saltin, 2006). Remaining active even in the absence of significant weight loss has also been shown to result in less depression, after accounting for potential confounders (Craig et al., 2006). Increased aerobic intensity with time may also be a factor, with training used as a supplement to medical treatment, except in the case of mild depression where it may be sufficient in itself.

Singh et al. (2005) investigated the effectiveness of two intensities of progressive resistance training exercise compared to normal GP care in patients with diagnosed clinical depression. Middle-aged patients with either major or minor depression were randomized into either high-intensity progressive resistance training (80% of 1 RM), low-intensity progressive resistance training (20% of 1 RM), or normal GP care. The exercise groups completed the exercises three days per week for eight weeks. Investigators found the high-intensity progressive resistance training was more effective than either low-intensity progressive resistance training or GP care in the treatment of these patients.

Further support for the benefits of progressive resistance training in patients with depression was reported by Motl et al. (2005), who studied the effects of six months of walking or resistance training exercise in older adults. They found depressive symptom scores significantly decreased following initiation of the exercise therapy and there was a sustained reduction for up to five years after both treatments.

OTHER CONDITIONS

As well as the diseases and disease risks mentioned herein, there are a number of common conditions where specific exercise prescription can improve or optimize health.

Guidelines for exercise in pregnancy, for example, are now available through a number of peak bodies, including the American College of Obstetrics and Gynecology (Artal et al., 2003). These include recommendations for protecting the fetus and health of the mother through the various stages of pregnancy, with a decrease in exercise volume and weight-supportive exercise during the last trimester. Particular attention needs to be paid to appropriate, rather than excessive, maternal weight gain and the specific needs of those at risk of or with gestational diabetes. All of the five S's are important, but should be achieved using particular precautions and individual specifications.

Studies with older people have shown the importance of emphasizing strength to reduce the sarcopenia (muscle wastage) that occurs with aging and to protect against falls. Suppleness or flexibility training is also recommended in this group, with less emphasis on stamina and the need for weight loss.

Perhaps counterintuitively, a graduated program of light to moderate aerobic exercise has therapeutic effects for chronic fatigue syndrome. Exercise and sleep are related, and recent research indicates a possible association between decreased sleep and obesity, with a possible link through increased inactivity (this is discussed more in Chapter 18).

TABLE 12.7 Common Problems and Possible Exercise Solutions

Problem	Possible Exercise Prescription
Cramping	Regular contractions (concentric and eccentric) of affected muscles, with and without resistance
Constipation	Endurance or extended aerobic exercise
Fatigue	Gentle endurance exercise in the mornings; increasing duration and intensity as tolerated
Incontinence	Kegel pelvic floor strengthening mobility improvement with endurance, balance and resistance training as needed
Insomnia	Endurance or resistance exercise in the early or midafternoon
Lower back pain	Resistance training to strengthen back extensor muscles (specific instruction required); strengthening of rectus abdominis and hip extensor muscles
Recurrent falls, balance gait and disorders	Lower extremity resistance training for hips, knees and ankles, including, for example, lateral movements, balance training, and tai chi
Weakness and sarcopenia	Moderate- to high-intensity resistance training for all muscle groups

Recently, attention has turned to exercise during menopause as an alternative to hormonal therapy. As well as protecting against osteoporosis, a structured stamina and strength-training program can assist in reducing menopausal symptoms (Fugate and Church, 2004). Appropriate exercise during the perimenopausal stage can also help ameliorate the weight gain that often occurs in menopause. Suppleness training is indicated for middle-aged people, as is an appropriate hypocaloric diet and exercise program to maintain body weight.

Some other common problems and possible exercise solutions are shown in Table 12.7.

SUMMARY

While a generic prescription for all patients to "move more" is appropriate in the modern environment, there are some conditions where more specific exercise prescription may provide added benefits. These include exercise regimens for rehabilitation, as well as prevention, where familial and biological risks indicate a propensity toward certain problems (e.g., dyslipidemia). The level of specialty required for this necessitates the use of a qualified exercise physiologist as part of the allied health professional team.

Practice Tips

Exercise Prescription for Specific Purposes

	Medical practitioner	Practice Nurse/Allied Health Professional
Assess	Whether a specific exercise prescription is appropriate. General fitness level.	Basic aspects of the five S's that may need attention: stamina, suppleness, strength, and size.
	Aspects of the five S's that may need attention.	
	Special precautions that may be needed/provided to exercise specialist.	
	Possible drug contraindications for exercise.	
Assist	Understanding of the importance of exercise.	By increasing motivation to undertake specific exercise.
	By obtaining regular updates on progress.	By providing feedback about interest in prescribed programs.
Arrange	Referral for specific exercise prescription advice where required.	Close consultation with an exercise physiologist to devise a program.
	List of other health professionals for a care plan (ensure exercise specialists have details about medical history and drug use).	Contact with gym or fitness center for ongoing exercise program if appropriate.

Key Points

Clinical Management

- Where specific exercise prescription and ongoing supervision is required, refer to an accredited exercise physiologist (see www.aaess.com.au).
- While basic exercise testing can be conducted in the clinical situation, this should only be used as the basis for more detailed testing and prescription carried out by a testing specialist.
- Regularly monitor exercise routines and ensure changes to exercise parameters as appropriate for the progression and goals of the program.
- Communicate regularly with an exercise specialist to ensure matching of medical and exercise goals.
- The exercise prescription must be individualized to ensure it is safe and clinically effective.

Key Points—cont'd

- Exercise intensity poses the greatest risk to any patient. Always ensure either the maximal exercise exertion (RPE on the Borg scale) or maximal exercise HR is set for each patient to ensure his or her safety during exercise. An accredited exercise physiologist can assist using the results of a recent graded exercise test.
- Telemetry heart rate monitors (i.e., Polar heart monitors, www.polar.fi/polar/entry.html) are particularly useful in helping patients ensure that they do not overexert themselves during exercise.
- It is good practice to require all cardiac patients to complete a hospital-conducted cardiac rehabilitation program prior to having them exercise in a community setting (i.e., gym or fitness center).

REFERENCES

ACSM, 2011. American College of Sports Medicine's Complete Guide to Health and Fitness. Human Kinetics, Champaign Ill.

ACSM (American College of Sports Medicine), Durstine, J.L., 2009. ACSM's Exercise Management for Persons with Chronic Diseases and Disabilities, third ed. Human Kinetics, Champ Ill.

ADA (American Diabetes Association), 2006. Standards of medical care in diabetes. Diabetes Care 29, S4–S42. Available from: http://care.diabetesjournals.org/cgi/content/full/29/suppl_1/s4.

Adams, J., Clin, M.J., Hubbard, M., et al., 2006. A new paradigm for post cardiac event resistance exercise guidelines. Am. J. Cardiol. 97, 281–286.

Albright, A., Franz, M., Hornsby, G., et al., 2000. American College of Sports Medicine position stand: exercise and type 2 diabetes. Med. Sci. Sports Exerc. 32, 1345–1360.

Artal, R., O'Toole, M., White, S., 2003. Guidelines of the American College of Obstetricians and Gynaecologists for exercise during pregnancy and the postpartum period. Br. J. Sports Med. 37, 6–12. Available from: http://bjsportmed.com/cgi/reprint/37/1/6.

Bouchard, C., 1983. Human adaptability may have a genetic basis. In: Landry, F. (Ed.), Health Risk Estimation, Risk Reduction and Health Promotion. Proceedings of the 18th Annual Meeting of the Society of Prospective Medicine. Canadian Public Health Association, Ottawa, pp. 463–476.

Bouchard, C., et al., January–February, 2015. Personalized preventive medicine: genetics and the response to regular exercise in preventive interventions. Prog. Cardiovasc. Dis. 57 (4), 337–346.

Boule, N.G., Kenny, G.P., Haddad, E., et al., 2003. Meta-analysis of the effect of structured exercise training on cardio-respiratory fitness in type 2 diabetes mellitus. Diabetologia 46, 1071–1081.

Brankston, G.N., Mitchell, B.F., Ryan, E.A., et al., 2004. Resistance exercise decreases the need for insulin in overweight women with gestational diabetes mellitus. Am. J. Obstet. Gynecol. 190, 188–193.

Brenes, G.A., Williamson, J.D., Messier, S.P., et al., 2007. Treatment of minor depression in older adults: a pilot study comparing sertraline and exercise. Ageing Ment. Health 11 (1), 61–68.

Craig, C.L., Bauman, A., Phongsavan, P., et al., 2006. Jolly, fit and fat: should we be singing the "Santa Too Fat Blues?" CMAJ 175 (12), 1563–1566.

Dempsey, J.C., Sorensen, T.K., Williams, M.A., et al., 2004. Prospective study of gestational diabetes risk in relation to maternal recreational physical activity before and during pregnancy. Am. J. Epidemiol. 159 (7), 663–670.

Drexel, H., 2006. Different Forms of Walking and Effects on Blood Chemistry. American Heart Association. Scientific Sessions abstract 3826.

Dunstan, D., Zimmet, P., Slade, R., et al., 2003. Diabetes and Physical Activity. International Diabetes Institute Position Paper. International Diabetes Institute, Melbourne.

Durstine, J.L., Moore, G.E., 2000. ACSM's Guidelines for Exercise Testing and Prescription. Human Kinetics, Champaign Ill.

Ehrman, J.K., Gordon, P.M., Visich, P.S., et al., 2003. Clinical Exercise Physiology. Human Kinetics, Champaign Ill.

Elhke, E., 2006. Resistance exercise for post-myocardial infarction patients: current guidelines and future considerations. Natl. Strength Cond. Assoc. J. 28 (6), 56–62.

Eves, N.D., Plotnikoff, R.C., 2006. Resistance training and type 2 diabetes. Diabetes Care 29, 1933–1941.

Fagard, R.H., 2006. Exercise is good for your blood pressure: effects of endurance training and resistance training. Clin. Exp. Pharmacol. Physiol. 33 (9), 853–856.

Fugate, S.E., Church, C.O., 2004. Nonestrogen treatment modalities for vasomotor symptoms associated with menopause. Ann. Pharmacother. 38 (9), 1482–1499.

Giannuzzi, P., Mezzani, A., Saner, H., et al., 2003. Physical activity for primary and secondary prevention. Position paper of the Working group on cardiac rehabilitation and exercise physiology of the European society of Cardiology. Eur. J. Cardiovasc. Prev. Rehab. 10 (5), 319–327.

Hootman, J., Bolen, J., Helmick, C., et al., 2006. Prevalence of doctor-diagnosed arthritis and arthritis attributable activity limitation—United States, 2003–2005. MMWR 55 (40), 1089–1092.

Jolliffe, J.A., Rees, K., Taylor, R.S., et al., 2000. Exercise-based rehabilitation for coronary heart disease. Cochrane Database Syst. Rev. 4, CD001800.

Kim, N.J., Lee, S.I., 2006. The effects of exercise type on cardiovascular risk factor index factors in male workers. J. Prev. Med. Pub Health 39 (6), 462–468.

Mathieu, P., Boulanger, M.C., Després, J.P., 2014. Ectopic visceral fat: a clinical and molecular perspective on the cardiometabolic risk. Rev. Endocr. Metab. Disord. 15 (4), 289–298.

Morgan, W.P., 2001. Prescription of physical activity: a paradigm shift. Quest 53, 366–382.

Motl, R.W., Konopack, J.F., McAuley, E., et al., 2005. Depressive symptoms among older adults: long-term reduction after a physical activity intervention. J. Behav. Med. 28 (4), 385–394.

NHMRC (National Health and Medical Research Council), 2005. National Physical Activity Guidelines. Canberra: DoHA. Available from: www.health.gov.au/internet/wcms/Publishing.nsf/Content.

Nieman, D.C., 1998. Exercise immunology: integration and regulation. Int. J. Sports Med. 19 (Suppl. 3), S171.

Paffenbarger Jr., R.S., Lee, I.M., Kampert, J.B., 1997. Physical activity in the prevention of non-insulin-dependent diabetes mellitus. World Rev. Nutr. Diet 82, 210–218.

Pedersen, B.K., Saltin, B., 2006. Evidence for prescribing exercise as therapy in chronic disease. Scand. J. Med. Sci. Sports 16 (Suppl. 1), 3–63.

Scrutinio, D., Bellotto, F., Lagioia, R., et al., 2005. Physical activity for coronary heart disease: cardioprotective mechanisms and effects on prognosis. Monaldi Arch. Chest Dis. 64 (2), 77–87.

Signal, R.J., Kenny, G.P., Wasserman, D.H., et al., 2004. Physical activity/exercise and type 2 diabetes. Diabetes Care 27 (10), 2518–2535.

Singh, N.A., Stavrinos, T.M., Scarbek, Y., et al., 2005. A randomized controlled trial of high versus low intensity weight training versus general practitioner care for clinical depression in older adults. J. Gerontol. Ser. A, Biol. Sci. Med. Sci. 60 (6), 768–776.

Snowling, N.J., Hopkins, W.G., 2006. Effects of different modes of exercise training on glucose control and risk factors for complications in type 2 diabetic patients: a meta-analysis. Diabetes Care 29 (11), 2518–2527.

Thomas, M.C., Zimmet, P., Shaw, J.E., 2006. Identification of obesity in patients with type 2 diabetes from Australian primary care: the NEFRON-5 study. Diabetes Care 29 (12), 2723–2725.

Thompson, P.D., Clarkson, P., Karas, R.H., 2003. Station-associated myopathy. JAMA 289 (13), 1681–1689.

Tsatsoulis, A., Fountoulakis, S., 2006. The protective role of exercise on stress system dysregulation and comorbidities. Ann. N. Y. Acad. Sci. 1083, 196–213.

US Surgeon General, 1996. Physical Activity and Health: a Report of the Surgeon General. Department of Health and Human Services Centres for Disease Control, Atlanta, GA.

Professional Resources

Indications for Deferring or Terminating Physical Activity

According to the Australian Association for Exercise and Sport Science, there are several indications for deferring activity and having a medical review:

- unstable angina
- uncontrolled cardiac failure
- severe aortic stenosis
- uncontrolled hypertension or grade 3 (severe) hypertension (e.g., blood pressure >180 mm Hg systolic or >110 mm Hg diastolic)
- symptomatic hypotension <90/60 mm Hg
- acute infection or fever or feeling unwell (including, but not limited to, acute myocarditis or pericarditis)
- resting tachycardia or arrhythmias
- diabetes with poor blood glucose control [e.g., blood glucose <6 mmol/L (108 mg/dL) or >15 mmol/L (270 mg/dL)]

There are also indications for terminating activity:

- squeezing, discomfort or typical pain in the center of the chest or behind the breastbone with or without spreading to the shoulders, neck, jaw, and/or arms, or symptoms reminiscent of previous myocardial ischemia
- dizziness, lightheadedness, or feeling faint
- difficulty breathing
- nausea
- uncharacteristic, excessive sweating
- palpitations associated with feeling unwell
- undue fatigue
- leg ache that curtails function
- physical inability to continue
- for people with diabetes: shakiness, tingling lips, hunger, weakness, palpitations

For more information, see the association's Website: www.aaess.com.au.

Chapter 13

Stress: Its Role in the S-AD Phenomenon

Garry Egger, Robert Reznik

> *The term 'stress' has simply come to mean so much that it actually means very little in real terms.*
>
> Dr. Albert Crum (2000)

INTRODUCTION: "STRESS" VERSUS "STRAIN"

Although they are quite often discussed independently, there are clear links, overlaps, and confusion between the process of stress and the states of anxiety and depression and their effects on human psychology and health. The connection is so close, at least in the public mind, that the urge to label them under the single acronym "SAD" is compelling. However, we have resisted this because of the need to differentiate the distinct operational differences between these terms as they are used in Lifestyle Medicine. Instead, we use the broken acronym S-AD in the following three chapters. This also avoids confusion with seasonal affective disorder, associated with obesity and depression during winter.

The confusion results partly from a problem of semantics. While anxiety and depression are outcome terms, stress is more often used as a process term, reflecting a possible cause of the former (according to the *Macquarie Dictionary (2016)*, stress is "the forces on a body which produce a deformation or strain"). In common parlance, people talk about being stressed as if it represents a medical condition, whereas the condition is actually the outcome of the stress, manifest in many different forms. According to American neuropsychiatrist Dr. Albert Crum (2000), "If we accept the dictionary definition, then stress is both a cause…and effect…The cause can be viewed as a good thing—a challenge or a push to do better—while the effect is not so good—a distress that may be felt as a momentary nervousness and free floating anxiety, or stark, paralyzing fear."

In like vein, Australian psychiatrist William Wilke suggested in 1990 that the term *stress* is increasingly misused as a synonym for distress, confusing both the response and the cause of nervous system overload. Canadian researcher Hans Selye first applied the term stress to human psychology in a 1946 journal article

Lifestyle Medicine. http://dx.doi.org/10.1016/B978-0-12-810401-9.00013-9

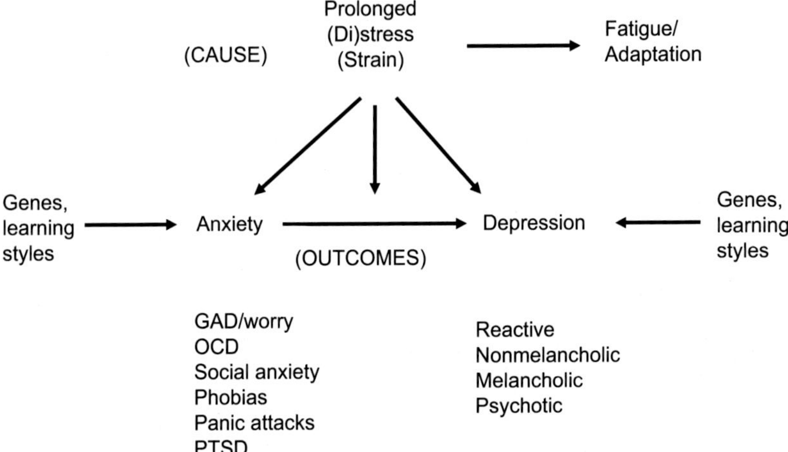

FIGURE 13.1 The links between stress/strain, anxiety, and depression. *GAD*, general anxiety disorder; *OCD*, obsessive compulsive disorder; *PTSD*, posttraumatic stress disorder.

(Selye, 1946). Before his death in the 1990s, however, Selye admitted that he may have popularized the wrong term and that the more appropriate expression for the outcome of stress would be the engineering term *strain*. Selye suggested that stress, in itself, is not a problem. "Stress," said Selye, "depends not on what happens to an individual, but upon the way he reacts." Using this, the links between stress, strain, anxiety, and depression become more obvious, as shown in Fig. 13.1.

In light of this, stress would more correctly be called the "stressed state," a "stress reaction," or "strain." However, given its popular use, we will continue to use the term stress.

As we will see, not all stress is bad. However, strain—when defined as the outcome of "distress" (in contrast to "eustress," or the feeling of euphoria associated with the right level of stimulation), as seen in Fig. 13.1—can have health-related consequences if it continues for long enough. This can lead to, or be caused by, anxiety, which if maintained at a high enough level for an extended period and accompanied by a sense of loss of control can lead to depression. American psychologist Dr. Martin Seligman (see Chapter 16) originally referred to this as learned helplessness (Seligman, 1972). As shown in Fig. 13.1, however, anxiety and depression can also develop spontaneously within an individual, not from any obvious strain from a stressor but as a result of a genetic predisposition toward these mood states.

So, having effectively complicated the S-AD phenomena, we will attempt to alleviate the confusion by untangling the phenomenon known as stress in this chapter and discussing the idiosyncrasies of anxiety and depression separately in the following chapters. We will take a more positive approach to the clinical management of mood states by examining happiness and the move toward positive psychology in Chapter 16.

WHAT IS STRESS?

In a comprehensive review of the relationship between stress and obesity, Greek endocrinologists Ionass Kyrou et al. (2006) defined stress as "a state of threatened homeostasis or disharmony caused by intrinsic or extrinsic adverse forces...counteracted by an intricate repertoire of physiologic and behavioral responses that aim to re-establish the challenged body composition."

The adaptive response referred to by Kyrou et al. and earlier by Selye, refers to an elaborate physiological system involving the hypothalamic–pituitary–adrenal (HPA) axis and the central and peripheral components of the autonomic nervous system (ANS). In situations of acute stress, the response of these systems restores homeostasis through behavioral and physical adaptations, which improves the individual's chances of survival. Behavioral adaptations include increased arousal, alertness, vigilance and cognition, and suppression of hunger, feeding, and sexual behavior. Physiological adaptations involve adaptive redirection of energy through increased respiratory rate, blood pressure, heart rate, cardiovascular tone, gluconeogenesis, and lipolysis. These place the body in an allostatic state, which is relieved by reactions to remove or evade the stressor. Where this does not occur, such as in a state of chronic stress, an allostatic overload results. This is caused by either a persistent stressor or a prolonged duration of the adaptive response.

Biological changes resulting from persistent changes in the HPA axis and ANS can affect a variety of physiological functions, resulting in symptoms defined as the metabolic syndrome, including increases in the most metabolically dangerous type of fat—visceral obesity (Mathieu et al., 2014). Chronic stress/strain is defined by Kryou et al. (2006) as a "pathological state of prolonged threat to homeostasis by persistent or frequently repeated stressors...considered a significant contributing factor in the pathophysiology of manifestations that characterize a wide range of diseases and syndromes."

Hence, while acute changes in homeostasis induced by a stressor can be stimulatory, chronic changes and the resultant allostatic state can have long-term health consequences. Chemical and hormonal changes that occur to assist the restoration of homeostasis through fight or flight are not corrected in the chronic state, and the body is left to "stew in its own juices." In readiness for fight or flight, the body prepares—the "general adaptation syndrome"—largely through the contraction of large preparatory muscles such as the trapezius (to raise the arms for "fight"), the abdominals (to avoid being "winded"), and the gluteals and lower back (for "running away"). If an adaptive response occurs, the muscles are relaxed and homeostasis reigns. However, if stress is extended, these muscles remain taut and become a source of pain when palpated.

Traditionally, nature's response to the allostatic state is to "escape" (fight or flight), either physically or mentally. This helps restore the psychological state of perceived control, which alleviates the stress reaction. It is these two concepts—escape and control—to which we will turn when looking at ways of dealing with stress. Before doing so, however, it is relevant to see why some people are affected more by particular stressors than other people and to examine the stages of strain through which progression can occur if stress reactions are not managed.

GENETICS, RESILIENCE, AND HARDINESS

Although we talk of stress as a single concept, there are two main influencing factors: the stressor, or the external factor(s) causing a stress response or strain in the individual, and the internal factors within any individual leading to a stress response.

There is little doubt that some people are more resistant to the outcome of stress than others (Tugade et al., 2004). This tends to run in families and new findings suggest that genes can actually be attributed to components such as worry. A general anxiety disorder (GAD) appears to exist in around 10% of the population and 60% of those suffering significant depression (Dolnak, 2006), and it may be that anxiety lies on a continuum, with severe phobias at one end and extreme hardiness or resilience at the other. Characteristics that appear to distinguish the hardy from the nonhardy—and which may or may not be "hard wired"—are being empowered or having a feeling of "control" rather than powerlessness; having the ability to see issues as a challenge rather than a threat; and having commitment as opposed to no commitment. But several other characteristics have also been identified (Steiger and Thaler, 2016). These provide the themes for management tips discussed here.

Learning styles can also contribute to a predisposition to develop anxiety or depression, although the extent to which this involves a genetic constituent is not known. Rumination, cyclical thinking and excessive self-based thinking, for example, can lead to a "downhill spiral," which could result in biochemical changes that "impress" a tendency to anxiety or depression on the brain and make this more difficult to then relieve (Dozois and Dobson, 2004).

STAGES OF STRESS/STRAIN

The response to stress can be conceptualized as passing through a number of "zones" and, although the transition through each may be imperceptible to the person being stressed, each zone has characteristic symptoms. These are shown in Fig. 13.2.

Distress can occur with too much or too little stress as shown in zones 1, 3, and 4 in Fig. 13.2. In the case of a stimulating stressor, such strain is minimal and in fact is even sought after by the individual as a feeling of excitement, enjoyment, or "flow." This is the stage known as "eustress" shown as stage 2.

	Zone	Symptoms	
High	4. Danger Zone	Insanity Breakdown Burnout*	Distress
	3. Warning Zone	"Brownout" Exhaustion Anxiety	↕
	2. Comfort Zone	Excitement "Flow" Enjoyment	Eustress ↕
Low	1. Boredom Zone	Disinterest Restlessness Immobilization	Distress

FIGURE 13.2 The stages of stress.

In the early stages of distress, a stage 1 (sensory level) perception only may have been made and hence symptoms such as exhaustion and anxiety can occur with little stage 2 awareness (intellectual level) of the cause. By the time this reaches stage 4, "burnout" becomes a common expression of symptoms, defined by the acronym DISINTEREST:

D Decreased sense of humor
I Increased physical problems (e.g., fatigue, infections)
S Social withdrawal
I Increased work load
N Not accomplishing
T Tension
E Emotional exhaustion
R Reduced sleep
E Easily taking offense
S Skipping meals and rest breaks
T Tranquilizer and/or alcohol use

According to prominent US psychologist Mihalyi Csikzentmihalyi (1975, 2008, 2013), the stages of stress and their outcomes on performance are influenced by an interaction between the severity of a stressor and the capacity of the individual being stressed (the "stressee"), as shown in Fig. 13.3.

Where a stressor is seen as insignificant by an individual capable of dealing with this (e.g., a very good mountain climber on a very easy slope), boredom is the likely outcome. Where the opposite occurs (an inexperienced climber on a very hard slope), anxiety is the likely outcome. Where both are matched, a feeling of "flow" or homeostasis is the likely outcome, putting the individual into the "comfort zone" (or state of eustress). This suggests that for any given task we have a "red line"—as seen in the top of Fig.

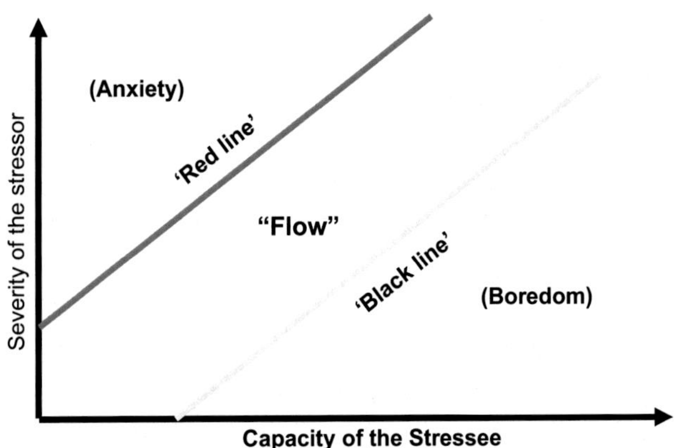

FIGURE 13.3 The relationship between stressor and stressee. *From Csikszentmihalyi, M., 1975. Beyond Boredom and Anxiety. Jossey Bass, San Francisco.*

13.3—above which the "heat becomes too hot." One solution is to lower the red line ("if you can't stand the heat, get out of the kitchen"). A second, "black line" represents the level below which stimulation is insufficient and boredom results.

Stress can becomes distress (leading to anxiety and other problems long term, as seen in Fig. 13.3) if the capacity of the stressee is genuinely overwhelmed by the severity of the stressor, or feels this to be the case, resulting in a feeling of loss of control. In cases of real stress, such as imminent danger, homeostasis can often only be achieved by attempting to change or eliminate the stressor. In many people, however, strain results from an inaccurate perception of stress. The classical reactions of fight or flight both lead to escape, in the broad sense of the term, either mental or physical. The concepts of control (or more correctly *perceived* control) and escape (both mental and physical) are core to the management of stress, and these can be achieved by changing the stressor and/or the stressee.

THE CONCEPTS OF CONTROL AND ESCAPE

A feeling of loss of control, whether this is real or perceived, is what differentiates stress from distress leading to anxiety and/or depression. Some people (the more hardy or resilient) can put up with extreme stressors without difficulty, whereas others crumble at the slightest pressure. Restoring a feeling of perceived control is therefore a prime goal of stress reduction, whether this is by removing the individual from a stressor, reducing its potency to the stressee (by building his or her psychological immunity), or changing the stressee's reaction to the stressor. In all cases, the medium of change is encompassed in the term "escape."

What Can Be Done in a Brief Consultation?

- ☐ Check whether stress is a possible influencing factor on health by asking about response patterns to chronic stress (e.g., overeating, self-medicating, passivity).
- ☐ Assess the extent of any existing stress response (i.e., stress "zone").
- ☐ Explain the concepts of "control" and "escape" in managing stress.
- ☐ Determine the patient's likely best management approach to stress: physical or mental (see Professional Resources).
- ☐ Where possible, identify and reduce the potency of stressor(s) by lowering the "red line."
- ☐ Help restore perceived control within the individual.

Interestingly, the modern interest in escape as a means of coping with stress arose from early research involving people who had little chance to put the real thing into practice. Sociologists Stanley Cohen and Taylor (1976) studied long-term prisoners—some of whom had life sentences—to find out about their mental processes. They found long-term inmates were no different to unconfined individuals. With no hope of physical escape the prisoners used a variety of mental escapes: fantasy, imagination, dreaming, meditation, writing, creativity, and even some strange hobbies. These helped to restore the feeling of control, which is vital for the management of stress.

AN APPROACH TO DEALING WITH STRESS: ACE (ANALYZE, CHANGE, EVALUATE)

The processes of stress management are described with a simple three-stage model: ACE (analyze, change, and evaluate) (Fig. 13.4).

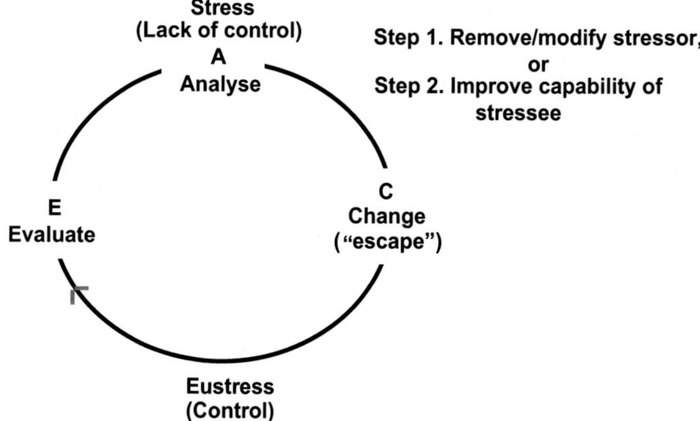

FIGURE 13.4 An ACE approach to dealing with stress (analyze, change, evaluate).

Analyze

This refers to establishing an awareness of the main source(s) of stress. As defined by Crum (2000), it involves moving from a stage 1 (sensory) to a stage 2 (intellectual) perception of the problem and the effects of stress are the results of particular external stressors or the individual's own coping mechanisms. Where a stress reaction is manifest with a less specific stressor or one that changes over time, a general anxiety disorder, or a higher order anxiety may be suspected. Where a stressor can be identified, and the stress reaction would be reduced or eliminated if this was made to disappear or reduced in potency, the obvious solution is to work on changing the stressor. Where removal of this would not reduce the reaction, it is more important to build immunity in the stressee. The exact situation determines the operational procedures of the next stage.

CHANGE (ESCAPE)

Changing the Stressor

Where an obvious stressor can be identified and is easily modified, changing the stressor is the obvious way to deal with the problem. A simple formula for this is the "R & D" method:

Reduce it Do it
Remove it Delegate it
Replace it Delay it
Rethink it Discuss it
Reframe it Drop it

However, because the causes of most stress reside in the home, work, or financial environments, and because these are situations that are often difficult to change, dealing with the stressor or the underlying cause of the problem is often not an option. As the prayer goes, "God grant me the serenity to accept the things I cannot change, the courage to change the things I can, and the wisdom to know the difference." Once a stressor is identified, these options can be considered. Most stress, however, comes from the individual's reaction and calls for changing the stressee.

Changing the Stressee

As discussed earlier, all methods of dealing with stress can be considered as involving some form of escape. Escapes can be either maladaptive (e.g., drug abuse, violence, blame, displacement) or adaptive. Maladaptive approaches work only in the short term; they do not solve a chronic problem and may even lead to its worsening. Alcohol and stimulant drugs such as caffeine, for example, may even lead to panic attacks (see Chapter 14). Medication can help patients change the situation but needs to be seen as an adjunct to a greater understanding and management of the problem, rather than treatment as such.

The adaptive approaches involve the classic two reactions—fight or flight. These still form the basis of modern stress management solutions. In the past, flight may have meant running away, but it is clear that escape can be either mental or physical (or both). This can mean removing oneself physically from a stressor (a holiday, quitting a stressful job, exercise) or using techniques of mental "escape." Similarly, tackling the situation head on physically through fight is now not always possible or advisable. However, other options, both physical and mental, are available and examples of these (as well as the flight options, divided into physical and mental approaches) are shown in Table 13.1. The examples shown are not meant to be exhaustive but to provide examples of the wide range of options available, depending on individual interests and capacity. The range is vast, and individual preferences probably explain why no one theory of stress management fits all.

In general, it can be said that some people escape best physically (by running long distances, sport, going to the gym, paddling a canoe, riding a bike). Others do it best mentally (by meditation, cognitive behavior therapy, prayer, tai chi, yoga, meditation, distractions such as reading or watching movies). A third group uses a combination of techniques. A brief test shown in the Professional Resources is one way of assessing the most appropriate approach for an individual.

TABLE 13.1 Somatic (Physical) and Cognitive (Mental) Escapes

Flight		Flight	
Physical	Mental	Physical	Mental
Exercise	Meditation	Confrontation	Cognitive
Sport	Muscle relaxation	Challenge	Restructuring
Warm baths	Pause/prayer	Preparation	Rational emotive
Holidays	Concentration	Social engineering	Therapy
Mini breaks	Counting to 10	Nutrition	Assertiveness
Hobbies	Art	Direct action	Psychotherapy
Games	Mind games	Expression	Planning
Yoga	Music	Talking	Mind control
Tai chi	Creativity	Meeting	Financial control
Breathing	Reading		Positive self-talk
Massage	Trance		Rehearsal
Sex	Biofeedback		Thought stopping
Sleep	Mind mapping		Brainstorming

Evaluate

Once change has been effected through some form of appropriate escape, setting stress-reduction goals and monitoring the patient's homeostatic state through standard measures such as blood pressure, blood glucose, and other biochemical measures, as well as through simple stress questionnaires, can test the outcome. If an appropriate outcome is not achieved, the process outlined in Fig. 13.4 may need to be instituted again, until a measure of control, and homeostasis, is restored.

SUMMARY

The term stress is often confused with the outcome of a chronic period of allostatic overload (strain), which can cause a variety of disorders—metabolic, physical, and psychological—often leading to more overt disease. There are major differences in hardiness (representing individual responses to a stressor), much of which seems to be genetically based and which, at one extreme, can lead to a range of phobic responses and even GAD. Stress management provides a range of mental or physical escapes appropriate to the individual, which help restore a sense of perceived control over the situation or one's life.

Practice Tips

Managing Stress

	Medical Practitioner	Practice Nurse/Allied Health Professional
Assess	• Whether stress is an issue likely to cause health problems • Effects of stress on health behaviors (e.g., eating, obesity) monitoring • Whether stress is a cause of greater morbidity (e.g., anxiety, depression)	• Stress as a factor in disease state • Stress levels with questionnaire (baseline and during monitoring) • Stress type (see Professional Resources) • Cardiovascular risk factors
Assist	• Patient to identify stressors where possible • By identifying goals • By providing (short-term) medication if needed • By explaining the control and escape concepts for stress management • Identification of "escapes" • By reviewing frequently • By using templates as appropriate...	• With general education about stress management, including suggesting or providing good literature

Practice Tips—cont'd

Managing Stress

	Medical Practitioner	Practice Nurse/Allied Health Professional
Arrange	• Referral for stress management if appropriate • List of other health professionals for a care plan	• Discussion of stress management tactics with other allied health professionals • Input from a psychologist or psychiatrist • Contact with self-help groups if desired

Key Points

Clinical Management

- Differentiate between stress and the outcome of chronic strain.
- Consider stress as a cause of other morbidities (e.g., anxiety, depression), not an outcome in itself.
- Offer techniques for reducing stress independent of treating other outcomes.
- Use medication as a window of opportunity for dealing with the lifestyle basis of the problem.
- Aim to empower the patient by restoring perceived control.
- Target the stressor or stressee according to where the problem lies.

Professional Resources

A Stress-Type Test

Patient instructions: Rate the degree to which you generally or typically experience each of the following symptoms when you feel anxious or stressed.

	Not at all				Very Much
A. I have difficulty concentrating because of uncontrollable thoughts.	1	2	3	4	5
B. My heart beats faster.	1	2	3	4	5
C. I worry too much over something that doesn't really matter.	1	2	3	4	5
D. I imagine terrifying scenes.	1	2	3	4	5
E. I feel jittery in my body.	1	2	3	4	5
F. I get diarrhea.	1	2	3	4	5
G. I can't keep anxiety-provoking pictures out of my mind.	1	2	3	4	5
H. I feel tense in the stomach.	1	2	3	4	5
I. Unimportant things bother me.	1	2	3	4	5

Continued

Professional Resources—cont'd

	Not at all				Very Much
J. I feel like I am losing out because I can't make up my mind quickly.	1	2	3	4	5
K. I nervously pace.	1	2	3	4	5
L. I become immobilized.	1	2	3	4	5
M. I perspire.	1	2	3	4	5
N. I can't keep anxiety-provoking thoughts out of my mind.	1	2	3	4	5

Scoring:

Add scores in the following way:

1. Physically oriented = B + E + F + H + K + L + M = Total score
2. Cognitively oriented = A + C + D + G + I + J + N = Total score

The results can indicate whether a respondent is best suited to a somatic (physical) or cognitive (mental) approach to managing stress.

REFERENCES

Cohen, S., Taylor, L., 1976. Escape Attempts. Allen Lane, London.

Crum, A., 2000. The 10-Step Method of Stress Relief: Decoding the Meaning and Significance of Stress. CRC Press, Boca Raton, FA.

Csikszentmihalyi, M., 1975. Beyond Boredom and Anxiety. Jossey Bass, San Francisco.

Csikszentmihalyi, M., 2008. Flow: The Psychology of Optimal Experience. Harper and Rowe, NY.

Csikszentmihalyi, M., 2013. Creativity: Flow and the Psychology of Discovery and Invention. Harper Perennial, NY.

Dozois, D.J.A., Dobson, K.S. (Eds.), 2004. The Prevention of Anxiety and Depression: Theory, Research, and Practice. American Psychological Association, Washington, D.C.

Dolnak, D.R., 2006. Treating patients for comorbid depression, anxiety disorders and somatic illnesses. J. Am. Osteopath. Ass. 106 (5 Suppl. 2), S9–S14.

Kyrou, J., Chrousos, G.P., Tsigos, C., 2006. Stress, visceral obesity and metabolic complications. Ann. N. Y. Acad. Sci. 1083, 77–110.

Macquarie Dictionary, 2016. Macquarie Library Pty Ltd., third ed. Macquarie University.

Mathieu, P., Boulanger, M.C., Després, J.P., 2014. Ectopic visceral fat: a clinical and molecular perspective on the cardiometabolic risk. Rev. Endocr. Metab. Disord. 15 (4), 289–298.

Seligman, M., 1972. Learned Helplessness. Random House, NY.

Selye, H., 1946. The general adaptation syndrome. J. Clin. Endocrinol. 6, 177.

Steiger, H., Thaler, L., 2016. Eating disorders, gene-environment interactions and the epigenome: Roles of stress exposures and nutritional status. Physiol. Behav. http://dx.doi.org/10.1016/j.physbeh.2016.01.041.

Tugade, M.M., Fredrickson, B.L., Barrett, L.F., 2004. Psychological resilience and positive emotional granularity: examining the benefits of positive emotions on coping and health. J. Pers. 72 (6), 1161–1190.

Chapter 14

Dealing With Worry and Anxiety

Andrew Binns, Garry Egger, Robert Reznik

Anxiety is a thin stream of fear trickling through the mind. If encouraged, it cuts a channel into which all other thoughts are drained.

Arthur Somers Roche

INTRODUCTION: ANXIETY AS "FEARED HELPLESSNESS"

If depression is "learned helplessness," anxiety could be thought of as "feared helplessness." Unlike depression, where the sufferer has all but "given up," the anxiety sufferer is still trying to cope, albeit in the presence of "an uncomfortable feeling of nervousness or worry about something that is happening or might happen in the future" (*Cambridge Dictionary*). The fear is recognized by individuals as being out of proportion to the event or situation but nevertheless they feel unable to control this. Psychodynamically, the anxiety and its associated signs and symptoms are often the displacement of conscious and unconscious stresses that are unresolved or seemingly insoluble.

Anxiety disorders follow a chronic course, often with relapse and remission, and although there is a significant genetic component involved in causality, lifestyle factors such as alcohol and drug use, caffeine, poor nutrition, smoking, and obesity can cause a significant exacerbation, or indeed initiate the symptoms (Coplan et al., 2015).

HEALTH EFFECTS OF ANXIETY

Everyone feels anxious at times and it can signal us to use our resources to minimize a threat to our wellbeing. It can stimulate us to step outside our comfort zone to achieve things of which we were not aware we were capable. However, excessive and prolonged anxiety itself can be debilitating; excessive anxiety can lead to an allostatic state of disequilibrium, which, over time, can result in a range of metabolic disorders up to and including heart disease (Celano et al., 2015). Pathologies can also include somatoform disorders—such as body dysmorphic syndrome, irritable bowel,

Lifestyle Medicine. http://dx.doi.org/10.1016/B978-0-12-810401-9.00014-0

hypochondriasis, and variants of fibromyalgia—where the focus of the anxiety becomes the body functioning. Recently, worry (as a subcomponent of anxiety) has also been shown to affect long-term memory and the risk of Alzheimer's disease.

TYPES OF ANXIETY DISORDERS

There are a number of separate classifications within the general category of anxiety disorders, ranging from generalized anxiety disorder (GAD), obsessive compulsive disorder (OCD), social anxiety disorder (SAD), panic disorder (PD), and specific phobias such as social phobia and agoraphobia through to posttraumatic stress disorder (PTSD). However, these can be comorbid with each other and with depression (Aina and Susman, 2006). For example, in a Swedish sample of over 3000 randomly selected adults, point prevalence of clinically significant depression and anxiety was around 17% and nearly 50% had comorbid disorders (Johansson et al., 2013).

Typically, an anxiety disorder will precede the development of depression (Kessler et al., 2003), although not all anxiety sufferers will progress to depression and not all depressive patients will suffer anxiety (Aina and Susman, 2006). Also, although each disorder has its own primary manifestation, symptoms of other disorders can be experienced in conjunction with the principal disorder. Indeed, the most common disorder of all (still not accepted as a disorder by Diagnostic and Statistical Manual of Mental Disorders, although it contains research criteria) appears to be a mixed anxiety/depression disorder (Johansson et al., 2013).

THE EXTENT OF THE PROBLEM

Anxiety disorders exist in up to 10% of the population at any one time but usually do not begin until the late teens. Studies have shown that up to 18% of over 18-year-olds may have an anxiety disorder in any one year. Over a lifetime, however, more than one in four adults will experience at least one anxiety disorder (Kessler et al., 2005, 2012). Based on epidemiological data, US data show that ~20% had at least one lifetime morbid anxiety risk disorder, 13.0% had PTSD, 10.1% had GAD, 8.7% had a PD, and 3.7% had OCD (Kessler et al., 2012)—and yet fewer than two in three sufferers reported receiving any treatment for their disorder. It has been estimated that anxiety accounts for around 30% of the total expenditure for mental illness (equivalent to about one-third the cost for heart disease) and that over half of this is accounted for by anxiety drug therapy (Kessler et al., 2012).

GAD is the most common of the anxiety disorders and makes up about 60% of anxiety cases. This is a pervasive problem, where patients have chronic daily anxiety that can interfere with their daily functioning and may require medication use. PD, on the other hand, produces discrete symptom attacks that arise suddenly, escalate over 10 to 15 minutes, then subside. We will look here only at the moderate end of the scale of anxiety, leaving severe panic attacks, PTSD, and OCD for specific attention from the appropriate allied health professionals. However, some of the treatment approaches discussed will be appropriate for a range of anxiety disorders.

Various websites, such as www.beyondblue.org and www.moodgym.anu.edu.au, give easy-to-follow descriptions of the symptoms of the different forms of anxiety disorder as well as brief self-tests for signs, symptoms, and potential risk factors. These and other websites quoted throughout this chapter also discuss treatment options and give details of where patients can get personalized help.

THE ROLE OF LIFESTYLE

While anxiety in its various forms has undoubtedly been with humans from time immemorial, certain modern lifestyle factors may have exacerbated the problem, and anxiety management needs to include a close analysis of lifestyle and thought patterns followed by appropriate modification.

Decreases in sleep time and quality, for example, can cause nervous system changes favoring increased parasympathetic arousal. This can be augmented by excessive caffeine, alcohol, and recreational drugs such as marijuana, cocaine, and amphetamines. Caffeine use can also raise anxiety levels in those predisposed individuals, particularly patients with GAD (Richards and Smith, 2015).

Inactivity, which fails to produce the standard modulatory hormonal effect on the sensory nervous system, is likely to have a similar effect (Tsatsoulis and Fountoulakis, 2006; Teychenne et al., 2015). Modern levels of obesity may increase anxiety as a result of social discrimination, workplace disadvantage, decreased sexual potency, and the resultant negative self-images that these create.

DIAGNOSIS

Unlike depression, anxiety may not be openly declared to a clinician. It can be picked up in a screening questionnaire such as the Hamilton Anxiety Scale or through questions about mood states (e.g., anxious mood, tension, difficulty getting to sleep, fears, worries, compulsions).

Somatic pain is often a presenting symptom, as with depression. Although anxiety may be comorbid with depression, the two symptoms have different internal mood states. Depression is characterized by feelings of helplessness, worthlessness, despair, and anger that reduce the patient's energy to carry out the activities of daily living. Anxiety, on the other hand, is an excess of energy generated by intense fear and panic, coupled with excessive worry. Anxiety sufferers typically have difficulty getting to sleep, whereas depressives typically have early morning wakening with excess sleeping at other times (however, the increased arousal associated with the depressed state can lead to a pervasive sense of exhaustion for the sufferer).

Diagnoses that may need to be ruled out include: thyroid disorders; other, less common medical disorders; use of a variety of medications including over-the-counter preparations and herbal remedies; substance abuse; and extreme overuse of caffeine, especially in those easily influenced by the effects of caffeine (such as the person who becomes jittery and hyperaroused after only one or two cups of strongly brewed coffee). It is also worth remembering that many substances contain caffeine, including all the so-called energy drinks, many diet drinks, and, especially, green tea. Care should also be taken to distinguish anxiety from bipolar disorder.

Aspects of each of the individual anxiety disorders are covered in detail in several specific texts (e.g., Bourne, 2015; Blashki et al., 2007) and popular publications (e.g., Pittman and Karle, 2015).

WORRY AS A UBIQUITOUS CHARACTERISTIC OF ANXIETY

Worry, in some form, is an underlying feature in all anxiety disorders (Newman et al., 2008). It also manifests as hypochondriasis in about 1% of the population (where displaced anxiety manifests as changes in somatic states that are interpreted by the person as a marker of significant or indeed catastrophic medical conditions) and up to 5% of those in primary care (Shearer and Gordon, 2006). Worry is a short-term response to uncertainty that can become self-perpetuating, with adverse long-term consequences.

However, worry differs in the forms it takes in different anxiety disorders. In PD, for example, worry is often specific to the cause, whereas those with GAD generally report a pervasive sense of worry about everything and almost any new stress is assumed to result in a negative outcome. Scores on worry tests correlate with the frequency of statements related to personal inadequacy, suggesting that this may be a characteristic of individuals who worry excessively. Older people (>65) worry more than the young, and the nature of the worry is usually different; over half the worries of older people are health related, whereas those of the young are more about psychological wellbeing and social evaluation (Barlow, 2002; Andrews et al., 2010).

Following evolutionary principles, worry (like anxiety) serves a purpose by diverting an individual's energies away from possible future life-threatening situations. People who worry chronically often believe that their worry serves positive functions. This is because the low probability of occurrence of feared events can often reward worry, shaping beliefs that this somehow has prevented bad things from happening. The reinforcement of the worried state can then, in extreme circumstances, be extended to "worrying about worrying" (meta-worry), taking the problem full circle (Barlow, 2002).

Ain't no use worryin' 'bout things you can't control, 'cause you can't control them. And ain't no use worryin' 'bout things you can control, 'cos you can control them. So just ain't no use worryin'.

Ed Moses (Olympic gold medalist 400 m hurdles 1976, 1984)

Management of excessive worry is covered in the general principles discussed later and primarily includes pharmacotherapy and cognitive behavioral therapy (CBT). Somatic and mental approaches such as relaxation training and meditation are likely to be less than fully effective because they do not specifically target the source of worry. Nevertheless, they do allow the person to develop a generalized lower level of arousal and to experience thoughts coming and going (with attendant tension and worry increasing and then dissipating). This process enhances the individual's sense of self-efficacy (i.e., control), a critical element for achieving proper homeostasis. Thought-stopping techniques are now out of favor, because, just as with telling someone not to think of white elephants, the deliberate effort to suppress some types of thought often just serves to promote these. Use of a "worry outcome diary" has been suggested to help decondition worry with a lack of reinforcement. This involves recording specific negative predictions (worries), then going back and assessing whether the outcomes justified such worry. It provides a form of extinction when positive reinforcement of the worry state is not forthcoming.

What Can Be Done in a Brief Consultation?

- ☐ Ask about ongoing worry if anxiety is suspected as a cause of ill-health (see Professional Resources).
- ☐ If necessary, administer a worry scale to quantify this (see Professional Resources).
- ☐ Suggest a worry outcome diary to be checked in retrospect to decondition worry.
- ☐ Determine the patient's likely best management approach to stress: physical or mental (see Professional Resources).
- ☐ Ask diagnostic questions to check on specific anxiety disorders (see Professional Resources).
- ☐ Offer pharmacotherapy if appropriate.
- ☐ Offer or refer for cognitive behavioral and other psychological treatments.

MANAGEMENT

Dealing with the extreme anxiety disorders generally requires a specialized psychologist or psychiatrist, but many of the principles of treatment are

TABLE 14.1 Management of Anxiety

	Pharmacotherapy	Psychological Treatment	Self-Help
General anxiety/ all anxiety disorders	SSRIs/SNRIs	CBT and other psychological processes RET Education	Reduced caffeine/ stimulants Support groups
GAD/ excessive worry	Anxiolytics TCAs	Positive psychology EMDR	Kava Exercise Relaxation Massage Yoga Support groups
Social anxiety/ panic disorder	Anxiolytics	Positive psychology Systematic desensitization Interpersonal therapy EMDR	Bibliotherapy Public speaking training Kava "Safe place" training
OCD	TCAs	Exposure and response Prevention Psychotherapy	Bibliotherapy
PTSD	Propranolol Anxiolytics	Counseling Psychotherapy	Support groups
Phobias	Anxiolytics	Exposure/flooding Psychotherapy	Support groups Bibliotherapy

EMDR, eye movement desensitization and reprocessing; *GAD*, general anxiety disorder; *OCD*, obsessive compulsive disorder; *PTSD*, posttraumatic stress disorder; *RET*, rational emotive therapy; *SSRI/SNRI*, selective serotonin reuptake inhibitor/selective noradrenaline reuptake inhibitor; *TCA*, tricyclic antidepressant Bibliotherapy means researching the problem; Kava is a soporific drink used by Pacific Islanders and is available as anxiolytic in Europe.
Adapted from Shearer, S., Gordon, L., 2006. The patient with excessive worry. Am. Fam. Physician 73 (6), 1049–1056.

common and are discussed in summary here (Dolnak, 2006). There are three main categories of treatment: pharmacotherapy, psychological processes, and self-help, including nutrition and exercise management and supplementation (see Table 14.1).

Pharmacotherapy

This consists largely of the selective serotonin reuptake inhibitors or serotonin and noradrenaline reuptake inhibitors as the first-line treatment. Tricyclic and bupropion antidepressants can also help in GAD and OCD. In some cases (GAD, SAD, PD, specific phobias), anxiolytics may also be indicated and a beta-blocker such as propranolol has been found to be effective in treating PTSD. Where a patient is accepting of this, pharmacotherapy is most effective when combined with psychological and self-help practices.

Psychological Processes

These are generally based around CBT, education, and other psychological processes, with specific counseling focusing on accepting uncertainty, curtailing reassurance seeking, thought stopping, rational emotional therapy, behavioral strategies (e.g., worry periods, worry recording), and mindfulness meditation. Psychological processes including CBT usually involve some means of changing internal thought processes. Several processes have been found to be effective even in the PTSD cases (Cusack et al., 2016).

An exception is eye movement desensitization and reprocessing. This involves focusing on a moving object while concentrating on the source of anxiety for a set period, followed by a period of relaxation or meditation. Meta-analyses show the effectiveness of this technique in many cases (i.e., PTSD), although the mechanisms of action are not known (Parnell, 2007; Shapiro, 2005).

Education and counseling that challenge thought processes (see Shearer and Gordon, 2006 for a list of strategies) are also effective in dealing with anxiety and form the mainstay of cognitive therapy (combined with the behavioral strategy of exposure leading to adaptation). For PTSD, websites such as www.acpmh.unimelb.edu.au and www.beyondblue.org provide some levels of self-help.

Self-Help Practices

These begin with the elimination of stimulant foods and drinks from the diet, including those containing caffeine (e.g., coffee, tea, chocolate, cocoa) and the inclusion of exercise sessions, particularly endurance or extended exercise to reduce somatic tension. Exercise is effective because anxiety is associated with increased sympathetic activity and vagal tone, and exercise tends to dampen this down through muscular contractions (de Souza Moura et al., 2015). Morning exercise may be expected to be best (although there is little current research on this) because it can have a somatic effect that lasts through the day and because it is more likely to prevent the buildup of anxiety. The most appropriate types of exercise are extended aerobic activities or resistance training (Powers et al., 2015). Progressive resistance training can be especially useful in the frail and elderly because it increases insulin sensitivity as well as decreasing anxiety and depression.

Self-help psychological processes such as developing a highly specific "safe place" in the mind for retreat and escape will help. For anxiety-provoking situations such as public speaking, training through groups such as Toastmasters will build confidence and allow for the minds adaptive capacity to reduce anxiety with exposure. Progressive muscular relaxation techniques can also help some patients who are willing to put in effort for treatment. Relaxation training, however, is less effective for inveterate worriers and thought-stopping techniques, as mentioned earlier, are no longer recommended for this group (Shearer and Gordon, 2006).

A review of self-help and complementary treatments has shown Kava, the soporific drink (or tablet) used by Pacific Islanders, to have the best evidence for management of GAD (Sarris et al., 2013; Jorn et al., 2004). However, although this is available as an anxiolytic in tablet form in Europe, it is not available (nor probably appropriate) in other countries. Other evidence-based self-help treatments are relaxation training and bibliotherapy (researching the problem). Common treatments with only limited evidence are acupuncture, music, inositol (a B vitamin derivative), and alcohol avoidance. Support groups and other self-help resources (see next) can also assist in management of the milder anxiety disorders.

Simple Approaches for Self-Management of Anxiety

Choose your battles wisely.
Be aware of the snowball effect of thinking.
Ask yourself, "Will this matter a year from now?"
Become aware of your moods—and don't be fooled by the low ones.
Practice random acts of kindness.
Do one thing at a time.
Think of what you have—not what you want.
Think of problems as potential teachers.
If someone throws you the ball, you don't have to catch it.
Keep asking the question, "What is really important?"
Be open to "what is," not "what should be."
Ask yourself "What do I lose by not doing what I fear?" if the consequence you fear is highly unlikely or far into the future.

From Carlson, R., 1997 Don't Sweat the Small Stuff. Hyperion, NY.

When to Refer

Referral to a psychiatrist, psychologist, or anxiety specialist should occur if:

- The patient fails to respond to simple treatments;
- Medical problems, such as medication use, complicate treatment or have not been ruled out as possible causes;
- Side effects occur from standard treatments; or
- The problem is escalating.

SUMMARY

At least one in four people have a lifetime risk of an anxiety disorder. This can range from a simple GAD to PTSD, panic attacks, or extreme phobia. Worry is a ubiquitous characteristic of all anxiety disorders and often needs to be managed with pharmacotherapy or cognitive behavioral techniques. The three methods of management of anxiety include pharmacotherapy, psychological techniques, and self-help processes, including relaxation therapy. Anxiety is a chronic condition with frequent recurrences, and a significant proportion of cases (around 50%) can progress to depression.

Practice Tips

Dealing With Anxiety

	Medical Practitioner	Practice Nurse/Allied Health Professional
Assess	• Worry as a health risk • Possible types of anxiety disorder • Physical effects of anxiety/worry	• Worry/anxiety (see Professional Resources for tests)
Assist	• By offering medication if appropriate • By discussing relaxation/meditation practices • By suggesting a worry outcome diary if appropriate	• General education about worry/anxiety • By providing good literature • By monitoring cardiovascular risk factors • Monitoring anxiety levels
Arrange	• Referral for CBT if appropriate • List of other health professionals for a care plan • Access to self-help websites	• Discussion of anxiety management tactics with other allied health professionals • Referral to a psychologist/psychiatrist • Contact with self-help groups

Key Points

Clinical Management

- Be aware of anxiety as an underlying cause of somatic problems.
- Consider specific aspects of "worry" in anxiety.
- Classify anxiety within accepted disorders.
- Use anxiolytic medication (such as a benzodiazepine, e.g., diazepam or alprazolam) only if needed and as a short-term option for GAD.
- Aim to empower the patient by restoring perceived control.

REFERENCES

Aina, Y., Susman, J.L., 2006. Understanding comorbidity with depression and anxiety disorders. J. Am. Osteopath. Assoc. 106 (Suppl. 2), S9–S14.

Andrews, G., Hobbs, M.J., Borkovec, T.D., Beesdo, K., Craske, M.G., Heimberg, R.G., Rapee, R.M., Ruscio, A.M., Stanley, M.A., 2010. Generalized worry disorder: a review of DSM-IV generalized anxiety disorder and options for DSM-V. Depress. Anxiety 27 (2), 134–147.

Barlow, D.H., 2002. Anxiety and its Disorders: The Nature and Treatment of Anxiety and Panic, second ed. Guilford, New York.

Blashki, G., Judd, F., Piterman, L., 2007. General Practice Psychiatry. McGraw–Hill, Sydney.

Bourne, E., 2015. The Anxiety and Phobia Worksbook, sixth ed. New Harbinger, Oaklands, CA.

Carlson, R., 1997. Don't Sweat the Small Stuff. Hyperion, NY.

Celano, C.M., Millstein, R.A., Bedoya, C.A., Healy, B.C., Roest, A.M., Huffman, J.C., 2015. Association between anxiety and mortality in patients with coronary artery disease: a meta-analysis. Am. Heart J. 170 (6), 1105–1115.

Coplan, J.D., Aaronson, C.J., Panthangi, V., Kim, Y., 2015. Treating comorbid anxiety and depression: Psychosocial and pharmacological approaches. World J. Psychiatry 5 (4), 366–378.

Cusack, K., Jonas, D.E., Forneris, C.A., Wines, C., Sonis, J., Middleton, J.C., Feltner, C., Brownley, K.A., Olmsted, K.R., Greenblatt, A., Weil, A., Gaynes, B.N., 2016. Psychological treatments for adults with posttraumatic stress disorder: a systematic review and meta-analysis. Clin. Psychol. Rev. 43, 128–141.

de Souza Moura, A.M., Lamego, M.K., Paes, F., Ferreira Rocha, N.B., Simoes-Silva, V., Rocha, S.A., de Sá Filho, A.S., Rimes, R., Manochio, J., Budde, H., Wegner, M., Mura, G., Arias-Carrión, O., Yuan, T.F., Nardi, A.E., Machado, S., 2015. Effects of aerobic exercise on anxiety disorders: a systematic review. CNS Neurol. Disord. Drug Targets 14 (9), 1184–1193.

Dolnak, D.R., 2006. Treating patients for comorbid depression, anxiety disorders and somatic illnesses. J. Am. Osteopath. Assoc. 106 (Suppl), S9–S14.

Fricchione, G., 2004. Clinical practice. Generalised anxiety disorder. N. Eng. J. Med. 351, 675–682.

Johansson, R., Carlbring, P., Heedman, Å., Paxling, B., Andersson, G., July 9, 2013. Depression, anxiety and their comorbidity in the Swedish general population: point prevalence and the effect on health- related quality of life. Peer J. (1), e98.

Jorn, A.F., Christensen, H., Griffiths, K.M., et al., 2004. Effectiveness of complementary and self help treatments for anxiety disorders. Med. J. Aust. 181 (7), S29–S46.

Kessler, R.C., Ormel, J., Demler, O., Stang, P.E., December 2003. Comorbid mental disorders account for the role impairment of commonly occurring chronic physical disorders: results from the National Comorbidity Survey. J. Occup. Environ. Med. 45 (12), 1257–1266.

Kessler, R.C., Berglund, P., Demler, O., et al., 2005. Lifetime prevalence and age-of-onset distributions of DSM-IV disorders in the National Comorbidity Survey Replication. Arch. Gen. Psychiatry 62, 593–602.

Kessler, R.C., Petukhova, M., Sampson, N.A., Zaslavsky, A.M., Wittchen, H.-U., 2012. Twelve-month and lifetime prevalence and lifetime morbid risk of anxiety and mood disorders in the United States. Int. J. Methods Psychiatr. Res. 21 (3), 169–184.

Newman, M.G., Llera, S.J., Erickson, T.M., Przeworski, A., Castonguay, L.G., 2008. Worry and generalized anxiety disorder: a review and theoretical synthesis of evidence on nature, etiology, mechanisms, and treatment. Annu. Rev. Clin. Psychol. 9, 275–297.

Parnell, L., 2007. A Therapist's Guide to EMDR. WW Norton and Co., NY.

Pittman, C.M., Karle, E.M., 2015. Rewire Your Anxious Brain: How to Use the Neuroscience of Fear to End Anxiety, Panic and Worry. New Harbinger, Oakland, CA.

Powers, M.B., Asmundson, G.J., Smits, J.A., 2015. Exercise for mood and anxiety disorders: the state-of-the science. Cogn. Behav. Ther. 44 (4), 237–239.

Richards, G., Smith, A., 2015. Caffeine consumption and self-assessed stress, anxiety, and depression in secondary school children. J. Psychopharmacol. 29 (12), 1236–1247.

Sarris, J., Stough, C., Bousman, C.A., Wahid, Z.T., Murray, G., Teschke, R., Savage, K.M., Dowell, A., Ng, C., Schweitzer, I., 2013. Kava in the treatment of generalized anxiety disorder: a double-blind, randomized, placebo-controlled study. J. Clin. Psychopharmacol. 33 (5), 643–648.

Shapiro, R. (Ed.), 2005. EMDR Solutions: Pathways to Healing. WW Norton and Co, NY.

Shearer, S., Gordon, L., 2006. The patient with excessive worry. Am. Fam. Physician 73 (6), 1049–1056.

Teychene, M., Costigan, S.A., Parher, K., 2015. The association between sedentary behaviour and risk of anxiety: a systematic review. BMC Public Health 15, 513. http://dx.doi.org/10.1186/s12889-0151843-x.

Tsatsoulis, A., Fountoulakis, S., 2006. The protective role of exercise on stress system dysregulation and comorbidities. Ann. N. Y. Acad. Sci. 1083, 196–213.

Professional Resources

Measuring Anxiety

Clinical Question

Approximately 90% of people with GAD will answer positively to the following question (Fricchione, 2004): During the past 4 weeks, have you been bothered by feeling worried, tense, or anxious most of the time?

Penn State Worry Questionnaire

Patient instructions: Enter the number that best describes how typical or characteristic each item is of you.

	Not at all Typical		Somewhat Typical		Very Typical
1. If I don't have enough time to do everything, I don't worry about it.	1	2	3	4	5
2. My worries overwhelm me.	1	2	3	4	5
3. I don't tend to worry about things.	1	2	3	4	5
4. Many situations make me worry.	1	2	3	4	5
5. I know I should not worry about things, but I just cannot help it.	1	2	3	4	5
6. When I am under pressure I worry a lot.	1	2	3	4	5
7. I am always worrying about something.	1	2	3	4	5
8. I find it easy to dismiss worrisome thoughts.	1	2	3	4	5
9. As soon as I finish one task, I start to worry about everything else I have to do.	1	2	3	4	5
10. I never worry about anything.	1	2	3	4	5

Continued

Professional Resources—cont'd

	Not at all Typical		Somewhat Typical		Very Typical
11. When there is nothing more I can do about a concern, I do not worry about it anymore.	1	2	3	4	5
12. I have been a worrier all my life.	1	2	3	4	5
13. I notice that I have been worrying about things.	1	2	3	4	5
14. Once I start worrying, I cannot stop.	1	2	3	4	5
15. I worry all the time.	1	2	3	4	5
16. I worry about projects until they are all done.	1	2	3	4	5

Practitioner instructions: Reverse score items 1, 3, 8, 10, and 11, then sum all 16 items. Possible scores range from 16 to 80. Means for groups with GAD range from 60 to 68.

Signs and Symptoms of Anxiety Disorders

For checklists for common types of anxiety disorders, see www.beyondblue.org.

Chapter 15

Depression

Robert Reznik, Andrew Binns, Garry Egger

He who has a why to live for can bear with almost any how.

Nietzche (Kraufman, 1954)

INTRODUCTION: DEPRESSION AS A WAY OF LIFE

Aldous Huxley in his 1932 novel *Brave New World* predicted a future of people "high" on mood-enhancing drugs. That future, it seems, is now. Antidepressants are among the most popularly prescribed medications in Western cultures. Up to one in five adults in Western, modern countries admit to suffering from anxiety and/or depression, and depression is a leading cause of disease burden worldwide (Ferrari et al., 2013). Is this a real phenomenon or just an artifact of modern living?

Indications from surveys carried out over 50 years ago suggest that there has been a genuine 10-fold increase in levels of depression over time (Seligman, 2006). Why this is so is a source of conjecture, but most analysts associate it with the advances in technology discussed in Chapter 2. While leading to improvements in living standards, these have resulted in a loss of the "meaning" gained from earlier struggles for survival. A more interesting theory links inactivity, obesity, and depression through neural changes in the brain (Ernst et al., 2006; Pereira et al., 2007), and this may explain the well-proven benefits of exercise in treating depression (discussed later).

In this chapter we look at the theory of, and management techniques for, depression. It specifically covers the less severe end of the scale. While reference is given to bipolar disorder and major depressive disorder, these require more detailed treatment than is given here. In Chapter 16, a more positive slant is given to the management of mild mood disorders through a developing discipline known as "positive psychology."

WHAT IS DEPRESSION?

Depression has been defined as "the common cold of psychopathology" (Tan and Ortberg, 1995). It is further defined as "an abnormal state characterized by exaggerated feelings of sadness, melancholy, dejection, worthlessness,

Lifestyle Medicine. http://dx.doi.org/10.1016/B978-0-12-810401-9.00015-2

emptiness and hopelessness that are inappropriate and out of proportion to reality." Up to 25% of women and 17% of men in Western countries admit to suffering depression at any one time—although men are less likely to admit it (Kessler et al., 2003, 2005; Kessler and Bromet, 2013). Those who have suffered one episode of depression have a 50% chance of a second episode if untreated (15% if treated); two episodes carry a 70% chance of relapse; and three episodes, a 100% chance (Parker, 2004).

Although it seems to be a useless emotion, McGuire in 1997 suggested that depression probably has an evolutionary function as the adaptive response to a loss of status or social holding power in a community. Depression reduces conflict and allows the social group to reach out and help the injured party and thus help maintain social cohesion within the group. According to Maguire, this is seen in many primate colonies following the successful coup of an alpha male, where serotonin levels reflect alpha male status. In an interesting experiment involving primate colonies (macaque monkeys), the alpha male when removed from the colony, who prior to removal had high serotonin levels, then had lower serotonin levels and only had them restored to high levels when reintegrated into the primate group and reachieved his alpha status.

DEPRESSION AND HEALTH

Although depression has its own pathological sequelae, including suicide (about 10% of depressed patients will attempt suicide; Kessler and Bromet, 2013), it is also associated with a range of other metabolic problems (Kessler et al., 2005; Yusuf et al., 2004). Depression can both follow and precede heart disease, type 2 diabetes, and possibly other chronic diseases, suggesting a sound basis for screening and treating those with recent disease for depression. Depression is now also routinely tested for and treated in cardiovascular disease (Dickens, 2015). The risk of further infarction after an initial myocardial infarct can be increased up to four times by depression. Somatic complaints are also higher in those with depression. In a study of almost 17,000 patients in Washington, somatic symptoms of four common medical disorders (diabetes, pulmonary disease, heart disease, arthritis) were at least as strongly associated with depression and anxiety as were objective physiologic measures (Katon et al., 2007). Improvement in depression outcome was associated with decreased somatic symptoms without improvement in physiologic measures.

DEPRESSIVE DISORDERS

As with anxiety, there are a number of depressive disorders, and these have been classified in a number of ways. The classification shown in Table 15.1, based on that by Professor Gordon Parker (2004) of the Black Dog Institute in Australia, is typical of most.

TABLE 15.1 Classification of Depressive Disorders

Minor depressive disorders: reactive (nonclinical) depression, nonmelancholic depression
Major depressive disorders: melancholic depression, psychotic depression/melancholia (including bipolar disorder)
From Parker, G., 2004. Dealing with Depression, second ed. Allen and Unwin, Sydney.

Reactive depression (also referred to in the Diagnostic Statistical Manual [DSM 5] as adjustment disorder and also sometimes as acute stress-related disorder) is a normal response to a major life event and does not usually last for more than about two weeks, although if the stressor is chronic, it may become a more significant depressive disorder, for example, those where symptoms of depressed mood, anhedonia (loss of joy in things that usually give you a kick-along) last for at least two weeks and affect daily functioning. Patients with "mild" depression probably vary most in the initial presentation, while those with major depression vary least.

Nonmelancholic depression lacks the psychomotor dysfunction of the more severe depressive categories, and sufferers usually have brief periods of cheerfulness with fewer lapses in concentration and memory. Often it is spontaneously remitting and different treatments—drugs, counseling, or psychotherapy—have similar effects.

Melancholic depression is more severe than its nonmelancholic counterpart. It is often described as more endogenous or biologically based and includes psychomotor disorders (both anergia—no energy—and agitation). This level of depression generally requires medication as the first-line therapy because nonmedical treatments alone are less effective. But even in such depression, a combined approach, when practical to initiate, is best.

Psychotic melancholia lies at the severe end of the scale and usually involves psychotic phenomena such as hallucinations and delusions. It includes bipolar disorder and responds only to medical treatments.

Postnatal depression is well described in new mothers, with increasing attention now being paid to early recognition and treatment. However, less has been reported on antenatal depression, which may occur in up to 20% of pregnant women (Bowen and Muhajarine, 2006) but which may be less well acknowledged because of the focus on maternal and fetal wellbeing and the attribution of complaints to the physical and hormonal changes associated with pregnancy. Moreover, many who suffer from postnatal depression will, on retrospective questioning, give a history of antenatal depression. Antidepressant medications are less advisable here, but lifestyle-based therapies including exercise, adequate nutrition, and adequate sleep are recommended (Bowen and Muhajarine, 2006). For more severe cases, clinical judgment must balance the reluctance of the mother to take medication, or the GP's reluctance to prescribe, against the dangers of inadequate treatment of her depression. The clinician must remember that maternal bonding (a critical phase in the development of the newborn) cannot be adequately provided by a depressed mother and poor bonding may lead to an increased risk of depression for the child in later life.

RISK FACTORS

The important signs to look for when screening for depression are a family history of depression and a previous episode. Some chronic illnesses—including stroke, Parkinson's disease, dementia, cancers and heart attack, as well as adverse life events in susceptible people—are risk factors for depression. Personality types such as the less hardy or resilient, or the pessimistic are at increased risk.

Risk factors for antenatal depression include a history of depression, lack of a partner, marital difficulties, lack of social support, poverty, family violence, increased life stress, substance abuse, history of previous abortions, unplanned pregnancy, ambivalence toward the pregnancy, and anxiety about the fetus (Bowen and Muhajarine, 2006).

It is now clear that depression has a biological as well as a psychological component, affecting neural connections and transmitters in different parts of the brain. A possible way in which the biological–psychological link works in minor depression is shown in Fig. 15.1, which suggests that continuation of adverse life events in susceptible people can lead to more severe depression through biological changes that require physical rather than psychological treatments (i.e., medication).

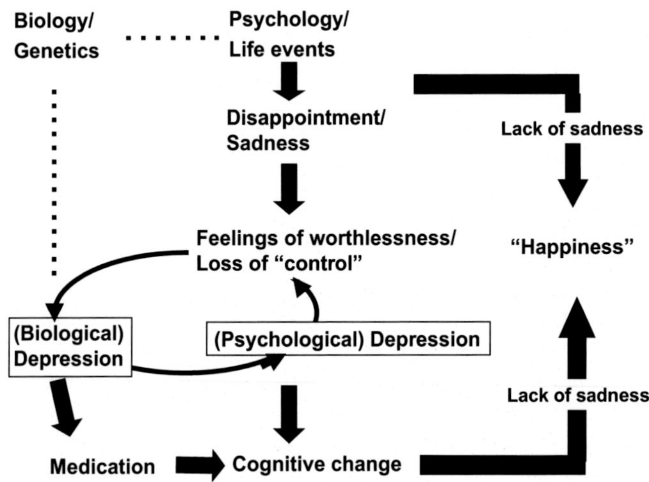

FIGURE 15.1 Suggested biological–psychological links in depression.

DIAGNOSIS

Several publications and websites provide lists of symptoms for diagnosing depression; however, some simple questions appear to have high validity in diagnosis (see Professional Resources). Symptoms can vary widely, but the acronym SIGECAPS (Mahendron and Yap, 2005) lists the defined DSM 5 symptoms:

S	Sleep increase/decrease
I	Interest in formally compelling or pleasurable activities diminished
G	Guilt, low self-esteem
E	Energy low
C	Concentration poor
A	Appetite increase/decrease
P	Psychomotor agitation or retardation
S	Suicidal thoughts

Aaron Beck, one of the pioneers of cognitive behavior therapy, defined a typical "thinking" triad in the depressed client (Beck, 1987). This involves:

- A negative view of the self;
- A tendency to interpret all external events as negative; and
- A pattern of viewing the future as negative.

The Beck Depression Inventory (Beck et al., 1996) is a licensed test for measuring depression, and it and the Hamilton Rating Scales for Depression can be used for detailed screening for depression as well as general anxiety disorder. Other measurement instruments such as the PHQ-9 (Hinz et al., 2016) and the simple question "are you depressed?" (Skoogh et al., 2010) are increasingly abbreviating the screening of depression with moderate validity.

MANAGEMENT

There are significant crossovers between treatments for anxiety and depression, with perhaps a greater emphasis on physical treatments (drugs) at the extremes of major depressive disorders. Table 15.2 lists possible options and indications for treatment.

Pharmacological Treatments

First-line pharmacological therapy today involves third-generation antidepressants known as SSRIs or SNRIs. These appear to have fewer side effects than earlier generation monoamine oxidase inhibitors or tricyclic or tetracyclic antidepressants and better long-term outcomes. However, side effects such as sexual dysfunction remain a problem, particularly in some middle-aged men. One of the more common side effects is a reduction in sensitivity of the glans penis and, in those more seriously affected, anorgasmia. The upside is less premature ejaculation, and for this reason antidepressants are commonly prescribed by

TABLE 15.2 Treatment Options for Depression

	Pharmacotherapy	Psychological Treatment	Nonpharmacotherapy (Self-Help)
General/all disorders	SSRIs/SNRIs Melatonergic agent TCAs MAOs	CBT Education Counseling	Exercise Support groups Bibliotherapy
Mild depressive disorder	St. John's wort Melatonergic agent SAMe	Psychological "immunization"	Positive psychology
Nonmelancholic depression	SSRIs/SNRIs Melatonergic agent St John's wort SAMe	Psychological "immunization"	Positive psychology
Melancholic depression	SSRIs/SNRIs Melatonergic agent TCAs MAOs	Psychotherapy Interpersonal therapy	Light therapy for SAD
Psychotic depression	Antipsychotic medication in addition to an antidepressant	Psychotherapy Interpersonal therapy	

CBT, cognitive behavior therapy; *MAO*, monoamine oxidase inhibitor; *SAD*, seasonal affective disorder; *SAMe*, *S*-adenosyl-L-methionine; *SSRI/SNRIs*, selective serotonin reuptake inhibitor/selective noradrenaline reuptake inhibitors. Melatonergic agent—agomelatine, TCA 5 tricyclic antidepressants.
Adapted from Parker, G., 2004. Dealing with Depression, second ed. Allen and Unwin, Sydney.

sex therapists for men with this problem. Side effects must be balanced against clinical improvement and the patient's preferences. For many, minor discomfort from side effects may be far outweighed by the benefits. An entirely new agent, agomelatine, has an entirely novel mode of action and, if it works, has the advantage of neither sexual side effects nor weight gain.

It is worth noting that St. John's wort's mechanism of action is that of an SSRI and thus the combined use of an SSRI and St. John's wort can boost the potential for unpleasant and potentially serious side effects.

Importantly, when one medication fails to give a good clinical response, a change in medication may be beneficial; consequently, a number of different medications may need to be trialed. A useful point to remember is that if a medication is going to work, 80% of responders will report an improvement (with up to a 50% decrease in depression) within 2 weeks. It is also worth stressing that, although patients may reach a good clinical response within 6–8 weeks of treatment with SSRIs, the improvement will continue, albeit at a reduced rate, for weeks and months later. A summary of the appropriate use of antidepressant medications for depression is given in Parker (2004, pp. 76, 82) and Nutt (2010) and by www.beyondblue.org. Possible limiting side effects of some of the major antidepressant medications are discussed in Chapter 24.

There are links between depression and obesity, which are complicated by some depression medications. Obesity can cause depression in some people, for obvious reasons. Seasonal affective disorder, although more common in high latitude countries with limited winter sunlight, is an increased negative energy balance and depression occurring in some individuals in winter. Depression, on the other hand, can also lead to obesity because of increased inertia, decreased physical activity, and increased comfort eating. Some SSRIs are obesogenic. People who eat more when depressed (typically women) may lose weight following treatment with antidepressants that do not cause weight gain, for example, fluoxetine or sertraline. People who eat less when depressed (typically men, who eat less but drink more) may eat more and gain weight following treatment.

While his attention was diverted it just seemed to pass him by.

Psychological Treatments

Psychological treatments are extensive and range from the simple “immunization” and self-help approaches discussed earlier to detailed psychodynamic approaches. A summary of the main approaches is shown in Table 15.3.

There is a good evidence base for treatments such as cognitive behavior therapy and its components, cognitive therapy and behavior therapy. Interpersonal therapy and psychodynamic approaches have also been shown to be effective under certain circumstances, including in combination with pharmacotherapy. More recently has been the so-called, third wave cognitive therapies, based on acceptance and commitment, that have an emphasis on actions (behavioral) that are aligned with the person’s values. There is less evidence for processes such as hypnosis, but these may have positive benefits in select individuals.

TABLE 15.3 The Main Psychological Treatments for Depression

	Description	Indications
Cognitive	Focuses on changing irrational and ruminating thoughts and images that affect patterns of behavior and social reinforcement. Includes awareness of thought distortion, rational emotive therapy, thought stopping, hypothesis testing, "homework," and behavioral techniques.	Internalizing, "ruminating" thinkers.
Behavior therapy	Based on learning principles of reinforcement. Involves self-monitoring, tracking moods, assertiveness training, problem solving, role playing, and behavioral rehearsal and relaxation.	Externalizers. Habit-influenced patients.
Interpersonal	Proposes that current functions are influenced by past relations. Designed to relieve relationship issues. Uses problem solving to resolve issues, improve communication, and form healthier relationships.	Where relationship difficulties are an obvious cause.
Psychodynamics	Uses a range of psychoanalytic Processes. Generally requires experienced psychotherapy training.	Complex early experience issues.

Other Treatments

As with anxiety, effective nonpharmacological interventions—such as exercise, relaxation training, meditation, research (bibliotherapy), and education—can have positive effects in some cases. Support groups, both in person and via the internet, can also be effective (see Professional Resources). Exercise, even without weight loss, is particularly effective in reducing depression (Danielsson et al., 2016). This is not only because it fulfills one of the main recommendations for depression (to "just do it") but because it may promote the release of "feel good" neurotransmitters and help re-form neural links in parts of the brain associated with depression (Yuan et al., 2015). As with anxiety, morning exercise is likely to be the most effective way to maintain mood throughout the day. It also reduces the problem of early wakening and the negative effects of rumination that can arise from lying awake.

In those who have suffered a depressive episode, psychological immunization and cognitive exercises can help prevent or reduce the severity of a recurrence (Dozois and Dobson, 2004). Self-help books (e.g., Carlson, 2003; Korbe, 2015) and research (bibliotherapy) often have the most profound effects on mild depressives who have tried everything else. Because depression and sleep are positively connected (and sleep and activity negatively connected), a process called wake therapy—which involves reducing sleep hours—has also been promoted, albeit with limited supporting evidence (Giedke, 2002).

Although many over-the-counter medications, such as ginseng, viridian, and lemon balm, are promoted for depression relief, only two formulations have reasonable supporting evidence. These are the herbal substance St. John's wort (Jorm et al., 2002. Linde et al., 2008) and *S*-denosyl-L-methionine, which is a naturally occurring amino acid and part of the natural dopamine–serotonin biochemical cycle associated with feelings of wellbeing (Papakostas et al., 2003; Sarris et al., 2014). One problem with herbal mixes such as St. John's wort is the dose of active ingredient, which may vary between manufacturer and tablets. There is little supporting evidence for other alternative therapies, such as massage, folate, or yoga, although some forms of meditation do appear to be more helpful for depression than for anxiety, where concentration is particularly difficult.

Although electroconvulsive therapy received bad press during the 1960s and 1970s, improvements in the process have led to it making a comeback—life saving in some cases—to the extent that it may be superior to pharmacotherapy, particularly in more severe cases of depression (Nordenskjoid, 2015).

What Can Be Done in a Standard Consultation?

- ☐ Check for anxiety (see Chapter 14).
- ☐ Check for a family history of depression, anxiety, and/or suicide.
- ☐ Determine the degree of depression and refer where appropriate.
- ☐ Check depression risk in pregnancy.
- ☐ Question life stages and events as triggers (e.g., heart problems, illness).
- ☐ Try to remove or reduce chronic stressor(s) that are obvious progenitors.
- ☐ Address the use of alcohol and psychoactive substances.
- ☐ Begin with simple steps (e.g., increased activity, activity planner) and add psychological and pharmacological treatments if required.

When to Refer

While suicidal thoughts are not necessarily likely to lead to completion of the act, obsessive and continual thoughts may require experienced help or referral to a psychologist or psychiatrist. Psychotic depression, including bipolar disorder, may be managed by the GP but also require ongoing care and advanced psychological input from a qualified therapist.

SUMMARY

Depression is a common and serious cause of misery for many people. It is important for the clinician and patient to recognize that this can be treated by a variety of methods. Whatever method is chosen, an objective measure such as a questionnaire or symptom checklist provides a useful baseline to enable both parties to evaluate treatment. Trying to remove or resolve a chronic stressor is a first line of attack; if this is not possible, consider reducing its significance. Education is critical, as is forming a partnership between the sufferer and clinical helper. As with most psychological problems, the earlier treatment is initiated, the better the outcome. Treatment begins with simple steps—such as increased activity levels or an activity planner—and, depending upon response, the addition of one of the forms of psychological treatment outlined earlier, with or without medication. However, no treatment for depression will succeed if comorbidity or other medical issues are not addressed, especially the use of psychoactive substances such as alcohol or illicit substances such as cocaine, amphetamines, or cannabis.

Practice Tips

Managing Depression

	Medical Practitioner	**Practice Nurse/Allied Health Professional**
Assess	• Depression, particularly in heart/diabetes patients • Possible comorbidities • Somatic problems • Suicidal ideation • The nature of any depressive disorder • Interest in self-help groups	• Depression: use depression scales on heart/diabetes patients if indicated • Comorbidity risk factors • The nature of any depressive disorder • Stressors and how these might be removed
Assist	• Understanding differences between anxiety and depression • By prescribing medication if indicated • By discussing the options for treatment • By referring to self-help books • Providing handouts • By developing prevention strategies with the patient • By providing ongoing support	• By offering alternative strategies for treatment/accompanying medication • By teaching management techniques (if able) • With self-help treatments • By referring to self-help books • By providing handouts • By referring to self-help groups/internet sites • By providing ongoing support

Practice Tips—cont,d

Managing Depression

	Medical Practitioner	Practice Nurse/Allied Health Professional
Arrange	• A list of other health professionals for a care plan • Referral to psychological/psychiatric care where appropriate • Referral for suicide counseling if appropriate	• A care plan • Contact with and coordination of other health professionals • Discussion of options with other professionals

Key Points

Clinical Management

- Identify the level of depression and extent of treatment required.
- Consider the role of worry and anxiety.
- Aim to reduce stressors where these are an obvious cause of concern.
- Use medication as a window of opportunity rather than a lifetime treatment (if possible).
- Encourage involvement in self-help and self-seeking.
- Make use of modern resources such as self-help websites (see Professional Resources).

REFERENCES

Beck, A.T., 1987. Cognitive models of depression. J. Cog Psychother. 1, 5–37.

Beck, A.T., Steer, R.A., Brown, G.K., 1996. Beck Depression Inventory, second ed. Psychological Corporation, San Antonio, TX.

Bowen, A., Muhajarine, N., 2006. Antenatal depression. Can. Nurse 102 (9), 26–30.

Carlson, R., 2003. Stop Thinking and Start Living. Harper Collins, London.

Danielsson, L., Kihlbom, B., Rosberg, S., 2016. "Crawling out of the cocoon": patients' experiences of a physical therapy exercise intervention in the treatment of major depression. Phys. Ther. (Epub ahead of print).

Dickens, C., 2015. Depression in people with coronary heart disease: prognostic significance and mechanisms. Curr. Cardiol. Rep. 17 (10), 83.

Dozois, D.J.A., Dobson, K.S. (Eds.), 2004. The Prevention of Anxiety and Depression: Theory, Research, and Practice. American Psychological Association, Washington, D.C.

Ernst, C., Olson, A.K., John, P.J., et al., 2006. Antidepressant effects of exercise: evidence for an adult-neurogenesis hypothesis? Rev. Pyschiat. Neurosci. 31 (2), 84–91.

Ferrari, A.J., Charlson, F.J., Norman, S.B., Freedman, G., Murray, C.J.L., Vos, T., Whiteford, H.A., 2013. Burden of depressive disorders by country, sex, age, and year: findings from the global burden of disease study 2010. PLoS Med. http://dx.doi.org/10.1371/journal.pmed.1001547.

Giedke, H., 2002. Therapeutic use of sleep deprivation in depression. Sleep Med. Rev. 6 (5), 77–361.

Hinz, A., Mehnert, A., Kocalevent, R.D., Brähler, E., Forkmann, T., Singer, S., Schulte, T., 2016. Assessment of depression severity with the PHQ-9 in cancer patients and in the population. BMC Psychiatry 16 (1), 22.

Jorm, A.F., Christensen, H., Griffiths, K.M., et al., 2002. Effectiveness of complimentary and self-help treatments for depression. Med. J. Aust. 176 (10 Suppl), S84–S96.

Katon, W., Lin, E.H., Kroenke, K., 2007. The association of depression and anxiety with medical symptom burden in patients with chronic medical illness. Gen. Hosp. Psychiatry 9 (2), 147–155.

Kessler, R.C., Berglund, P., Demler, O., et al., 2003. The epidemiology of major depressive disorder: results from the National Comorbidity Survey Replication (NCS-R). JAMA 289, 3095–3105.

Kessler, R.C., Berglund, P., Demler, O., et al., 2005. Lifetime prevalence and age-of-onset distributions of DSM-IV disorders in the National Comorbidity Survey Replication. Arch. Gen. Psychiatry 62, 593–602.

Kessler, R.C., Bromet, E.J., 2013. The epidemiology of depression across countries. Annu. Rev. Public Health 34, 119–138.

Korbe, A., 2015. The Upward Spiral: Using Neuroscience to Reverse the Course of Depression, One Small Change at a Time. New Harbinger, Oakland, CA.

Linde, K.1, Berner, M.M., Kriston, L., October 8, 2008. St John's Wort for major depression. Cochrane Database Syst. Rev. (4), CD000448. http://dx.doi.org/10.1002/14651858.CD000448.pub3.

Lovibond, S.H., Lovibond, P.F., 1995. Manual for the Depression Anxiety Stress Scales, second ed. Psychology Foundation, Sydney.

Mahendran, R., Yap, H.L., 2005. Clinical practice guidelines for depression. Singap. Med. J. 46 (11), 610–615.

Mcguire, M.T., 1997. Darwinian Psychiatry. Harvard University Press, Cambridge, MA.

Nutt, D.J., 2010. Rationale for, barriers to, and appropriate medication for the long-term treatment of depression. J. Clin. Psychiatry 71 (Suppl. E1), e02.

Nordenskjöld, A., November 2015. ECT is superior to pharmacotherapy for the short-term treatment of medication-resistant inpatients with bipolar depression. Evid. Based Ment. Health 18 (4), 11.

Papakostas, G.I., Alpert, J.E., Fava, M., 2003. S-adenosyl-methionine in depression: a comprehensive review of the literature. Curr. Psychiatry Rep. 5 (6), 460–466.

Parker, G., 2004. Dealing with Depression, second ed. Allen and Unwin, Sydney.

Pereira, A.C., Huddleston, D.E., Brickman, A.M., et al., 2007. An in vivo correlate of exercise-induced neurogenesis in the adult dentate gyrus. PNAS 104, 5638–5643.

Sarris, J., Papakostas, G.I., Vitolo, O., Fava, M., Mischoulon, D., 2014. S-adenosyl methionine (SAMe) versus escitalopram and placebo in major depression RCT: efficacy and effects of histamine and carnitine as moderators of response. J. Affect. Disord. 164, 76–81. http://dx.doi.org/10.1016/j.jad.2014.03.041. Epub.

Seligman, M., 2006. In: Paper Presented to Australian Psychological Society Meeting, Sydney, August 2006.

Skoogh, J., Ylitalo, N., Larsson Omerόv, P., Hauksdόttir, A., Nyberg, U., Wilderäng, U., Johansson, B., Gatz, M., Steineck, G., Swedish-Norwegian Testicular Cancer Group, 2010. "A no means no"—measuring depression using a single-item question versus Hospital Anxiety and Depression Scale (HADS-D). Ann. Oncol. 21 (9), 1905–1909.

Tan, S.Y., Ortberg, J., 1995. Coping with Depression: The Common Cold of the Emotional Life. Baker Publishing Group, Grand Rapids, MI.

Yuan, T.F., Paes, F., Arias-Carriόn, O., Ferreira Rocha, N.B., de Sá Filho, A.S., Machado, S., 2015. Neural mechanisms of exercise: anti-depression, neurogenesis, and serotonin signaling. CNS Neurol. Disord. Drug Targets 14 (10), 1307–1311.

Yusuf, S., Hawken, S., Ounpuu, S., Interheart Study Investigators, et al., 2004. Effect of potentially modifiable risk factors associated with myocardial infarction in 52 countries (the INTERHEART study): case-control study. Lancet 364 (9438), 937–952.

Professional Resources

Measuring Depression

Clinical questions

During the past month have you often been bothered by feeling down, depressed, or hopeless?

During the past month have you often had little interest or pleasure in doing things?

The $DASS_{21}$ (Depression, Anxiety, and Stress Scale—short version)

This is a valid Australian single test measuring stress, anxiety, and depression. It was developed by Lovibond and Lovibond (1995). It can be downloaded from www.psy.unsw.edu.au/groups/dass and a manual purchased if desired.

Patient Instructions

Please read each statement and circle a number 0, 1, 2, or 3 to indicate how much the statement applied to you over the past week. There are no right or wrong answers. Do not spend too much time on any statement.

The applicability rating scale is:

0 = Did not apply to me at all
1 = Applied to me to some degree or some of the time
2 = Applied to me to a considerable degree or a good part of the time
3 = Applied to me very much or most of the time

	Question	Score				
1.	I found it hard to wind down.	0	1	2	3	**S**
2.	I was aware of dryness of my mouth.	0	1	2	3	**A**
3.	I couldn't seem to experience any positive at all.	0	1	2	3	**D**
4.	I experienced breathing difficulty (e.g., excessively breathlessness in the absence of physical exertion).	0	1	2	3	**A**
5.	I found it difficult to work up the initiative to do things.	0	1	2	3	**D**
6.	I tended to overreact to situations.	0	1	2	3	**S**
7.	I experienced trembling (e.g., in the hands).	0	1	2	3	**A**
8.	I felt that I was using a lot of nervous energy.	0	1	2	3	**S**
9.	I was worried about situations in which I might panic and make a fool of myself.	0	1	2	3	**A**
10.	I felt that I had nothing to look forward to.	0	1	2	3	**D**
11.	I found myself getting agitated.	0	1	2	3	**S**
12.	I found it difficult to relax.	0	1	2	3	**S**

Continued

Professional Resources—cont'd

	Question	Score				
13.	I felt downhearted and blue.	0	1	2	3	**D**
14.	I was intolerant of anything that kept me from getting on with what I was doing.	0	1	2	3	**S**
15.	I felt I was close to panic.	0	1	2	3	**A**
16.	I was unable to become enthusiastic about anything.	0	1	2	3	**D**
17.	I felt I wasn't worth much as a person.	0	1	2	3	**D**
18.	I felt that I was rather touchy.	0	1	2	3	**S**
19.	I was aware of the action of my heart in the absence of physical exertion (e.g., sense of heart rate increase, heart missing a beat).	0	1	2	3	**A**
20.	I felt scared without any good reason.	0	1	2	3	**A**
21.	I felt that life was meaningless.	0	1	2	3	**D**

Practitioner Instructions

Apply the template to both sides of the sheet and sum the scores for each scale. For the short (21 item) version, multiply the sum by two. Bolded letters should be printed on a plastic template for scoring: S = stress, A = anxiety, D = depression. The total score for each of the four subscales is evaluated using the severity-rating index given in Table 15.4. Normative data are available on a number of samples. From a sample of 2914 adults the means (and standard deviations) were 6.34 (6.97), 4.7 (4.91), and 10.11 (7.91) for the depression, anxiety, and stress scales, respectively. A clinical sample reported means (and standard deviations) of 10.65 (9.3), 10.90 (8.12), and 21.1 (11.15) for the three measures.

TABLE 15.4 Severity-Rating Index for the $DASS_{21}$

	Depression	Anxiety	Stress
Normal	0–9	0–7	0–14
Mild	10–13	8–9	15–18
Moderate	14–20	10–14	19–25
Severe	21–27	15–19	26–33
Extremely severe	28+	20+	34+

OTHER RESOURCES

www.adaa.org
Anxiety Disorders Association of America; has excellent self-tests for different anxiety disorders.
www.anxieties.com
Free self-help for anxiety sufferers; includes a good free newsletter.
www.beyondblue.org
Australian site with a range of excellent information, free advice, and referrals for anxiety and depression.
www.blackdoginstitute.com.au
Australian site concentrating on the management of different levels of depression.

Chapter 16

Happiness and Mental Health: The Flip Side of S-AD

Julia Anwar-McHenry, Rob Donovan, Garry Egger

> *"Hans Selye is wrong; it is not stress that kills us. It is effective adaptation to stress that permits us to live."*
>
> George Vaillant (1995)

INTRODUCTION

Psychiatrist George Vaillant was three years old when the long-term prospective study of Harvard alumni commenced. He was in the third decade of his life when he joined the study and began to forge his career based on a detailed understanding of how humans use ego *mechanisms of defense* to adapt, with varying degrees of success, to life's trials and tribulations.

Vaillant's message, based on five decades of observation of growth to maturity, is a simple one: "soundness is a way of reacting to problems, not an absence of them." Of all the 287 men studied, none had only "clear sailing through life." What interested Vaillant were the adaptation mechanisms used to cope with the inevitable conflicts that occurred. Vaillant and his team identified four main types of defense mechanisms, from psychotic (in people typically managing least effectively in their adaptation to life) to mature (those who manage most effectively). The levels are shown in Table 16.1.

The Harvard findings represent an early awareness of mental "health" in contrast to mental illness. This has since been adopted by the "positive psychology" movement, which has developed practices for capitalizing on this to promote if not happiness, then a form of positive affect with potential benefits for health.

DEPRESSION AND HAPPINESS

It goes without saying that nobody likes to be "bitten by the black dog" of depression; but it seems as though more people than ever before are suffering from depression and anxiety disorders. True depression is undoubtedly a serious

Lifestyle Medicine. http://dx.doi.org/10.1016/B978-0-12-810401-9.00016-4

TABLE 16.1 Adaptive Mechanisms of (Ego) Defense

Level 1: Psychotic mechanisms (common in psychosis, dreams, and childhood)	Denial (of external reality) Distortion Delusional projection
Level 2: Immature mechanisms (common in severe depression, personality disorders, and adolescents)	Fantasy Projection Hypochondriasis Passive–aggressive behavior (e.g., masochism, turning against the self)
Level 3: Neurotic mechanisms (common in everyone)	Intellectualization (isolation, obsessive behavior, rationalization) Repression Reaction formation Displacement (conversion, phobias) Dissociation (neurotic denial)
Level 4: Mature mechanisms (common in "healthy" adults)	Sublimation Altruism Suppression Anticipation Humor

From Vaillant, G., 1995. Adaptation to Life, second ed. Harvard University Press, London and Vaillant, G., 2015. Triumphs of Experience: The Men of the Harvard Grant Study, NY Belknap Press.

and apparently growing problem in modern societies, as we saw in Chapter 15. But does feeling unhappy from time to time justify reaching for the latest soma pill? Are there other ways of dealing with mood fluctuations? Can we use George Vaillant's mechanisms of defense or develop other more positive ways of preventing mental illnesses (which Vaillant describes as a continuum, rather than a discrete entity), or lessening their severity or duration when they do strike?

LIFESTYLE MEDICINE AND MENTAL HEALTH PROMOTION

Lifestyle Medicine has largely been focused on physical illnesses and chronic disease, with the emphasis on lifestyle factors such as physical activity, smoking, alcohol consumption, diet, and obesity, with the occasional inclusion of stress (e.g., Rohrer et al., 2005; Duaso and Cheung, 2002) and some interest in depression and anxiety (American College of Preventive Medicine, 2009).

Lifestyle interventions in the mental health area have tended to target only people with a defined mental illness, and, apart from looking at the impact of physical activity on illnesses such as depression or anxiety (O'Connor et al., 2010; Herring et al., 2013), have focused on improving those individuals' physical health, not their mental health (Brown and Chan, 2006). Improving mental health through lifestyle changes such as physical activity and diet have been proven effective in the treatment of mental health problems, and are sometimes preferred for their lack of associated stigma and fewer side effects and complications when compared to psychotherapy and pharmacotherapy (Amminger et al., 2010; Borgonovi, 2008).

Overall though, as far as we are aware, apart from the Act-Belong-Commit framework described later in this chapter, there is currently no available overall framework for a mental health promotion lifestyle intervention in clinical practice.

WHAT IS (GOOD) MENTAL HEALTH?

Mental health has been defined by the World Health Organization as "a state of wellbeing in which the individual realises his or her own abilities, can cope with the normal stresses of life, can work productively and fruitfully, and is able to make a contribution to his or her community" (WHO, 2001). In lay terms, being mentally healthy means being alert, socially competent, emotionally stable, enthusiastic, and energetic (Donovan et al., 2003).

In this chapter we look at these positive aspects of mental health, with a view to encouraging patients to take up activities that build and maintain good mental health. As most people associate mental illness with the term mental health, it is suggested that in dealing with patients, mental health issues should be discussed in a context of "keeping mentally healthy." Being "mentally healthy" has positive connotations and a mentally healthy individual is generally described by laypeople as being content with who they are and what they have; being in control of their lives; being mentally competent and emotionally stable; generally feeling happy most of the time; being able to cope with problems and crises in life; and being interested and involved in things in their lives (Donovan et al., 2003).

Defining Happiness

It has been said that the search for happiness is one of the main sources of unhappiness in the world. Defining happiness, on the other hand, is like trying to describe a color or wrestle with a column of smoke. Happiness has been variously defined as:

- An activity of the soul expressing virtue (Aristotle);
- The absence of desire (Epicurus;
- Unity with one's nature (John F. Schumaker);
- A way station between too little and too much (Channing Pollock); and
- A way of travel, not a destination (Roy M. Goodman).

Contemporary society tends to focus on the affective nature of happiness and on overall enjoyment with one's life as a whole: "feeling good—enjoying life and wanting that feeling to be maintained" (Layard, 2005). Of most importance to our discussion here is that happiness is more than just affect and more than just the absence of feeling sad.

APPROACHES TO MENTAL HEALTH

There are a number of overlapping approaches to the promotion of good mental health. One stream, called "positive psychology," grew out of mainstream psychology and hence tends to be individual focused. Positive psychology is differentiated from "traditional" clinical psychology because of its emphasis on individual strengths rather than weaknesses, and by putting equal emphasis on building mental health and happiness in the absence of pathology. A second stream is more concerned with the factors that contribute to or detract from "wellbeing" or "satisfaction with life" as measured by structured items in large-scale population surveys (Eckersley, 2004; Heady and Wearing, 1992). Where positive psychologists are concerned with individual applications of their principles, wellbeing researchers see their data as also applicable at the population level. A third stream looks at mental health from a more comprehensive and proactive view. While drawing on these approaches, it takes a broader view of the factors that lead to poor mental health, provide resilience to stressors, and strengthen mental health. This approach is concerned more with the neglected area of mental health promotion; that is, increasing proactivity about positively building good mental health (Donovan et al., 2006).

POSITIVE PSYCHOLOGY?

Dr. Martin Seligman, a former head of the American Psychological Society, became famous in the 1970s for "shocking the living daylights" out of rats. This was not because he was a sadist, but rather to prove a point. Seligman found that if the rats were warned by a light that they were going to be shocked through the iron grills of a cage floor, and were able to learn a task to avoid that shock, they stayed happy and healthy (in rat terms, of course). However, if the ability to escape was taken away from them, they gradually became agitated, then catatonic, then just gave up trying until they eventually died. Seligman called this "learned helplessness," and this was the title of his seminal book (Seligman, 1976). His thesis was that stress per se is not such a bad thing—in fact, it can even be invigorating. It is an organism's ability to react to that stressor which is important.

Fast forward to the 1990s and Dr. Seligman has a road to Damascus moment. He realizes that rats and people are different. He also appreciates that studying helplessness, anxiety, and depression only helps one relieve these symptoms—it does not do the more positive things that make people happy. Hence he was converted (along with many other psychologists of the time) to positive psychology, or the study of human happiness.

Seligman's Positive Psychology Center (www.ppc.sas.upenn.edu) defines positive psychology as "the scientific study of the strengths and virtues that enable individuals and communities to thrive." Positive psychology focuses on positive affect (contentment with the past; happiness in the present; hope for the future) and individual strengths and virtues such as the capacity for love and work, courage, compassion, creativity, curiosity, resilience, and integrity. It also aims to create positive families, communities, schools, and other social institutions.

Positive psychology has three main characteristics that distinguish it from other forms of psychopathology:

- It works on an individual's underlying strengths and virtues, rather than concentrating on weaknesses;
- It is aimed mainly at untroubled or mildly troubled people, not the pathological; and
- It works on making people happy, not just on making them less miserable.

There are a number of practical applications involved in doing this (see www.authentichappiness.com). These include:

Finding and working on signature strengths: A questionnaire designed to identify signature strengths and, based on the handbook of signature strengths (Pearson and Seligman, 2004), it offers a short cut to identifying an individual's five most prominent strengths. Strengths include such things as the ability to find humor, summon enthusiasm, appreciate beauty, be curious, and love learning. The idea of the exercise is that using one's "signature" strengths may be a way to become engaged in satisfying activities. One or more strengths is then applied each day in a different way.

Counting blessings: At the end of each day, patients are advised to think of three good things that have happened that day and analyze why they have occurred. This enables them to focus on the good things that happen, which might otherwise be forgotten because of daily disappointments.

Expressing appreciation: Patients are advised to find someone who has done something helpful at some stage, and for whom proper appreciation was never given, and to thank them for this. This increases attention to good relationships and the good things that have happened in life, in contrast to the bad.

Other practical techniques include:

- Savoring the pleasing things in life, such as a warm shower or a good breakfast;
- Writing down what one may want to be remembered for, to help bring daily activities in line with what is really important;
- Regularly practicing acts of kindness for strangers; and
- Thinking about the happiest day in one's life over and over again, without analyzing it.

The general idea is to improve self-image and promote good interactions with others. Participants who perform a variety of acts, rather than repeating the same ones, have been shown in published research to have an increase in happiness, even a month after the experiment ended. Those who kept on doing the acts on their own did better than those who did not.

LIVES LED

Positive psychology assumes that there are different levels of life that can be led on the road to happiness.

The pleasant life: Seligman defines this as the superficially happy life espoused by modern celebrity worship, which is generally artificial and based around material possessions.

The engaged life (interest and involvement): This involves more engagement in activities that include a feeling of belonging and interest, such as joining a sports or interest group.

The meaningful life (giving/commitment): At this level, a sense of purpose and meaning in what has been committed to provides an inner satisfaction. This may or may not involve religious belief and can include such things as political, social, or environmental action or commitment.

As discussed later, Seligman's lives led can be transposed into action at a public health as well as a clinical level.

IS POSITIVE PSYCHOLOGY THE WAY FOR EVERYBODY?

Most psychologists working in the area of mental health acknowledge the organic nature of much mental illness. At the extreme ends, this requires pharmacological as well as psychological intervention. However, much of the so-called normal population is burdened with what have been called neuroses, often encouraged by a happiness-fixated media, resulting in an inability to thrive and achieve full potential. As pointed out by Schock (2006), "somewhere between Plato and Prozac, happiness stopped being a lofty achievement and became an entitlement." Positive psychology aims to assist people in achieving a level of happiness by focusing on core strengths and deemphasizing neurotic failings.

MENTAL HEALTH PROMOTION, ILLNESS PREVENTION, AND EARLY INTERVENTION

Mental health promotion is any action taken to maximize mental health and human wellbeing that focuses on improving environments that affect mental health and the coping capacity of communities and individuals (Australian Government Department of Health and Aged Care, 2000). Mental health promotion aims to improve the wellbeing of all people regardless of whether or not they have a mental illness. Prevention refers to interventions that prevent the development of a disorder by targeting known risk factors. Early intervention involves actions that specifically target people displaying the early signs and symptoms of a mental disorder (Australian Government Department of Health and Aged Care, 2000). As for health in general, such interventions need to occur at the policy and structural level (e.g., housing, education, employment, physical environment, discrimination, etc.) (Herrman et al., 2005).

In this chapter we are concerned primarily with mental health promotion and illness prevention, although as noted next, the Act-Belong-Commit framework is also applicable as part of an early intervention process.

The Act-Belong-Commit Approach to Mental Health Promotion

Mentally Healthy WA's Act-Belong-Commit campaign is a whole-of-population intervention, developed from qualitative and quantitative research. The Act-Belong-Commit message provides a simple mnemonic of behavioral concepts that are relevant at the clinical as well as the public health level. A primary communication objective for the campaign is to reframe individuals' perceptions of mental health as the absence of mental illness, to the belief that an individual can (and should) act proactively to protect and strengthen his or her mental health.

Act-Belong-Commit is the world's first and only comprehensive, population-wide mental health promotion program, as distinct from mental illness prevention or early intervention initiatives. While a number of school and worksite interventions aimed at building positive mental health exist, most community-wide campaigns have aimed at one or more of: increasing awareness of specific mental illnesses; education about stress reduction and coping strategies; encouraging help-seeking; the early detection and treatment of mental problems; and the destigmatization of mental illness (Patterson, 2009; Barry et al., 2005; Saxena and Garrison, 2004).

In line with the principles of the Ottawa Charter for Health Promotion (WHO, 2004) and the seventh principle of the Perth Charter for the Promotion of Mental Health and Wellbeing (Anwar-McHenry and Donovan, 2013) (i.e., "mental health promotion must take place at the individual and societal levels"), Act-Belong-Commit seeks to influence individual behavior and to create supportive environments for fostering and maintaining mental health and wellbeing. That is, the campaign targets individuals to be proactive about their mental health and engage in mentally healthy activities while at the same time supporting and encouraging organizations that offer mentally healthy activities to promote, and increase participation in, their activities.

Formative Research Findings

Participants in formative research identified a number of factors that they considered impact negatively on mental health, ranging from economic and sociocultural factors to individual personality and lifestyle factors. These included: unemployment or job insecurity; early childhood negative experiences and coercive parenting; exposure to violence; alcohol and drug abuse; and being subject to discrimination (racial, age, gender, sexual orientation, disability).

On the other hand, participants suggested a number of factors that increased resilience and ability to cope with stressors, including: positive parenting, educational and workplace practices; having access to good support networks; good self-esteem; and feelings of self-efficacy. There was near universal agreement that keeping oneself active (physically, socially, spiritually, and mentally), having good friends, being a member of various groups in the community, and feeling in control of one's circumstances were necessary for good mental health. There was also widespread agreement that having opportunities for achievable challenges at home, school or work, or in hobbies, sports or the arts, were important for a positive sense of self. Helping others (including volunteering, coaching, mentoring) was also frequently mentioned as a source of feeling good about oneself and satisfaction, as well as providing a source of activity and involvement with others (Donovan et al., 2003, 2007).

The Act-Belong-Commit Slogan

The three verbs "act," "belong," and "commit" were chosen as the campaign "branding" because they not only provide a colloquial "ABC for mental health," but also represent the three major behavioral domains that both the literature and people in general consider contribute to good mental health (Donovan et al., 2006). In the tradition of Aristotle's "virtue is cultivated by practice," Act-Belong-Commit is focused on getting people to engage in behaviors known to improve and maintain good mental health. According to Aristotle "we become just by doing just acts, temperate by doing temperate acts, brave by doing brave acts." The Act-Belong-Commit "philosophy" similarly states that "we become mentally healthy by engaging in mentally healthy activities." The Act-Belong-Commit brand is therefore a simple message to act on (Fig. 16.1), and is articulated as follows:

ACT	Keep alert and engaged by keeping mentally, socially, spiritually, and physically active.
BELONG	Develop a strong sense of identity and belonging by keeping up family relationships, friendships, joining groups, and participating in community activities.
COMMIT	Do things that provide meaning and purpose in life, such as taking up challenges, supporting causes, and helping others

For example, at the basic physical and cognitive levels, individuals can be encouraged to take a walk, read a book, do a crossword puzzle, tidy the garden, take a correspondence course, visit a museum, and so on. At a basic social level, individuals are encouraged to interact with salespeople while shopping, talk to their

FIGURE 16.1 Act-Belong-Commit brand/logo.

neighbors, and maintain contact with family and friends. Many activities can be done alone or as a member of a group (e.g., read a book vs join a book club; go for a walk alone or join a walking group; play solitaire or bridge games). In some cases there are synergistic effects: belonging to a book club not only adds a connectedness dimension but is likely to expand the cognitive activity involved; joining a walking group can expand the physical activity while adding a social dimension.

Commitment can be to a cause or organization that benefits the group or wider community, or to achieve a personal goal. The Feeling Blue? Act Green advertisement (Box 16.1) aimed at increasing interest in nature, captures the notion of joint rewards from physical activity that is part of a cause.

Religious belief is one form of commitment, but meaning can be obtained from short-term as well as long-term commitment to an ideal, as pointed out by Frankl in developing "Logotherapy" while incarcerated in a Nazi concentration camp (see Chapter 20). Volunteering and undertaking activities to benefit the community at large, especially where these involve the disadvantaged, have special returns for feeling good about oneself and overall mental health benefits, particularly in the retired elderly (Vaananen et al., 2005). Volunteering and greater participation in community activities and organizations also have substantial implications for community cohesion and social capital, and hence quality of life (Borgonovi, 2008; ESRC, 2004).

Overall, the three domains can be seen as a hierarchy for increasing levels of involvement, and thus a deeper contribution to wellbeing. For example, an individual may choose to be physically active by going for a run, increase their sense of belonging by joining a running group, and derive a sense of challenge and meaning by setting goals for participation in a fun run to raise funds for a charitable cause.

Box 16.1 Act-Belong-Commit Message in Nature Context

Feeling **blue?** Act **green!**

It seems that watching wildlife shows, exploring parks and gardens, looking at fabulous mountain and ocean views, and getting away from it all to the bush and Pacific island beaches are not only pleasurable, but are actually good for us!

Eminent biologists, psychologists and health professionals are showing that contact with nature – whether through parks, natural bush, pets or farm animals – helps us recover from stress and mental fatigue, helps us relax and puts us in a good frame of mind.

Of course, most of us know this intuitively and it's probably why we are drawn to nature instinctively. We all know that a walk on the beach, down a bush track or in a park is good to clear the head when we feel a little tired or stressed.

So, next time you are feeling like a lift, 'act green': do some gardening, pet the cat or dog, take a walk around the park or head down to the water for some time out.

Better still, don't wait until you're tired or feeling flat. Act green more often. Being in touch with nature makes us feel good, builds good mental health and helps beat the blues. And it's as easy as A-B-C

Act – do some gardening; take a walk around the local park; watch a wildlife documentary; take time to watch the sun set; spend time with pets ...

Belong – get a group together for a picnic in a natural setting; visit a wildlife sanctuary with friends; join a hiking group ...

Commit – become a 'civic environmentalist'; join a tree planting group; volunteer to keep your local parks & gardens clean; take up orienteering; learn more about ecology; offer to take a home-bound person out to a park ...

Being active, having a sense of belonging, and having a purpose in life all contribute to happiness and good mental health.

If you want to know more, visit www.mentallyhealthywa.org.au
Phone Professor Rob Donovan on 9266 4598
or email r.donovan@curtin.edu.au

www.mentallyhealthywa.org.au

Similarly, one can read a book, join a book club, and books can be selected that are challenging to read and require concentration and new learnings.

Overall, the Act-Belong-Commit message encourages people to be physically, spiritually, socially, and mentally active in ways that increase their sense of belonging to the communities in which they live, work, play, and recover and that involve commitments to causes or challenges that provide meaning and purpose in their lives. There is substantial scientific evidence that these three behavioral domains contribute to increasing levels of positive mental health and wellbeing (and, in fact, to physical health) (Donovan and Anwar-McHenry, 2014; Patterson, 2009; Barry et al., 2005). Furthermore, although different groups may articulate the domains differently and place different emphases on each, these three domains appear universal across different cultures (e.g., Koushede et al., 2015; Takenaka et al., 2012).

The Act-Belong-Commit framework also has broader implications for suicide prevention and civic engagement. According to Joiner (2005) the desire or motivation to suicide is driven by two main factors: low or "thwarted" belongingness and perceived burdensomeness. Given that Belong is about building and maintaining connections with others, including community and civic organizations and institutions, and that Commit involves doing things that provide meaning and purpose in life, including taking up causes and volunteering that help society and other individuals, the Act-Belong-Commit campaign can be viewed as strengthening two major protective factors for suicide. Furthermore, by encouraging participation in public events by people with different demographic and ethnic backgrounds, the campaign also contributes to greater understanding between groups and to what Aristotle called "civic virtue"; that is, greater feelings of obligation and responsibilities toward communities to which people have a greater sense of belonging (Sandel, 2012).

The campaign has a number of resources, including a self-help guide ("A Great Way to Live Life: the Act-Belong-Commit Guide to Keeping Mentally Healthy"; Fig. 16.2), which not only provides individuals with a tool for enhancing their mental health, but also provides the clinician with a helpful tool in the clinical setting. The Guide is downloadable from the website or can be completed online. Other resources include a mobile phone app, a search tool to find clubs and organizations in one's areas of interest, various fact sheets, curriculum materials for schools and worksites, and print and video advertisements (visit www.actbelongcommit.org.au).

As the self-help guide provides the basis for implementation in the clinical setting, an overview of the guide's contents is provided in Box 16.2. The guide can be downloaded from the website for paper-and-pencil completion (www.actbelongcommit.org.au) or can be completed interactively online (www.actbelongcommit.org.au). Readers are encouraged to visit the website and complete the guide online.

Act-Belong-Commit conducts continuous and ad hoc evaluation via annual population impact surveys, surveys of partners/franchisees, and evaluation of specific projects (Anwar-McHenry et al., 2012; Jalleh et al., 2013). Of most

FIGURE 16.2 Self-help guide: "A Great Way to Live Life."

interest here is that population surveys show individuals with a diagnosed mental illness or experience of a mental health problem are significantly more likely than others in the population to try to do something for their mental health as a result of campaign exposure.

Furthermore, a limited pilot study showed that discharged psychiatric patients who completed and acted on the self-help guide were significantly more likely to report enhanced wellbeing in recovery than those who did not (Robinson et al., 2015).

HEALTH PROFESSIONAL GOALS IN APPLYING ACT-BELONG-COMMIT IN THE CLINIC

Regardless of the level of involvement of the clinical practice overall in implementing the Act-Belong-Commit framework (see next section), the individual health professional's main aims should be to:

Box 16.2 The self-help guide: "A Great Way to Live Life! The Act-Belong-Commit Guide to Keeping Mentally Healthy"

The Self-help Guide: " A Great Way to Live Life! The Act-Belong- Commit Guide to Keeping Mentally Healthy"

The self-help guide contains the following sections. These can be viewed and completed online or via a downloaded hard copy.

Introductory Sections
The guide's introductory sections:

* define what it means to be mentally healthy;
* describes what is meant by each of Act, Belong, and Commit;
* refers to the evidence for these behaviours in building resilience, good mental health, and wellbeing; and
* outlines how to use the guide.

Wellbeing Questionnaire
Users complete an overall well-being questionnaire (The Warwick-Edinburgh Mental Wellbeing Scale, NHS Scotland, 2006) and are able to compare their score with population scores.

Act-Belong-Commit Sections
Each of the Act, Belong, and Commit domains are dealt with in turn in the following format:

* elaboration on what the domain means,
* a brief self-assessment questionnaire to measure how much the individual is active in that domain.

For Act, separate questions assess levels of physical, social, spiritual, and mental activity;
For Belong, the questions assess interactions with friends and family, the local community, specific interest groups, and attendance at large public events;
The Commit items measure involvement in activities for personal challenges and goals, formal roles in organizations and groups, involvement in causes, volunteering, and helping activities.

The person's score in each domain is used to provide 3 overall recommendations:

* "definitely" need to increase participation level,
* "could do more," or
* maintain current level.

Where increased activity is recommended in a domain, users are asked to look through the individual questionnaire items to identify where they could increase their activity levels.

Tips and activities
Each Act-Belong-Commit section contains tips and activities for increasing activity levels in each of the separate areas covered in each questionnaire.

Goals
Users are asked to set goals in each of the Act, Belong, and Commit domains in terms of what they intend to keep on doing and what they intend to try to do and to repeat the self-assessment questionnaires in the future to check their progress.

Additional information
The guide contains links to additional information and resources in the three domains as well as links to nutrition, sleep, alcohol, illicit drugs, coping strategies, positive psychology, and support services for those with mental health problems or a mental illness. All these links can be readily adapted to local, state, or national areas.

- Increase patients' understanding that maintaining good mental health is just as important as maintaining good physical health;
- Increase patients' awareness of things that they can and should do to build and maintain good mental health (i.e., the three behavioral domains under the Act-Belong-Commit framework); and

- Encourage patients to take up Act-Belong-Commit activities where they appear to lack "sufficient" activity in one or other of these behavioral domains.

The self-help guide is a useful clinical tool to apply these aims.

How the Clinical Practice Can Implement Act-Belong-Commit

At a baseline level, a clinical practice can simply promote the campaign messages by having posters and pamphlets in the waiting room. The clinic could also provide paper and pencil self-assessment questionnaires, along with an invitation to complete them, and/or promote a link to a website where patients can download or interact online with the self-assessment questionnaire and the self-help guide. This level would necessarily require staff to have a general understanding of the campaign messages and materials, with one or more individuals being able to answer patient queries. It is desirable that all staff become familiar with the self-help guide by working through it themselves.

At a more intensive level, the clinic can adopt a more proactive approach by actively seeking their patients' involvement. This would involve the clinic deciding whether to target all patients or selected groups: that is, targeting all patients, regardless of their mental health status; only targeting patients identified as at risk for mental illness or experiencing difficulties; and/or only targeting patients with a known diagnosis of mental illness. However, all patients, regardless of mental health status, can benefit from an increased awareness of how they can strengthen and maintain good mental health.

Targeting all Patients (Primary Prevention)

The clinic may choose to actively target all patients by encouraging them to read an introductory pamphlet (or watch an introductory video) in the waiting room, complete the self-assessment questionnaire (either via paper and pencil or via touch screen kiosk), and visit a website and work through the guide. In this mass approach, waiting room posters could also encourage patients to "ask their doctor" (or a designated staff member in the clinic) about "keeping mentally healthy" via the Act-Belong-Commit message.

Selective Targeting (Secondary Prevention)

This approach focuses on identifying patients showing early signs of depression or anxiety disorder, or simply languishing, and where adoption of the Act-Belong-Commit behaviors could improve their quality of life, alleviate current symptoms, or prevent further decline. This would include patients identified as at risk because of recently experienced or ongoing trauma or stress, such as recent job loss, divorce or separation, bereavement, or dealing with chronic illness or pain. Other indications would be provided by a patient's symptoms—for example, those expressing feelings of excess stress or an inability to relax, those feeling bored or expressing a lack of energy or enthusiasm for general activities, and those lacking social support.

In these cases, the clinician actively encourages (or requests) that the patient do the self-assessment questionnaires, discusses the results with the patient, and encourages the patient to work through the guide and return for a session to discuss the patient's goals. It would be helpful if, at this level, the clinician was able to provide a directory of organizations and activities in the local area, and, where necessary, details of contact persons who could introduce the patient to an organization they would like to join.

Diagnosed Patients (Tertiary Prevention)

In this approach, the clinic includes Act-Belong-Commit as a low-intensity mental illness management treatment/intervention and promotes self-management of mental health and wellbeing for patients currently being treated for illnesses such as anxiety and depression and for those with a known diagnosis (such as bipolar disorder). In this case, the Act-Belong-Commit intervention becomes a formal lifestyle add-on to any medication or psychotherapy that these patients are receiving. As earlier, the clinician discusses the results of the patient's self-assessment questionnaires with the patient and encourages the patient to work through the guide and to return for a session to discuss the patient's goals. However, these cases might require the clinician to work through parts of the guide with the patient. Furthermore, the patient's self-assessment results should be monitored on an ongoing basis as part of their treatment.

Overall, a typical consultation would involve:

- **First:** Establish via the self-assessment questionnaire (or by the clinician asking the questions in the questionnaire) to what extent the patient is: (A) physically, mentally, spiritually, and socially active; (B) an active member of groups and actively participates in community activities or events; and (C) engages in activities, hobbies, or interests that provide meaning and purpose in his or her life.
- **Second:** Using the previous answers, establish in which of the Act, Belong, or Commit domains, if any, the individual has low levels of activity.
- **Third:** Using tips and suggestions from the self-help guide, or simply direct questioning, establish what activities the individual is interested in and capable of doing in each of the domains where they are low, and facilitate the uptake of those activities.

Additional Suggestions for Analyzing and Identifying Activities

Some patients may have "social overload" through their jobs or family commitments, but are not getting enough physical or mental activity. Others may get plenty of mental stimulation, but make little time for socializing or physical activity. Others may be active individually but appear socially isolated. Some may have few or no real commitments to causes or challenges.

If an individual is lacking in all three behavioral domains, the aim would be to get him or her to join some group activity that involves physical and/or mental or spiritual activity along with the group membership. The first step would be to establish the activities he or she might like doing and then provide a contact name (preferably) or organization providing such activities. Uncovering past hobbies, activities, and interests is a step to identifying possible interests. If the patient is too shy to personally make contact, try to provide a contact name.

Similar principles apply to increasing levels for each of the Act, Belong, and Commit domains where they are deemed deficient: identify activities he or she may be interested in and capable of doing. Older people are quite receptive to the benefits of increasing their levels of mental and physical activity and appreciate the sense of satisfaction from volunteering. Younger people might be more interested in participating in activities requiring some challenge and learning new things. Many individuals find spiritual relaxation in activities such as Tai Chi or spending time in natural environments. For Commitment, depending on time and other demands, people can be encouraged to volunteer for causes they are interested in, or, for personal goals, to look at local skills providers, whether for a qualification certificate or simply personal development. Learning new skills can provide cognitive and social activity as well as building self-esteem through achievements.

Minimal social activity can be maintained by encouraging the individual to just "say hello" to neighbors and customer service staff, and by maintaining contacts with friends and family, preferably in person if able, or via telephone or social media. There are also wellbeing benefits by simply being exposed to other people in public gatherings.

SUMMARY

This chapter demonstrates how clinicians can use the Act-Belong-Commit framework as a Lifestyle Medicine approach to promote good mental health in the clinical setting. Act-Belong-Commit is a positive message that seeks to build and maintain good mental health by encouraging patients to participate in activities that keep them mentally healthy. Consistent with what both the literature and laypeople consider to contribute to good mental health, this simple and versatile message is universally applicable across cultures and can easily be adapted to an individual's personal circumstances, abilities, and interests, regardless of their mental health status. The campaign's supportive materials and resources, such as the self-help guide and self-assessment tool, make the adoption and integration of the framework into the clinic an even easier transition.

Practice Tips

Encouraging Mentally Healthy Activities

	Medical Practitioner	Practice Nurse
Assess	• Assess for mild depression • Check Act-Belong-Commit scores • Check physical and mental activities; social contacts • Check your own Act-Belong-Commit status	• Check more detail in Act-Belong-Commit format • Check belonging to "causes" • Check on time available and commitment—overload or underload • Check your own Act-Belong-Commit status
Assist	• Provide medication for risk if appropriate • Provide website for self-assessment • Discuss actions from self-assessment scores • Suggest appropriate activities	• Provide list of available organizations • Check places for being volunteers • Suggest options for daily actions • Provide reading materials
Arrange	• Identify relevant local organizations • Refer to self-help websites	• Help contact relevant local organizations • Connect with self-help groups

Key Points

Clinical Management

- Consider mental health and not just mental illness.
- Regard everyone as a potential target for good mental health.
- Practice good personal mental health tactics.
- Encourage the patient to do personal research (reading, searching websites).

REFERENCES

Australian Government Department of Health and Aged Care, 2000. Promotion, Prevention and Early Intervention for Mental Health—a Monograph. Mental Health and Special Programs Branch, Department of Health and Aged Care, Canberra.

American College of Preventive Medicine, 2009. Lifestyle Medicine—Evidence Review. www.acpm.org/resource/resmgr/lmi-files/lifestylemedicine- literature.pdf American College of Preventive Medicine.

Amminger, G.P., Schafer, M.R., Papageorgiou, M., Klier, C.M., Cotton, S.M., Harrigan, S.M., MacKinnon McGorry, P.D., Berger, G.E., 2010. Long-chain omega-3 fatty acids for indicated prevention of psychotic disorders: a randomised, placebo-controlled trial. Arch. Gen. Psychiatry 67 (2), 146–154.

Anwar-McHenry, J., Donovan, R.J., 2013. The development of the Perth Charter for the promotion of mental health and wellbeing. Int. J. Ment. Health Prom. http://dx.doi.org/10.1080/14623730.2013.810402.

Anwar-McHenry, J., Donovan, R.J., Jalleh, G., Laws, A., 2012. Impact evaluation of the Act-Belong- Commit mental health promotion campaign. J. Public Ment. Health 11 (4), 186–195.

Barry, M.M., Domitrovich, C., Lara, M.A., 2005. The implementation of mental health promotion programmes. IUHPE - Promot. Educ. (Suppl. 2), 30–35.

Borgonovi, F., 2008. Doing well by doing good. The relationship between formal volunteering and self-reported health and happiness. Soc. Sci. Med. 66 (11), 2321–2334.

Brown, S., Chan, K., 2006. A randomised controlled trial of a brief health promotion intervention in a population with series mental illness. J. Ment. Health 15 (5), 543–549.

Donovan, R., Henley, N., Jalleh, G., et al., 2007. People's beliefs about factors contributing to mental health: implications for mental health promotion. Health Prom J. Aust. 18 (1), 50–56.

Donovan, R.J., James, R., Jalleh, G., et al., 2006. Implementing mental health promotion: the 'act-belong-commit' mentally healthy WA campaign in Western Australia. Int. J. Ment. Health Prom 8 (1), 29–38.

Donovan, R.J., Watson, N., Henley, N., et al., 2003. Report to Healthway: Mental Health Promotion Scoping Project. Curtin University, Centre for Developmental Health, Perth, WA.

Donovan, R.J., Anwar-McHenry, J., 2014. Act-belong-commit: lifestyle medicine for keeping mentally healthy. Am. J. Lifestyle Med. 8 (1), 33–42.

Duaso, M.J., Cheung, P., 2002. Health promotion and lifestyle advice in a general practice: what do patients think? J. Adv. Nurs. 39 (5), 472–479.

Eckersley, R., 2004. Well & Good. The Text Publishing Company, Melbourne.

ESRC (Economic Research Council), 2004. The Art of Happiness… Is Volunteering the Blueprint for Bliss? ESRC Press Release. Available from: www.esrc.ac.uk/esrccontent/news/september04-2.asp.

Heady, B., Wearing, A., 1992. Understanding Happiness: A Theory of Subjective Well-Being. Longman Cheshire, Melbourne.

Herrman, H., Saxena, S., Moodie, R. (Eds.), 2005. Promoting Mental Health: Concepts, Emerging Evidence, Practice. World Health Organization, Geneva.

Herring, M.P., Lindheimer, J.B., O'Connor, P.J., 2013. The effects of exercise training on anxiety. Am. J. Lifestyle Med. http://dx.doi.org/10.1177/1559827613508542.

Jalleh, G., Anwar-McHenry, J., Donovan, R., Laws, A., 2013. Impact on community organisations that partnered with the Act-Belong-Commit mental health promotion campaign. Health Prom. J. Aust. 24 (1), 44–48.

Joiner, T.E., 2005. Why People Die by Suicide. Harvard University Press, Cambridge.

Koushede, V., Nielsen, L., Meilstrup, C., Donovan, R.J., 2015. From rhetoric to action: Adapting the Act-Belong-Commit Mental Health Promotion Programme to a Danish context. Int. J. Ment. Health Prom. http://dx.doi.org/10.1080/14623730.2014.995449.

Layard, R., 2005. Happiness: Lessons from a New Science. Penguin, London.

O'Connor, P.J., Herring, M.P., Caravalho, A., 2010. Mental health benefits of strength training in adults. Am. J. Lifestyle Med. 4 (5), 377–396.

Patterson, A., 2009. Building the Foundations for Mental Health and Wellbeing: Review of Australian and International Mental Health Promotion, Prevention and Early Intervention Policy. Department of Health and Human Services, Hobart, Australia.

Pearson, C., Seligman, M., 2004. Character Strengths and Virtues. Oxford University Press, USA.

Robinson, K., Donovan, R.J., Jalleh, G., Lin, C., 2015. Act-belong-commit in Recovery: Train the Trainer Pilot and Intensive Pilot Study Final Report. Perth: Mentally Healthy WA, Curtin University and the Mental Health Commission, Government of Western Australia.

Rohrer, J.E., Pierce, J.R., Blackburn, C., 2005. Lifestyle and mental health. Prev. Med. 40, 438–443.

Sandel, M.J., 2012. What Money Can't Buy: The Moral Limits of Markets. Farrar, Straus, and Giroux, New York.

Saxena, S., Garrison, P.J., 2004. Mental Health Promotion: Case Studies from Countries. World Health Organization and World Federation for Mental Health, Geneva.

Schock, R., 2006. The Secrets of Happiness. Profile Books, London.

Seligman, M., 1976. Learned Helplessness. Random House, NY.

Takenaka, K., Bao, H., Shimazaki, T., Lee, Y.H., Konuma, K., 2012. Mental health promotion contributing to resilience for children after Tsunami disaster in Japan. Poster Presentation. In: The Seventh World Conference on the Promotion of Mental Health and the Prevention of Mental and Behavioural Disorders Perth, Australia. http://www.papersearch.net/view/detail.asp?detail_key=20212446.

Vaananen, A., Buunk, B.P., Kivimaki, M., et al., 2005. When it is better to give than to receive: long- term health effects of perceived reciprocity in support exchange. J. Pers. Soc. Psychol. 89 (2), 176–193.

Vaillant, G., 1995. Adaptation to Life, second ed. Harvard University Press, London.

Vaillant, G., 2015. Triumphs of Experience: The Men of the Harvard Grant Study. Belknap Press, NY.

WHO, 2004. Prevention of Mental Disorders: Effective Interventions and Policy Options A Report of the WHO and the Prevention Research Centre, Universities of Nijmegen and Maastricht. WHO, Geneva.

WHO, 2001. Strengthening Mental Health Promotion. World Health Organization (Fact Sheet No. 220). WHO, Geneva.

Chapter 17

Technology-Induced Pathology: Watch (This) Space

John Stevens, Garry Egger

Life is like riding a bicycle. To keep your balance, you have to keep moving.
Albert Einstein

INTRODUCTION: THE GOOD AND THE BAD OF TECHNOLOGY

In considering the lifestyle determinants of disease, clinicians often stop at nutrition and exercise (which together have been called the "penicillin of Lifestyle Medicine"). The more insightful might add stress, smoking, and even excessive alcohol intake as an afterthought. But this is often where it ends.

As we have shown in Chapter 4 and elsewhere (Egger et al., 2011), there are up to 15 determinants, generically labeled "anthropogens" with an evidence base linking these to chronic disease under the acronym NASTIE MAL ODOURS (Chapter 4). Many are detailed in individual chapters. But one that is not often considered are the diseases induced by the very advances in technology that are recognized (and rightly so) for improving health and living standards. These are labeled in Chapter 4 as "Technology-Induced Pathologies." Some fit into the category of injury. Others have more chronic effects that are also systemic. Yet despite their ill-defined nature, the category is destined to expand with further likely expansions of technology.

THE CHANGING NATURE OF HEALTH

Human health is paved with evolving challenges: the nomadic existence, for example, predisposed humans to the elements and injury, agrarianism to zoonotic diseases, and urbanization to the spread of pestilence-related ailments. In recent times, lifestyle and modern environment-driving lifestyles have become the dominant etiology. It's not just what we eat and do, but the vagaries of modern times that pose new, albeit less obvious, challenges.

Lifestyle Medicine. http://dx.doi.org/10.1016/B978-0-12-810401-9.00017-6

Unquestionably, poor nutrition, inactivity and stress (considered in Chapters 8–15) can all contribute to ill health. But the association of disease with certain modern forms of technology is often overlooked. Problems can vary from death or chronic pain from motor vehicle or machine injuries, to hearing problems from amplified music (Henderson et al., 2011). At the extremes, they can range from death and disability from firearms and high-tech weapons used in warfare, to apparently obscure problems like "screen dermatitis" (Ghasri and Feldman, 2010) and other skin disorders (Wintzen and van Zuuren, 2003), impaired vision, and repetitive strain injury from excessive computer and small screen use (Agarwal et al., 2013). Skin damage from sunbeds (Tierney et al., 2015) and cell disturbances from endocrine disrupting chemicals (like phthalates from plastics; Song et al., 2015) are what 19th century philosopher John Ruskin called "Illth," or the effects of materials and services that have economic benefits but human and environmental costs.

Other adverse health outcomes (death, injury) can occur, for example, while focusing on the new technology of social media (e.g., texting, tweeting) while carrying out other activities, like driving (Llerena et al., 2015; O'Connor et al., 2013). Because of its immediacy, social media bullying and intimidation can lead to mental health issues and even suicide among prone youth. Strange as it may seem, "Facebook depression" has already hit the medical literature (O'Keefe and Clarke-Pearson, 2011). Social contagion effects on disease are also amplified through social media as shown in the association between social networks and disease risks like obesity and smoking (Kramer et al., 2014; Christakis and Fowler, 2013).

Another of the vicarious outcomes of our modern technology is that we are staying indoors more and more. This has led to a growing phenomenon called nature deficit disorder (Louv, 2010). Especially noticeable in children, according to Louv (2010) it mimics the spectrum of attention deficit disorder. The increasing use of technology can also alter other behaviors such as nutrition (snacking) and reduced physical activity (TV, computer usage).

TECHNOLOGY'S "HORMETIC" EFFECT

Most advances in humanity have been related to the discovery or introduction of new innovations and/or technology. There should be no suggestion that all technology is bad or that major advances in health, science, and well-being have not come from industrial and technological advances. But just like the invention of the wheel would have no doubt caused people to be run over and modern life-saving medicines can have side effects, modern technology can have a downside, which needs to be recognized.

One set of technologies that is having a big impact on humanity at the moment is the personalization of computers, communication devices, and the internet (i.e., small screen technology). The uptake and impact of these are pivotal in redefining the way we work, shop, seek entertainment, communicate, socialize, reproduce, interact with the environment, and seek information—i.e., the way we live. As with the introduction of all new technologies there is a corresponding adaption of human behavior and lifestyle with positive and negative outcomes for some that accompanies the progress. Some potential pathological outcomes of this type of technology are:

1. Adolescent obesity and physical well-being

   In general the literature suggests that screen time for many adolescents especially is related to many of the determinants of ill health we have discussed in this book. This includes: a reduction in physical activity (Streb et al., 2015; Barnett et al., 2010; Hardy et al., 2010) and an increased intake of calorie-dense food both of which can be determinants of obesity, metabolic disorder, and chronic disease. A World Health Organization project supported these reports and also found a decreased intake of vegetables and fruit among adolescent screen users suggesting they will be at an increased risk for cancer and other specific nutritional deficit disorders (Börnhorst et al., 2015).

   Research also indicates that screen time can increase sleep disruption, which is a determinant of obesity, chronic disease, and poor mental health by: (1) reducing hours available for sleep of an increasing number of adolescent late night users (Falbe et al., 2015), and (2) interfering with melatonin production. Melatonin is the hormone produced in the brain that determines the quality and quantity of one's sleep (Cajochen et al., 2011). Decreased screen time in this same population was shown to improve melatonin production and sleep (Cajochen et al., 2011). Bener et al. (2011) also found that there is a relationship between increased screen time obesity and poorer vision for adolescent screen users.
2. Mental health

   There are increasing reports of mental health in adolescents being associated with cyberbullying, which is being viewed as a new form of violence expressed through electronic media. A systematic review by Bottino et al. (2015) reported that the prevalence of cyberbullying ranged from 6.5% to 35.4% of all social media users. Cyberbullying is associated with moderate to severe depressive symptoms, substance use, ideation, and suicide attempts. Reducing screen time also reduces cyberbullying (Bottino et al., 2015; Carli et al., 2013). As mentioned previously, studies identifying "Facebook depression" have also reported an association between the use of social media and self-esteem, body image, and depression and anxiety (Carli et al., 2013; Richards et al., 2015).
3. Addictive characteristics

   These are conceptualized as impulse-control disorders stemming from pathological internet use (PIU). Research has indicated a potential link between PIU and psychopathology, such as depression and symptoms of attention deficit hyperactivity disorder (Carli et al., 2013). O'Connor et al. (2013) have shown that motor vehicle injuries are associated with heightened anticipation about incoming phone calls or messages, which they use as an example of PIU.

Similarly, according to Kuss and Griffiths (2010) internet gaming addiction is linked to behavioral problems largely because of lack of sleep and inactivity. The tabloid literature is reporting deaths related to gaming addiction through exhaustion, dehydration, and embolism. One report is of a 17-year-old boy dying after internet gaming for 22 consecutive days without a break (save for toilets and limited sleep) (http://www.news.com.au/world/europe/russian-teenager-dies-after-playing-online-computer-game-defense-of-the-ancients-for-22-days-in-a-row/news-story/7f178341c80c5896a9c8c315b8e5c9b6).

4. "Phantom vibration" or "phantom ringing"
   Phantom ringing or phantom vibration is described as a sense that a phone is ringing or vibrating when it is not. Men who carry mobile phones in their pockets, for example, describe a real vibration felt in their quadriceps that mimics their telephone signaling an incoming call. It is thought to be a form of involuntarily activated muscle memory related to ongoing exposure to the actual physical stimulus of the phone ringing (or vibrating) and potentially the electromagnetic radiation emitted by the device. In a systematic review of the field, Deb (2015) expects phantom ringing to increase with potential implications for people with mental illness issues.
5. Human reproduction
   At a base level, small screen devices may also affect human reproduction. A systematic review and metaanalysis by Adams et al. (2014) concluded that pooled results from in vitro and in vivo studies suggest that mobile phone exposure *might* negatively affect sperm motility, viability, and concentration. Further study is required to determine the full clinical implications for both subfertile men and the general population.

What Can Be Done in a Brief Consultation?

- ☐ Consider the possibility of a technology-induced pathology.
- ☐ Where obvious, recommend a break from suspected technology to see if the problems persist.
- ☐ Explain the difference between exercise for fitness and movement for fat loss.

THE FUTURE OF RESEARCHING TECHNOPATHOLOGY

Disease related to technopathology, like much modern chronic disease, is unlikely to respond to sophisticated medical intervention (except for palliation). Proper management requires vigilance and awareness and use of the legislative/regulative arms of public health and health promotion, as has been effective in aspects of injury prevention to date (seat belt use, pool fences, bike helmets, random breath testing, etc.). Avoiding allegations of promoting a "nanny state" will tend to temper these future health actions. Still, there will be a price to pay for advances in economic wealth to make this genuine human "wellth," if Ruskin's ideas of "illth" are to be taken seriously. Given that modern technology is mostly corporatized and the bigger corporations who develop and market the technology are politically powerful organizations, health professionals need to remain vigilant to the future veracity of research that reveals disease—or the lack of it—related to the deployment and uptake of new technologies. Without scientific vigilance, truth itself may indeed become a technology-induced pathology.

The scholarship of technology-induced pathology is a green field. There will undoubtedly be more to come in the future. Watch this space.

SUMMARY

It may not be unexpected that lifestyle-related problems can arise from the overuse of modern technology and that this is likely to increase with the technological revolution. Dealing with such problems may simply involve desisting using offending technology; however, this may be easier in theory than in practice in a society that has become dependent on all forms of technology within employment and recreation.

Key Points

Clinical Management

- Keep up to date with potential problems arising from the use and overuse of technology.
- Consider the possibility of these in any full diagnosis.

REFERENCES

Adams, J.A., Galloway, T.S., Mondal, D., Esteves, S.C., September 2014. Effect of mobile telephones on sperm quality: a systematic review and meta-analysis. Environ. Int. 70, 106–112.

Agarwal, S., Goel, D., Sharma, A., 2013. Evaluation of the factors which contribute to the ocular complaints in computer users. J. Clin. Diagnos Res. 7 (2), 331–335.

Barnett, T.A., O'Loughlin, J., Sabiston, C.M., Karp, I., Bélanger, M., Van Hulst, A., Lambert, M., Augest 1, 2010. Teens and screens: the influence of screen time on adiposity in adolescents. Am. J. Epidemiol. 172 (3), 255–262.

Bener, A., Al-Mahdi, H.S., Ali, A.I., Al-Nufal, M., Vachhani, P.J., Tewfik, I., 2011. Obesity and low vision as a result of excessive Internet use and television viewing. Int. J. Food Sci. Nutr. 62 (1), 60–62.

Börnhorst, C., Wijnhoven, T.M., Kunešová, M., Yngve, A., Lissner, L., Duleva, V., Petrauskiene, A., Breda, J., 2015. WHO European childhood obesity surveillance initiative: associations between sleep duration, screen time and food consumption frequencies. BMC Public Health 15, 442.

Bottino, S.M., Bottino, C.M., Regina, C.G., Correia, A.V., Ribeiro, W.S., 2015. Cyberbullying and adolescent mental health: systematic review. Cad. Saúde Pública 31 (3), 463–475.

Carli, V., Durkee, T., Wasserman, D., Hadlaczky, G., Despalins, R., Kramarz, E., Wasserman, C., Sarchiapone, M., Hoven, C.W., Brunner, R., Kaess, M., 2013. The association between pathological internet use and comorbid psychopathology: a systematic review. Psychopathology 46 (1), 1–13.

Cajochen, C., Frey, S., Anders, D., Späti, J., Bues, M., Pross, A., Mager, R., Wirz-Justice, A., Stefani, O., 2011. Evening exposure to a light-emitting diodes (LED)-backlit computer screen affects circadian physiology and cognitive performance. J. Appl. Physiol. 110 (5), 1432–1438.

Christakis, N.A., Fowler, J.H., 2013. Social contagion theory: examining dynamic social networks and human behavior. Stat. Med. 32 (4), 556–557.

Deb, A., 2015. Phantom vibration and phantom ringing among mobile phone users: a systematic review of literature. Asia Pac. Psychiatry 7 (3), 231–239.

Egger, G., Binns, A., Rossner, S. (Eds.), 2011. Lifestyle Medicine: Managing Disease of Lifestyle in the 21st Century, second ed. McGraw-Hill, Sydney.

Falbe, T., Davidson, K., Franckle, R., Ganti, C., Gotmaker, S., Smith, C., Land, T., Taveras, E., 2015. Sleep duration, restfulness, and screens in the sleep environment. Pediatrics 135 (2), 367–375.

Ghasri, P., Feldman, S.R., 2010. Frictional lichenified dermatosis from prolonged use of a computer mouse: case report and review of the literature of computer-related dermatoses. Dermatol. Online J. 16 (12), 3.

Hardy, L.L., Denney-Wilson, E., Thrift, A.P., Okely, A.D., Baur, L.A., 2010. Screen time and metabolic risk factors among adolescents. Arch. Paed. Adoles. Med. 164 (7), 643–649.

Henderson, E., Testa, M.A., Hartnick, C., 2011. Prevalence of noise-induced hearing-threshold shifts and hearing loss among US youths. Pediatrics 127 (1), e39–e46.

Kramer, A.D., Guillory, J.E., Hancock, J.T., 2014. Experimental evidence of massive-scale emotional contagion through social networks. Proc. Natl. Acad. Sci. USA 111 (24), 8788–8790.

Kuss, D.J., Griffiths, M.D., 2010. Internet and gaming addiction: a systematic literature review of neuroimaging studies. Brain Sci. 2 (3), 347–374.

Llerena, L.E., Aronow, K.V., Macleod, J., Bard, M., Salzman, S., Greene, W., Haider, A., Schupper, A., 2015. J. Trauma Acute Care Surg. 78 (1), 147–152.

Louv, R., 2010. Do our kid have a nature deficit disorder? Educ. Leadership (4), 24–30.

O'Connor, S., Whitehill, J., King, J., Ebel, E.B., 2013. Compulsive cell phone use and history of motor vehicle crash. J. Adoles. Health 53 (4), 512–519.

O'Keefe, G.S., Clarke-Pearson, K., 2011. Council on communications and media. The impact of social media on children, adolescents, and families. Paediatrics 127 (4), 800–804.

Richards, D., Caldwell, P.H., Go, H., 2015. Impact of social media on the health of children and young people. J. Paediatr. Child Health. http://dx.doi.org/10.1111/jpc.13023 November 26. Date of Electronic Publication (Epub ahead of print).

Song, Y., Chou, E.L., Baecker, A., You, N.Y., Song, Y., Sun, Q., Liu, S., 2015. Endocrine-disrupting Chemicals, risk of type 2 diabetes, and diabetes-related metabolic traits: a systematic review and meta-analysis. J. Diabetes. http://dx.doi.org/10.1111/1753-0407.12325.

Streb, J., Kammer, T., Spitzer, M., Hille, K., 2015. Extremely reduced motion in front of screens: investigating real-world physical activity of adolescents by accelerometry and electronic diary. PLoS One 10 (5), e0126722.

Tierney, P., de Gruijl, F.R., Ibbotson, S., Moseley, H., 2015. Predicted increased risk of squamous cell carcinoma induction associated with sunbed exposure habits. Br. J. Dermatol. http://dx.doi.org/10.1111/bjd.13714.

Wintzen, M., van Zuuren, E.J., 2003. Computer-related skin diseases. Contact Dermatitis 48 (5), 241–243.

Chapter 18

To Sleep, Perchance to ... Get Everything Else Right

Caroline West, Garry Egger

> *Sleep touches on nearly every aspect of our physiology and psychology, of our interaction with the world and with others.*
>
> William C. Dement (2000)

INTRODUCTION: THE VALUE OF SLEEP RESEARCH

Sleep research, such as was carried out by some of the pioneers in the area in the 1960s and 1970s might, in a different setting have been classified as torture. Researchers kept subjects awake for extended periods and, just when they tended to nod off, would wake them every minute or so throughout an entire night. They then evaluated their alertness the next day. Not surprisingly, this was found to be less than optimal. But the important thing was how and why daytime functioning was affected by sleep deprivation and what this tells us about the functions of, and reasons for, sleep.

Humans, like most mammals, sleep for about a third of their lives. Anyone living to the current average age (for women in the Western world) of around 82 will have slept for a staggering 27 years. Yet, given the amount of time invested in sleep, it is surprising how little time is spent prioritizing it—until something goes wrong. In many ways, healthy sleep is the anchor for a healthy life. Healthy sleep not only helps boost mood and IQ and improve memory and concentration but also decreases the risks of heart disease, diabetes, depression, and even obesity (Irish et al., 2015).

Sleep disorders are also on the rise, as shown initially in the classic sleep text by Dement (2000), and more recently with colleagues in an update on the *Principles and Practice of Sleep Medicine* (Kryger et al., 2015). As many as 80% of people will suffer from a sleep problem at some stage in their life ranging from mild intransient insomnia to severe and crippling narcolepsy. The burden is even heavier in lower socioeconomic groups, where obesity, sleep-disordered breathing, and cardiovascular risk are all greater (See et al., 2006). At any given time as many as 30–50% of people will have difficulty sleeping.

Lifestyle Medicine. http://dx.doi.org/10.1016/B978-0-12-810401-9.00018-8

Modern lifestyles are often in direct competition with sleep, so much so that it could be argued that the majority of modern sleep problems have a basis in lifestyle choices. Inactivity, poor nutrition, anxiety, depression, and stress—all by-products of modernity—have an impact on sleep. Obesity, another by-product of modernity, has been linked both with too little and too much sleep (Marshall et al., 2008; Vargas, 2016). The good news, however, is that sleep can often be dramatically improved with a healthy approach to lifestyle and a simple "sleep hygiene" approach to sleeping patterns. While we will consider the best ways of achieving this here, this is not intended to be an in-depth discussion of medically related sleep problems. Our main concerns are those sleep issues that are related to, and can be modified by, lifestyle factors and the lifestyle changes that can make sleep more functional for ongoing well-being.

THE REASONS FOR SLEEP

Ironically, despite the importance of sleep in human lives and functioning, there is still no agreed understanding of why it is necessary (Vyazovskiy, 2015). The notion of giving the brain and/or body a "rest" seems facile in light of the fact that body organs such as the heart, liver, and kidneys function 24 hours a day. Other suggestions are that sleep serves an energy conservation function, a body tissue restitution function, or a temperature downregulation function.

Regardless of the reason(s), it is clear that sleep plays an important role in human health and well-being. The fact that the world record for avoiding sleep is only a matter of days in a lifespan that lasts scores of years suggests that the psychophysiological rationale of sleep, albeit illusive, is undoubtedly significant.

HOW MUCH SLEEP DO WE NEED?

Humans have large variations in their individual needs for sleep, ranging from 1–2 to 12–14h a day. Variations occur with age, life stages, health conditions, lifestyle patterns, and other conditions. On average, most adults need 7–8h a night to wake refreshed and function well during the day, and individuals reporting both an increased (>8h/day) or reduced (<7h/day) sleep duration are at moderately increased risk of all-cause mortality, cardiovascular disease, and symptomatic diabetes (Spiegel et al., 2005). Unfortunately, though, the number of hours slept per day appears to be decreasing, with increasingly busy lifestyles. The average of 8–9h in previous years has now dropped to about 7h per night, according to the National Sleep Foundation (2011). The proportion of young adults (19–29 years) getting less than 7h sleep per night has more than doubled from 15.6% in 1960 to 24% in 2011.

What are the consequences? Is sleep deprivation really a problem, or can humans simply be conditioned to sleep less? Accumulating evidence suggests that sleeping less than 5h a night over an extended period can significantly raise the risks of all those health problems referred to earlier. There is also evidence that chronic sleep deprivation of as little as an hour or more a day can lead to a situation of sleep debt (Kryger et al., 2015). This can have an enormous impact on the risk of fatigue-related accidents (Bhattacharyya, 2015). In other words, if sleep is not taken seriously, there is a price to pay.

SLEEP CYCLES

Prior to the invention of the electroencephalogram (EEG) in 1929 by Hans Berger, very little was known about sleep. Up until that point, sleep was often considered to be a blank state, an unconscious wasteland. However, the EEG revealed that sleep was dynamic, varied, and complex. It was soon discovered that sleep could be divided into five stages, each with its own distinct brain wave patterns. And sleep was far from a linear journey. While on average each cycle or journey of sleep lasts around 90 minutes, each can be slightly different. For example, later in the evening, people may drift from the deep sleep of stage 4 back to the light sleep of stage 2 before going into stage 5 rapid eye movement (REM) or dream sleep.

The Sleep Cycle: Stages of Sleep

Stage 1	Very light sleep, theta brain waves. During this phase we may experience fleeting visual images called "hypnagogic hallucinations."
Stage 2	Light/transitional sleep characterized by theta waves with K complex waves and sleep spindles.
Stage 3	Deep sleep with a combination of theta and delta waves.
Stage 4	Deepest sleep, theta waves disappear and there are only delta waves.
Stage 5	REM (dream) sleep.

Individuals usually have about five sleep cycles per night. The earlier part of the night may be characterized by more time spent in deep sleep, and often as the night progresses more time is spent in REM sleep. We know that REM/dream sleep is important for preparing the mind for peak daytime performance and improving memory retention and recall (Suzuki et al., 2012). Too little sleep can certainly come at a cost. Missing out on the longer periods of REM sleep that come near morning can also have serious consequences for thinking, memory, and daytime performance.

THE CONCEPT OF SLEEP DEBT

One of the most accepted axioms among sleep researchers is the concept of "sleep debt." Sleep debt accumulates during periods of wakefulness. In fact, according to many sleep researchers all periods of wakefulness are sleep deprivation and add to sleep debt (Kryger et al., 2015). This is usually paid back after a good night's sleep. However, if this does not occur, and the debt is left unpaid, it accumulates progressively until eventually it has to be paid off in another form, for example, with sickness, mood change, or ongoing fatigue (Haack and Mullington, 2005). According to McEwen (2006), sleep deprivation is a chronic stressor that causes cognitive problems, which can exacerbate pathways that lead to disease, possibly through inflammation (Koren et al., 2015). Researchers who woke sleepers regularly found that while 1-min sleep periods followed by forced awakenings (leading to sleep deprivation) caused major deficits in alertness the next day, longer periods of sleep (10–15 min) had much less of an effect. This suggests that there are minimal units of restorative sleep.

As Kryger et al. (2015) put it, "It's as if the bank that keeps track of sleep debt doesn't accept small deposits." Siestas, and naps, on the other hand, seem to play a significant role in restoring debt. Heart disease rates are lower in countries where siestas are part of daily life (Naska et al., 2007). Paradoxically, short daytime naps (<30 min) also appear to be more effective than longer naps in maintaining alertness (Dhand and Sohal, 2006).

In terms of performance, vigilance, alertness, decision-making, and reaction times, sleep loss can been compared to the effects of blood alcohol. This has particular relevance for shift workers (and indeed hospital doctors), who are at risk of fatigue-related accidents and mistakes. Dawson and Reid (1997) showed that 17 hours of wakefulness had an equivalent effect on performance to a blood alcohol level (BAL) of 0.05%; 24 hours without sleep was the equivalent of a BAL of 0.1%.

While accountants deal with monetary debts and bankers call in bad loans, the body deals with sleep debt, and clinicians are left to deal with the unpaid loans by treating sleep-related disorders.

FATIGUE AND EXCESSIVE DAYTIME SLEEPINESS

As might be expected, poor or inadequate sleep can lead to fatigue (tiredness without increased sleep propensity) and excessive daytime sleepiness (EDS), the latter of which has been found to be quite common (National Sleep Foundation, 2011). This is often attributed to periods of wakefulness caused by obstructive sleep apnea, which in turn is causally related to obesity. One review, however, showed that obese individuals without sleep apnea can suffer EDS and fatigue and those with sleep apnea may not (Vgontzas et al., 2006). This led to the suggestion that other factors, such as depression, diabetes, and inactivity in the obese, may be risk factors for daytime fatigue and EDS, independent of nighttime sleep quality. In turn, fatigue and EDS may themselves cause disease via a link to cytokines and the stress system. Vgontzas et al. (2006) suggest that fatigue is associated more with psychological distress and EDS with metabolic factors such as insulin resistance and diabetes. This suggests the need to distinguish between both conditions in treating the problem. Apnea should also not be discounted as a cause.

RISK FACTORS ASSOCIATED WITH POOR SLEEP

As well as the common disorders that occur in sleep (see Table 18.1), there is a range of pathological conditions associated with inadequate or poor quality sleep. These include:

- depression/anxiety
- obesity
- heart disease
- decreased cognition/performance/memory
- impaired decision-making
- impaired immunity
- fatigue and work-related accidents
- car accidents
- substance abuse
- suicide risk

Poor sleep can also include too much sleep. Although this is controversial, some researchers suggest that sleeping for prolonged periods may adversely affect health, although perhaps not nearly to the same extent as short sleeping (Youngstedt and Kripke, 2004). The deleterious effects of sleep duration less than 8 h a day appear to be related to depression and low-socioeconomic status (Patel et al., 2006). Hence, it could be argued that the ideal amount of sleep (as with many lifestyle issues) may be like a U-shaped curve, with too little or too much placing an individual at risk of health problems.

SLEEP AND LIFESTYLE INTERACTIONS

As with most topics discussed in this book, there are significant interactions between other components of lifestyle, sleep, and disease. The process is often cyclical, with lifestyle-related problems causing poor sleep patterns, which then make the initial problems worse. Treating lifestyles behavior is therefore necessary to break this cycle and reduce the chances of further ill health (Nam et al., 2015). Research suggests there are two-way interactions between sleep problems and a range of health problems that include obesity, alcohol misuse, and diabetes, for example.

Sleep and Obesity

The findings on sleep and body weight are interesting and challenging. On the one hand, a lack of sleep may be an important trigger for obesity (Bayon et al., 2014). Sleep deprivation can lead to a surge in appetite. In one study (Gangwisch and Heymsfield, 2005), people who sleep only 4 h a night were shown to be 73% more likely to be obese, and those sleeping 5 h have a 50% higher risk of obesity. In another (Cappuccio et al., 2007), those who slept 6 h a night (about 1 h less than the current 7 h average) were 23% more likely to be very overweight.

The reasons for this may be complex. There may be a mechanism humans have evolved to protect themselves against famine in winter, when food is in shorter supply. Sleeping less could act as a trigger to eat more. Sleep deprivation also affects hormone levels associated with appetite and satiation. Leptin is decreased (making it harder to sense satiety) and ghrelin, the stomach hormone related to hunger, is increased. Cortisol levels are also increased with sleep deprivation, and this can encourage weight gain (Spiegel et al., 2005).

Quality and length of sleep may also play an important role in childhood obesity, with short sleep duration adversely influencing weight in children, possibly through the medium of poor diet (Cespedes et al., 2016). Obesity also decreases sleep quality by increasing the risk of sleep apnea and snoring (see following). For these reasons, a sleep review is vital when managing other lifestyle-related health conditions, such as overweight or obesity. Improving sleep quality can help prevent weight gain while adding benefit to other weight loss initiatives.

What Can Be Done in a Standard Consultation?

- ☐ Ask about sleep quality, as well as daytime fatigue and excessive daytime sleepiness.
- ☐ Ask if patients have ever been told they snore.
- ☐ Test for diabetes in cases of unexplained fatigue.
- ☐ Consider psychological factors as a cause of fatigue and insulin resistance as a cause of EDS.
- ☐ If a sleep problem is suspected, get the patient to complete a sleep diary (see professional resources).
- ☐ Explain the link between obesity, inactivity, depression, anxiety, and sleep.
- ☐ Order a home sleep study if apnea is suspected.

Sleep and Alcohol

Alcohol and sleep are not the best bedfellows. Although a nightcap is often seen as a relaxant that will help with sleep, the reality is that alcohol (along with caffeine and nicotine) is a sleep thief. While alcohol may help to induce sleep a little faster (reducing sleep latency), it has a definite downside in that quality of sleep can be adversely affected. Having an alcoholic drink in the hour before bedtime may disrupt the second part of the sleep cycle. Surprisingly, even having a drink much earlier in the evening at "happy hour" (about 6 h before retiring) can still interrupt an evening's sleep. This suggests that alcohol, even after it has been recently metabolized, can exert lasting influences over sleep regulation (Park et al., 2015). After a few hours of sleep, those who have been drinking even modest amounts can suffer rebound wakefulness that leads to difficulty getting back to sleep. The depth of sleep is also adversely affected, leading to fatigue the next day (Thakkar et al., 2015).

Certainly the extent of sleep disturbance is often related to how much alcohol is consumed, but if sleep problems are a concern, even small amounts of alcohol consumption need to be considered. A small amount of alcohol can also uncover accumulated sleep debt, which may make it dangerously easy to lapse into sleep or microsleeps (Kryger et al., 2015). When combined with sleep debt, alcohol becomes a potent sedative that can have dangerous repercussions, for example, in traffic and work injuries (Taylor and Dorn, 2006). A large sleep debt and even a small amount of alcohol can cause fatal fatigue.

Alcohol-dependent sleep disorder is diagnosed when a patient needs alcohol to get to sleep on at least 30 consecutive days. The problem can lead to early wakening, poor sleep quality, and greater alcohol tolerance, leading to more and more alcohol being required to get to sleep. Specialist care is required (Kryger et al., 2015).

Sleep, Exercise, and Inactivity

Initial research on exercise and sleep showed the complexity of this relationship. Sleep following exercise was found to be disturbed. However, the results were influenced by the fact that the subjects used in the research did not exercise regularly. When regular exercisers (joggers) were tested, sleep quality generally improved. Exercise just before bedtime, however, may decrease alertness and decrease sleep capacity; however, once again this may be less so in highly fit individuals who, incidentally, are less likely to present to a clinician with a sleep problem (Kline, 2014).

Data from the US National Health and Nutrition Survey (NHANES III) showed that people with insufficient physical activity were two to four times more likely to report feeling tired or exhausted compared to the group that reported feeling fresh (Resnick, 2006). Exercise in fit individuals has also been found to increase the deepest stages of sleep (Kline, 2014). Excessive exercise, on the other hand, can lead to disturbed and shortened sleep. Patients should be advised that regular exercise will improve sleep as fitness increases but that this should not be carried out in excess and usually not just before retiring.

Sleep and the New Media

Since the publication of the first edition of this book, social media has become much more ubiquitous, so much so that the last National Sleep Survey carried out by the US National Sleep Foundation (2011) has shown the following recent developments that have yet to be fully appreciated in terms of their effects on sleep function:

- Around 40% of Americans (with 72% of 13–18-year-olds and 67% of 19–29-year-olds) use cell phones in the hours and minutes before trying to get to sleep.

- Around 21% of young Americans text almost every night before going to sleep, and those who do text:
  - Are less likely to get a good night's sleep
  - Are more likely to wake up feeling unrefreshed
  - Are more likely to admit to driving drowsy
- The two main responses to drowsiness amongst 17–29-year-olds is drinking caffeine (67% drink an average of 3.1 caffeinated drinks a day) and napping during the day (44%).

The consequences of these relatively new lifestyle patterns have yet to be realized, i.e., in driving injuries and long-term disease patterns.

Sleep and Diabetes

Type 2 diabetes can be a cause of disturbed sleep. More importantly, however, it could result from sleep loss. This association has been shown epidemiologically (Al-Delaimy et al., 2002), but there are also physiological explanations related to the modulatory effect of sleep on glucose metabolism and the molecular mechanisms for the interaction between sleeping and feeding (Spiegel et al., 2005). Restless legs syndrome, a relatively common sleep disorder, is also increased in diabetic patients and those with impaired glucose control, possibly through the activation of small neural fibers associated with peripheral neuropathy (Gemignani et al., 2007). Decreased glucose tolerance and insulin sensitivity have been found with increased sleep deprivation and hunger-related hormones such as leptin and ghrelin change in a way conducive to greater food intake.

Obesity and sleep loss are related and sleep-disordered breathing is linked to obesity; this suggests a feedforward cascade of negative events generated by sleep problems, which are likely to exacerbate the severity of metabolic problems.

Sleep, Bedding, and Air-Conditioning

As discussed in Chapter 28, bedding can have a significant impact on skin, as well as sleep patterns. Overheating under a continental quilt or with an electric blanket, for example, can cause disturbed sleep, sleep debt, consequent daytime sleepiness, and mood changes.

Sleep Disorders

Sleep disorders represent the extreme of sleep-related problems. Over 80 different types of such disorders have now been identified and described in the *International Classification of Sleep Disorders, Diagnostic and Coding Manual* published by the American Sleep Disorders Association. The major classifications are shown in Table 18.1.

Those disorders associated with lifestyle are mainly dyssomnias. The two most common problems considered here are insomnia and obstructive sleep apnea.

TABLE 18.1 Classification of Sleep Disorders

Major Category	Subcategories	Examples
Dyssomnias (Disorders that cause difficulty initiating or maintaining sleep or excessive sleepiness)	Intrinsic sleep disorders (originating in the body)	Obstructive sleep apnea Insomnia Narcolepsy Restless legs syndrome
	Extrinsic sleep disorders (originating outside the body)	Alcohol-dependent sleep disorder Altitude insomnia Inadequate sleep hygiene
	Circadian rhythm sleep disorders	"Jet lag" syndrome Shiftwork sleep disorder Delayed sleep phase syndrome
Parasomnias (disorders of partial arousal or those that interfere with sleep stage transitions)	Arousal disorders	Sleepwalking Night terrors Confusional arousal
	Sleep-wake transition disorders	Sleep talking Nocturnal leg cramps Rhythmic movement disorders
	Parasomnias usually associated with REM sleep	Nightmares Sleep paralysis REM sleep behavior disorder
	Other parasomnias	Sleep bruxism Enuresis Abnormal swallowing syndrome SIDS
Medical and psychiatric sleep disorders	Sleep disorders associated with medical disorders	Alcoholism Nocturnal cardiac ischemia Chronic obstructive pulmonary disease
	Sleep problems associated with neurological disorders	Degenerative brain disorders Sleep-related epilepsy Sleep-related headache
	Sleep disorders associated with psychiatric disorders	Psychoses Mood disorders Anxiety disorders Panic disorders
Proposed sleep disorders (not yet formally categorized as discrete sleep disorders)		Short sleeper Long sleeper Menstrual-associated disorder Sleep choking syndrome

WHAT IS INSOMNIA?

Insomnia is not so much a condition as a symptom. It is defined as difficulty getting to sleep or staying asleep, waking through the night, waking up earlier in the morning, or not getting the amount or quality of sleep needed (Wong and Ng, 2015). Short-term or transient insomnia lasting days or a couple of weeks may be triggered by stress, temporary illness, excitement before an upcoming event, travel, or environmental factors, such as light, heat, or noise. Chronic insomnia lasts 2–3 weeks or more. It can be related to lifestyle issues, beliefs or attitudes about sleep, or an underlying medical problem. The list shown in Table 18.2 contains just some of the factors to consider when someone presents with chronic insomnia.

In order to establish the source of insomnia, it is vital that a full medical history is taken, paying attention to any potential underlying medical conditions, recreational drugs, stimulants, prescription medications, beliefs about sleep, and lifestyle factors. Getting to the root of the problem is vital, as this will govern management of the problem. It is important to determine the pattern of and beliefs about sleep and to work out the impact of sleep deprivation on the

TABLE 18.2 Triggers and Medical Conditions That Can Cause Insomnia

Triggers	Medical Conditions
Attitudes and beliefs about sleep	Sleep-related epilepsy
Poor sleep habits	Cerebral degenerative disorders
Feelings of loss of control about sleep	Thyroid disease
Inadequate exercise	Anxiety/depression
Trying to control sleep	Alcohol abuse
Stress/worrying	Parkinson disease
Lying in bed tense and frustrated	Obstructive sleep apnea
Too much light or noise in the bedroom	Pain disorders
Excess caffeine or stimulant intake	Gastrointestinal diseases
Drugs and alcohol	Hormone imbalances
	Medication side effects
	Urological conditions
	Itching/allergies
	Fibromyalgia

individual. For example, exercise before sleep at the time of arrival at a destination (if this is during the daylight) is a way of dealing with insomnia from jet lag.

For chronic insomnia relating to lifestyle issues, sleeping medications are usually inappropriate because they fail to deal with the underlying triggers for poor sleep. However, for transient insomnia relating to travel or a change in life circumstances, short-term sleeping medication use may be an appropriate option along with lifestyle/sleep habit modification. Careful review, dispelling myths about sleep and providing practical healthy sleep tips, can be vital in improving sleep. Lifestyle change prescriptions such as increasing daytime physical activity, winding down in the evening, eating less, watching caffeine and alcohol intake, trying to avoid accumulating a sleep debt, and using sleep hygiene practices (see Practice Tips) can all help to dramatically improve sleep for many patients.

THE VALUE OF SLEEP DIARIES IN DETERMINING SLEEP PATTERNS

Just as patients are regularly asked about their diet and exercise habits as part of lifestyle medicine, it is also appropriate to review sleep patterns. One of the most effective ways of doing this is to get patients to keep a sleep diary (mental recall alone is notoriously inaccurate when trying to assess a prior week's sleep patterns). Ideally this should include:

- daytime lifestyle factors (exercise, caffeine intake, alcohol and drugs, wind down routine in the hour before bed)
- getting into bed time
- sleep time
- any nighttime wakening
- time awake in the morning
- time out of bed in the morning
- mood on waking (e.g., refreshed, tired, exhausted)

Patterns quickly emerge that can then be targeted with healthy lifestyle changes, counseling, and cognitive behavior therapy. Maintaining the sleep diary can then help chart progress and improvements (see Professional Resources for an example of a sleep diary).

SNORING AND OBSTRUCTIVE SLEEP APNEA

Apnea is a classical Greek word for absence of breath. With sleep apnea, a person stops breathing repeatedly during sleep for at least 10 seconds and sometimes for a minute or more. The most common type of sleep apnea, which has a strong lifestyle link, is obstructive sleep apnea (OSA). With this condition, the soft tissue at the back of the throat collapses and obstructs the airway, which stops or impedes airflow. The sufferer usually snores, then stops breathing. This is followed by arousal, a gasp for breath, and resumed breathing. This exhausting cycle can happen hundreds of times a night, leading to fractured, poor quality sleep along with fatigue and sleepiness the next day.

The exact number of people affected by OSA remains controversial, although some research suggests that up to 20% of people could have clinically significant apnea (National Sleep Foundation, 2011). What is agreed, however, is that the incidence is on the rise, coincident with the increase of one of its most prominent causes, obesity. Because of its link with disease, sleep apnea has even been defined as a manifestation of the metabolic syndrome (Vgontzas et al., 2005).

Sleep apnea is known to increase blood pressure, with decreases associated with continuous positive airway pressure (CPAP) treatment (Yang et al., 2015). However, a more immediate danger is the large sleep debt accumulated by sufferers from their brief (usually unrecognized) wakening. This increases risks from driving or using machinery and is the reason apnea sufferers are excluded from becoming professional drivers in some countries. Other problems common to people with apnea are reflux, nocturnal urination, sweating at night, morning headaches, raspy voice, personality changes, loss of hearing, and even male impotence and reduction of sex drive in men and women (Ting and Malhotra, 2005). Risk factors for apnea identified from the prospective Busselton study in Western Australia are being male, obesity, and weight gain (Knuiman et al., 2006). Development of asthma over time (also associated with obesity and taking up smoking) may also play a role.

Snoring is highly associated with sleep apnea, although not all snorers have OSA (Knuiman et al., 2006). Treatment for OSA and snoring can require surgery, but opening of the airways through CPAP machines has become standard treatment since their invention by Dr. Colin Sullivan at the Royal Prince Alfred Hospital in Sydney in the 1980s (Kaparianos et al., 2006). Another technique, which is cheaper and useful in some cases, involves dental devices to move the lower jaw forward. While highly successful, these techniques should be seen as treating the symptoms, not the cause. Weight loss, a moderate amount of physical activity, and reduction of alcohol before retiring are the primary lifestyle prescriptions in conjunction with these devices where they assist in reducing sleep debt and hence daytime sleepiness (which otherwise reduces the desire to be active during the day). If this does not occur, a vicious circle can be created, leading to greater weight gain and more severe apnea and disease risk.

SLEEP HYGIENE

The following are the key points in an effective patient sleep hygiene program. All should be undertaken together for best effect.

- **Establish a routine**. Babies and children love sleeping routines, but adults also benefit from regular bedtime and wake-up times, including weekends. Sleeping in on the weekend can reset the body clock and lead to a type of weekend jet lag. Sleeping in on a Sunday can mean going to bed later on Sunday night, making Monday morning particularly tough.

- **Wind down before bed**. Particularly in the hour before bed, it is useful to engage in soothing, calming activities. Avoid activities that stimulate, for example, watching TV, searching the Internet, reading emails or newspapers, paying bills, or exercising. Instead, focus on enjoying a warm shower or bath, listening to calming music or relaxing.
- **Keep the bedroom for sleep and sex only**. Avoid working, eating, or other activities in bed and try to keep computers and TV out of the bedroom. Reading, however, may be helpful to help wind down.
- **Get comfortable**. Sleep on a comfortable mattress and pillow. Make sure the bedroom is dark, quiet, and cool. Eyeshades and earplugs may be useful additions.
- **Control caffeine (coffee, tea, cola, and energy drinks) and limit alcohol before bed**. Keep in mind that caffeine can still create a buzz in some people up to 12 hours after consumption. Avoiding caffeine at least in the 8h prior to bed can help improve sleep quality. Alcohol should be limited to one to two drinks for women and three to four drinks for men, in line with other recommendations for alcohol consumption.
- **Get regular physical activity**. Regular daily exercise promotes sleep. Avoid exercise in the few hours prior to bed, however, because the adrenaline released increases alertness.
- **Finish dinner at least a couple of hours before bed**. Eating too late or eating spicy foods (that trigger heartburn) or heavy meals can interfere with sleep.
- **Avoid nicotine**. Nicotine is a stimulant, so smoking close to bedtime can make it more difficult to fall asleep.
- **Avoid long daytime naps if insomnia is a problem**. Napping for more than 20–30min may disrupt evening sleep. Shorter naps, however, may be advantageous.
- **Develop healthy thinking about sleep**. Controlling worry and "racing thoughts" through relaxation practices can facilitate sleep.

SUMMARY

While the reasons for our physiological need for sleep are unknown, the need for a consistent amount is unquestioned. And while there are big individual differences in sleep requirements, less than around 7 h or more than around 8.5 h appear to have negative health outcomes for the majority of the population. Sleep problems can begin as mild forms of insomnia or hypersomnia, often associated with lifestyle habits, but can also include serious disorders or signs of other problems, such as snoring and sleep apnea, which are linked to heart disease. At the even more serious end of the spectrum are major disturbances such as narcolepsy or night terrors. Sleep (or, more usually, lack of it) is thus a significant cause of modern lifestyle-related ill health and should be included in any considered lifestyle analysis.

Practice Tips

Improving Sleep Quality

	Medical Practitioner	Practice Nurse/Allied Health Professional
Assess	• sleep as part of a routine checkup • underlying lifestyle factors including exercise, eating patterns, and stress levels • underlying medical conditions such as depression or thyroid disease • use of sleeping medications including over-the-counter preparations • intake of drugs, alcohol, stimulants, and caffeine	• sleep as part of a routine checkup • underlying lifestyle factors including exercise, eating patterns, and stress levels
Assist	• by recommending a sleep diary • by providing tips for healthy sleep habits (see below) • by reviewing sleeping medication and reducing/eliminating it where appropriate	• by explaining and assisting in completion of a seven-day sleep diary
Arrange	• referral for treatment of comorbidities where appropriate • list of other health professionals for a care plan • sleep specialist review if further assessment required • sleep laboratory or home monitoring investigations if a disorder such as sleep apnea is suspected • follow-up to monitor sleep improvements (first follow-up session with the seven-day sleep diary completed) and a timetable for ongoing review • psychologist referral for cognitive behavior therapy if appropriate	• a discussion of tactics with other involved health professionals • a sleep therapist, psychologist, or life coach if appropriate • contact with self-help groups

Key Points

Clinical Management

- Consider sleep as an issue in other lifestyle-related illnesses, for example, obesity, diabetes and fatigue.
- Discuss sleep-related problems such as snoring, early wakening, late onset to sleep, and regular wakening.
- Relate excessive regular daytime sleepiness with sleep problems/patterns.
- Consider bedding, air-conditioning, and other environmental factors as potential contributors to poor sleep.

REFERENCES

Al-Delaimy, W.K., Manson, J.E., Willett, W.C., et al., 2002. Snoring as a risk factor for type II diabetes mellitus: a prospective study. Am. J. Epidem. 5, 387–393.

Bhattacharyya, N., 2015. Abnormal sleep duration is associated with a higher risk of accidental injury. Otolaryngol. Head Neck Surg. 153 (6), 962–965.

Bayon, V., Leger, D., Gomez-Merino, D., Vecchierini, M.F., Chennaoui, M., 2014. Sleep debt and obesity. Ann. Med. 46 (5), 264–272.

Cappuccio, F., Taggart, F., Kandala, N.-B., et al., 2007. Short sleep duration and obesity: meta-analysis of studies world-wide. In: Paper Presented to 5th Congress of World Federation of Sleep Research, Cairns, Australia, 2007.

Cespedes, E.M., Hu, F.B., Redline, S., Rosner, B., Gillman, M.W., Rifas-Shiman, S.L., Taveras, E.M., 2016. Chronic insufficient sleep and diet quality: contributors to childhood obesity. Obesity (Silver Springs) 24 (1), 284–290.

Dawson, D., Reid, K., 1997. Fatigue, alcohol and performance impairment. Nature 388 (6639), 235.

Dement, W.C., 2000. The Promise of Sleep. Delacort Press, NY.

Dhand, R., Sohal, H., 2006. Good sleep, bad sleep! the role of daytime naps in healthy adults. Curr. Opin. Pulm. Med. 12 (6), 379–382.

Gangwisch, P., Heymsfield, 2005. Columbia University's Mailman School of Public Health and the Obesity Research Centre, Analysis of National Health and Nutrition Examination Survey 1. NHANES 1 28 (10), 1289–1296.

Gemignani, F., Brindani, F., Vitetta, F., et al., 2007. Restless legs syndrome in diabetic neuropathy: a frequent manifestation of small fiber neuropathy. J. Periph. Nerv. Syst. 12 (1), 50–53.

Haack, M., Mullington, J.M., 2005. Sustained sleep restriction reduces emotional and physical well-being. Pain 119 (1–3), 56–64.

Irish, L.A., Kline, C.E., Gunn, H.E., Buysse, D.J., Hall, M.H., 2015. The role of sleep hygiene in promoting public health: a review of empirical evidence. Sleep Med. Rev. 22, 23–36. http://dx.doi.org/10.1016/j.smrv.2014.10.001.

Kaparianos, A., Sampsonas, F., Karkoulias, K., et al., 2006. The metabolic aspects and hormonal derangements in obstructive sleep apnea syndrome and the role of CPAP therapy. Eur. Rev. Med. Pharmacol. Sci. 10 (6), 319–326.

Kline, C.E., 2014. The bidirectional relationship between exercise and sleep: implications for exercise adherence and sleep improvement. Am. J. Lifestyle Med. 8 (6), 375–379.

Knuiman, M., James, A., Divitini, M., et al., 2006. Longitudinal study of risk factors for habitual snoring in a general adult population: the Busselton health study. Chest 130 (6), 1779–1783.

Koren, K., O'Sullivan, K.L., Mokhlesi, B., 2015. Metabolic and glycemic sequelae of sleep disturbances in children and adults. Curr. Diab. Rep. 15 (1):562.

Kryger, M.H., Roth, T., Dement, W.C., 2015. Principles and Practice of Sleep Medicine, sixth ed. Elsevier, Pennsylvannia, PA.

Marshall, N.S., Glozier, N., Grunstein, R.R., 2008. Is sleep duration related to obesity? A critical review of the epidemiological evidence. Sleep Med. Rev. 12 (4), 289–298.

McEwan, B.S., 2006. Sleep deprivation as a neurobiological and physiologic stressor: allostasis and allostatic load. Metabolism 55 (10 Suppl. 2), S20–S23.

Nam, S., Stewart, K.J., Dobrosielski, D.A., 2015. Lifestyle intervention for sleep disturbances among overweight or obese individuals. Behav. Sleep Med. Sep 16, 1–8.

Naska, A., Oikonomou, E., Trichopoulou, A., et al., 2007. Siesta in healthy adults and coronary mortality in the general population. Arch. Intern. Med. 167, 296–301.

National Sleep Foundation, 2011. 'Sleep in America' Poll. National Sleep Foundation, Washington, DC.

Park, S.Y., Park, S.Y., Oh, M.K., Lee, B.S., Kim, H.G., Lee, W.J., Lee, J.H., Lim, J.T., Kim, J.Y., 2015. The effects of alcohol on quality of sleep. Korean J. Fam. Med. 36 (6), 294–299.

Patel, S.R., Malhotra, A., Gottlieb, D.J., et al., 2006. Correlates of long sleep duration. Sleep 29 (7), 881–889.

Resnick, H.E., 2006. Cross-sectional relationship of reported fatigue to obesity, diet and physical activity: results from the Third National Health and Nutrition Examination Survey. J. Clin. Sleep Med. 2, 163–169.

See, C.Q., Mensah, E., Olopade, C.O., 2006. Obesity, ethnicity, and sleep-disordered breathing: medical and health policy implications. Clin. Chest Med. 27 (3), 521–533.

Spiegal, K., Knutson, K., Leprault, R., et al., 2005. Sleep loss: a novel risk factor for insulin resistance and Type 2 diabetes. J. Appl. Physiol. 99 (5), 2008–2019.

Suzuki, H., Uchiyama, M., Aritake, S., Kuriyama, K., Kuga, R., Enomoto, M., Mishima, K., 2012. Alpha activity during rem sleep contributes to overnight improvement in performance on a visual discrimination task. Percep. Mot. Skills 115 (2), 337–348.

Taylor, A.H., Dorn, L., 2006. Stress, fatigue, health, and the risk of road traffic accidents among professional drivers: the contribution of physical inactivity. Annu. Rev. Public. Health 27, 371–379.

Thakkar, M.M., Sharma, R., Sahota, P., 2015. Alcohol disrupts sleep homeostasis. Alcohol 49 (4), 299–310.

Ting, L., Malhotra, A., 2005. Disorders of sleep: an overview. Prim. Care 32 (2), 305–318.

Vargas, P.A., February 13, 2016. The link between inadequate sleep and obesity in young adults. Curr. Obes. Rep. 5 (1).

Vgontzas, A.N., Bixler, E.O., Chrousos, G.P., 2005. Sleep apnoea is a manifestation of the metabolic syndrome. Sleep Med. Rev. 9 (3), 211–224.

Vgontzas, A.N., Bixler, E.O., Chrousos, G.P., 2006. Obesity-related sleepiness and fatigue: the role of the stress system and cytokines. Ann. N.Y. Acad. Sci. 1083, 329–344.

Vyazovskiy, V.V., 2015. Sleep, recovery, and metaregulation: explaining the benefits of sleep. Nat. Sci. Sleep 7, 174–184.

Wong, S.H., Ng, B.Y., 2015. Review of sleep studies of patients with chronic insomnia at a sleep disorder unit. Singapore Med. J. 56 (6), 317–323.

Yang, M.C., Huang, Y.C., Lan, C.C., Wu, Y.K., Huang, K.F., 2015. Beneficial effects of long-term CPAP treatment on sleep quality and blood pressure in adherent subjects with obstructive sleep apnea. Respir. Care 60 (12), 1810–1818.

Youngstedt, S.D., Kripke, D.F., 2004. Long sleep and mortality: rational for sleep restriction. Sleep Med. Rev. 8 (3), 159–174.

Other Useful Resources

National Sleep Foundation, USA (www.sleepfoundation.org) a nonprofit independent organization dedicated to research and providing information to the general public; sleep diaries can be downloaded; online questionnaires regarding sleepiness and general information are available.

University of Newcastle Sleep Disorders Centre (www.newcastle.edu.au).

American Academy of Sleep Medicine (www.aasmnet.org).

Professional Resources

Measuring Sleep

Assessing Sleep Problems

To work out how successful a patient is at meeting his or her sleep requirements, try asking these simple questions:

- How many hours a night do you sleep throughout the week?
- Do you fall asleep the instant your head hits the pillow?
- Do you struggle to get out of the bed in the morning?
- Do your eyelids droop (or do you even drop off) during a boring presentation or meeting at work?
- Do you sleep extra on weekends to catch up?
- Do you rely on coffee to stay alert?
- Do you need an alarm to wake up in the mornings?
- Do you regularly use the snooze function?
- Do you often feel drowsy while driving?
- Do you have trouble concentrating or remembering?

If the answer to some of these questions is yes, the patient may be getting insufficient sleep or have a sleep debt. While nodding off during a meeting may not be life threatening, falling asleep at the wheel definitely is—heeding warning signs of sleepiness and fatigue is vital.

Briefly Assessing for Sleep Apnea

Two useful questions to determine whether someone should be tested for sleep apnea are:

- Are you excessively tired during the daytime?
- Have you been told that you snore?

A Sample Sleep Diary

A sleep dairy is an excellent way of monitoring sleep patterns and charting progress. To begin, the diary should be kept for a minimum of two weeks to establish initial sleep patterns. Maintaining the diary helps keep track of progress.

	Monday	**Tuesday**	**Wednesday**	**Thursday**	**Friday**	**Saturday**	**Sunday**
What time did you go to bed?							
How long did it take to fall asleep?							
How many times through the night did you wake up?							

Continued

Professional Resources—cont'd

	Monday	Tuesday	Wednesday	Thursday	Friday	Saturday	Sunday
What time did you get up?							
How many hours did you sleep in total?							
Caffeine (number of drinks and time consumed)							
Alcohol (number of drinks and time consumed)							

Chapter 19

Health and the Environment: Clinical Implications for Lifestyle Medicine

Garry Egger, Andrew Binns, Stephan Rössner

Anything else you're interested in is not going to happen if you can't breathe the air and drink the water.

Carl Sagan

"Environment" (Gk, *en*, in: L, *viron*, circle). All of the many factors, as physical and psychologic, that influence or affect the life and survival of a person. See also **biome**, **climate**.

INTRODUCTION: ENVIRONMENTAL INFLUENCES ON HEALTH

The influence of the environment is paramount in public health, yet is often downplayed at the clinical level. The classic epidemiological triad (Fig. 19.1), for example, considers host, vector, and environment as vital to the assessment of disease outbreaks, whether of infectious or chronic disease. And while intervention strategies are generally focused on public health, the interaction with host strategies cannot be ignored at the clinical level. As shown in Fig. 19.1, education, behavior change, and clinical practices deal predominantly with the host, although multidisciplinary practices such as shared care and shared medical appointments might also be expected to impact on other corners of the triad. Technology deals best with the agent of disease (e.g., energy imbalance in the case of obesity) and its vectors (excessive energy intake and/or inadequate energy expenditure), and policy and social change are needed to cope fully with the environment.

Politics and the Environment

In the context of classical political dialog, influences on lifestyle-related disease are often considered from either a distinctly right- or left-wing perspective.

Lifestyle Medicine. http://dx.doi.org/10.1016/B978-0-12-810401-9.00019-X

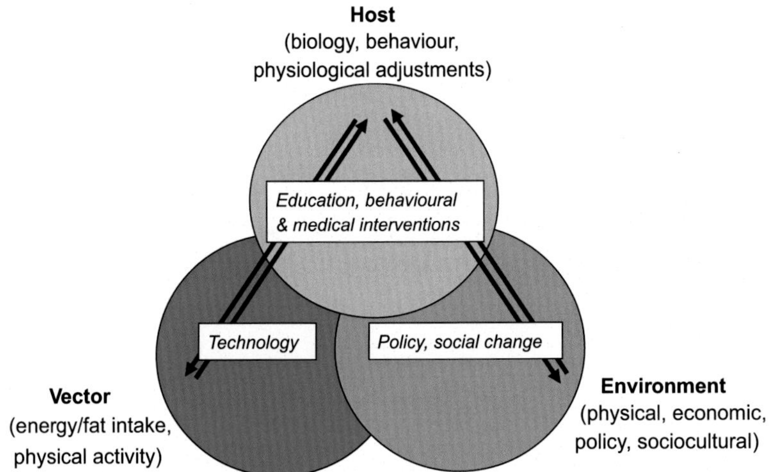

FIGURE 19.1 The epidemiological triad and approaches to intervention. *From Egger, G., Swinburn, B., Rossner, S., 2003. Dusting off the epidemiological triad: Could it apply to obesity. Obes. Rev. 4(2), 115–120.*

From the right, the responsibility of the individual is paramount, irrespective of the influences of the environment, policy, government action, or social and economic status. From the left, it is these latter factors that need to be held to account with individual responsibility of little consequence in the weight of such external pressures, making individual choice under the circumstances difficult, if not impossible.

In reality, both views have potency, although the relative influence of each can be dynamic. While ultimate charge for an individual's actions falls on the individual, or host, such actions are strongly dependent on the power of both the agent and vectors driving these, and the environments in which they exist. In the case of type 2 diabetes, for example, it is ultimately up to the individual, irrespective of his or her genetic burden, to resist the drive to overuse tasty fattening foods and effort-saving transport devices, both of which can be determinants of the disorder. However, if the environment is conducive or even obligatory toward behaviors antithetical to this, social responsibility—at least at the population level—is difficult to ensure.

More telling even is the dynamism between these extremes. As was once pointed out by Winston Churchill: "…first, we change the environment. Then the environment changes us." This has been particularly evident since the advent of the industrial, and now the technological, revolutions, which have widened the abundance of choice, while paradoxically restricting the need for restraint.

So where does this leave the clinician in managing environmentally induced lifestyle-related disease? In the first instance, we need to define the types of environments that may or may not influence unhealthy behaviors, then consider the influence a clinician might have on any of these, and ultimately potential disease outcomes.

DEFINING ENVIRONMENTS

There are many different types of environments and environmental influences on health and/or health-related behavior. Swinburn et al. (1999) and Egger et al. (2003) differentiated environments relating to obesity within a model called ANGELO (Analysis Grid for Environments Leading to Obesity). They elaborated four influential environmental "types" or sectors and four environmental sizes or "settings," as shown in Table 19.1.

Individuals interact with the environment in multiple micro (local) environments, or settings, including schools, workplaces, homes, and neighborhoods. Microenvironmental settings, in turn, are influenced by the broader macroenvironments, or sectors (such as the education and health systems, all levels of government, the food industry, and society's attitudes and beliefs) and these tend to be less amenable to the influence of individuals.

TABLE 19.1 Sizes and Types of Environments With Potential Influence on Human Health

Environment Size	Microenvironment (Settings)	Macroenvironment (Sectors)
Environment Type		
Physical	What is or is not available?	
Policy	What are the rules?	
Sociocultural	What are the attitudes, beliefs, perceptions, and values?	
Economic	What are the financial factors?	

As this text is primarily about the clinical aspects of Lifestyle Medicine, broader public health initiatives are not included. In relation to the environment this leaves little scope for action on the part of the individual clinician. On the other hand, it should encourage the clinician to the view, as expressed by Rose (1992), that "all health is politics," and that involvement in public health, at some level, should also be within the ambit of the medical and allied health professional practitioner. Elements of the environment that are particularly relevant include the following.

The Physical Environment

The physical macroenvironment ("what is or is not available?") includes not only the visible world (food, physical activity, behavioral choices, etc.) but also less tangible factors such as the availability of educational training opportunities, nutrition and exercise expertise, technological innovations, information, and food labels. Some factors such as the weather, climate change, or terrain may be important determinants of behavior but because they are not amenable to personal influence they are of little value to clinicians.

Endocrine disrupting chemicals (EDCs) are defined as exogenous chemicals or a mixture of chemicals that can interfere with any aspect of hormone action (Zoeller et al., 2012). EDCs are demonstrated contributors to infertility, premature puberty, endometriosis, and other disorders and have been implicated in metabolic syndrome and obesity (Casals-Casas and Desvergne, 2011). Because adipose (fat) tissue is a true endocrine organ, and is highly susceptible to disturbances by EDCs (e.g., hunger, metabolism), a subset of these have been labeled "Obesogens" with a proposed direct influence on body weight (Janesick and Blumberg, 2016). Other potential chemical contributors to disease in the modern environment are shown in Table 19.2.

EDCs like those shown in Table 19.2 may not be able to be influenced by individuals, but the effects can often be measured and certain lifestyle actions taken to minimize the effects in the body (Genuis, 2011, 2012; Sears and Genuis, 2012).

TABLE 19.2 Known Endocrine Disrupting Chemical (EDC) Contributors to Chronic Disease

- Toxic elements (e.g., arsenic, cadmium, lead, mercury)
- Naturally occurring substances (e.g., molds, animal/plant/food allergens)
- Pesticides (e.g., pesticides, herbicides, fungicides)
- Persistent organic pollutants (such as dioxins, polychlorinated biphenyls, dichlorodiphenyltrichloroethane)
- Volatile organic compounds (solvents, petrol, fragrances, benzene)
- Plastics (generally inert unless dissolved or burnt to release, e.g., bisphenols)

From Casals-Casas, C., Desvergne, B., 2011. Endocrine disruptors: from endocrine to metabolic disruption. Ann. Rev. Physiol. 73, 135–162.

Prescriptive initiatives encompass risk recognition and chemical assessment, then exposure reduction, remediation, monitoring, and avoidance. Applicable strategies include nutrition and supplements to counter toxic effects and to support metabolism, as well as exercise and sweating, and possibly medication to enhance excretion (Table 19.3). Early exposure to EDCs can lead to later dysregulated inflammation or "metaflammation" that can influence long-term chronic disease (Dietert, 2012).

The physical microenvironment includes foods, dust and other irritants, and pollutants available in the household, schools, workplaces, and immediate neighborhood. It can also include existence and positioning of unhealthy foods and chemicals, availability of effort-saving devices and access to passive leisure time activities such as TVs, videos, and small screen technology, which, if used excessively, can reduce physical activity levels and encourage obesity. Clinical advice on purchasing and placement of unhealthy and energy-dense foods, and the prescription "…if you can't stand the heat get out of the kitchen" in a workplace inducing stress, are examples of possible clinical influences at this level.

The Invironment

At an even more micro level is the recently discovered influence of the gut microbiome within an individual (sometimes called the invironment) as shaped by the external (food, stress, exercise) environment (Konkel, 2013). Although largely influenced by those lifestyle factors previously enumerated here, dysmetabolism of the microbiome, which is often a result of a modern Western

TABLE 19.3 A Seven Step Clinical Approach to Reduction of Endocrine Disrupting Chemicals (EDCs)

1. A detailed environmental history and exposure inventory, followed by avoidance
2. Specific toxicological testing if indicated according to the patient history
3. Remediation of abnormal biochemistry, as identified by laboratory investigations
4. A combination of optimal diet and supplemental nutrients, e.g.:
    a. Glutathione and sulfur-containing foods (eggs and alliums)
    b. Folate (for arsenic metabolism and excretion)
    c. Lipoic acid therapy as protection against mold toxins
    d. Adequate minerals (calcium, iron, zinc) to reduce absorption of heavy metals
    e. Fiber to reduce absorption and facilitate elimination
5. Regular sweating to facilitate transdermal excretion
6. Daily exercise
7. Directed therapies (as determined by lab tests) for retained toxicants

From Sears, M.E., Genuis, S.L., 2012. Environmental determinants of chronic disease and medical approaches: recognition, avoidance, supportive therapy, and detoxification. J. Environ. Public Health. http://dx.doi.org/10.1155/2012/356798.

lifestyle (Clemente et al., 2015), can mediate and maybe even perpetuate illness associated with such lifestyles (Lam et al., 2011). Correction is usually through lifestyle change (and particularly diet, although some supplements [e.g., pre- and probiotics] may offer some assistance, despite the "shotgun" approach to this and the current lack of significant evidence for it). Fecal microbial transplants have shown promise with some gut-related diseases and, interestingly, with reduction of an obesogenic microbiome profile in overweight animals (Duca et al., 2014), which may, in the long run, help reduce obesity in humans.

What Can Be Done in a Brief Consultation?

- ☐ Determine and delineate any environmental influences on health.
- ☐ Discuss potential (micro) environmental influences with patients and parents.
- ☐ Consider and discuss potential influences on obesity of the home and local environments.
- ☐ Consider an assay of the gut microbiome if dysmetabolism is suspected.

Other Environments Relevant to Clinical Practice

Economics at the macro level plays a huge and often unrecognized part in human health, from the overconsumption required (beyond a point) to maintain economic growth (Egger and Swinburn, 2010), to the policies instituted by governments to provide incentives and disincentives for healthy and unhealthy behaviors. Few of these, however, have practical implications at the clinical level for Lifestyle Medicine practitioners outside their public health implications. Similarly, the microeconomic environment, which refers to the costs related to products and activities that may affect health, such as food, physical activity, leisure, etc., offer opportunities for public policy aimed at behavior change, while having little prescriptive relevance for the clinician.

The *sociocultural environment* principally refers to a community or society's attitudes, beliefs, and values related to health influences such as food and physical activity. These social and cultural norms, which are influenced by gender, age, ethnicity, traditions, religion, and subgroup affiliations, have a powerful effect on the behavior of individual members of the community group. While difficult to modify at the individual level, some cultural mores, such as the attraction of feasting and the effects of obesity among Pacific Islanders or the taboo on women walking in public within some religious groups, or any form of physical activity in pregnancy, may need to be challenged under certain conditions.

Policy environments relate to the "rules" determined with specific populations (at the macro level) or families, schools, or neighborhoods at the micro level. Family rules for children, such as consumption and overconsumption of certain foods, overuse of small screen technology, and drug and alcohol use, can be discussed with parents in clinical consultations for possible implementation in the home.

REFERENCES

Casals-Casas, C., Desvergne, B., 2011. Endocrine disruptors: from endocrine to metabolic disruption. Ann. Rev. Physiol. 73, 135–162.

Clemente, J.C., Perhrsson, E.V., Blaser, M.J., et al., April 3, 2015. The microbiome of uncontacted Amerindians. Sci. Adv. 1 (3), pii: e1500183.

Dietert, R.R., 2012. Misregulated inflammation as an outcome of early-life exposure to endocrine-disrupting chemicals. Rev. Environ. Health 27 (2–3), 117–131.

Duca, F.A., Sakar, Y., Lepage, P., Devime, F., Langelier, B., Dore, J., Covasa, M., 2014. Replication of obesity and associated signaling pathways through transfer of microbiota from obese-prone rats. Diabetes 63 (5), 1624–1636.

Egger, G., Swinburn, B., Rossner, S., 2003. Dusting off the epidemiological triad: could it apply to obesity. Obes. Rev. 4 (2), 115–120.

Egger, G., Swinburn, B., 2010. Planet Obesity: How we are eating ourselves and the planet to death. Allen & Unwin, Sydney.

Genuis, S.J., 2011. Elimination of persistent toxicants from the human body. Hum. Exp. Toxicol. 30 (1), 3–18.

Genuis, S.J., 2012. What's out there making us sick? J. Environ. Public Health. Article ID 605137, 10 pages. http://dx.doi.org/10.1155/2012/605137.

Janesick, A.S., Blumberg, B., 2016. Obesogens: an emerging threat to public health. Am. J. Obstet. Gynecol. http://dx.doi.org/10.1016/j.ajog.2016.01.182.

Konkel, L., 2013. The environment within: exploring the role of the gut microbiome in health and disease. Environ. Healh Perspect. 121 (9), 81.

Lam, Y., Mitchell, A., Holmes, A., Denyer, G., Gummesson, A., Caterson, I., Hunt, N., Storlien, L., 2011. Role of the gut in visceral fat inflammation and metabolic disorders. Obesity 19 (11), 2112–2113.

Rose, G., 1992. A Strategy for Public Health. Oxford University Press, Oxford, UK.

Sears, M.E., Genuis, S.L., 2012. Environmental determinants of chronic disease and medical approaches: recognition, avoidance, supportive therapy, and detoxification. J. Environ. Public Health. http://dx.doi.org/10.1155/2012/356798.

Swinburn, B., Egger, G., Raza, F., 1999. Dissecting obesogenic environments: the development and application of a framework for identifying and prioritizing environmental interventions for obesity. Prev. Med. 29, 563–570.

Zoeller, R.T., Brown, T.R., Doan, L.L., et al., 2012. Endocrine-disrupting chemicals and public health protection: a statement of principles from the Endocrine Society. Endocrinology 153, 4097–4110.

Chapter 20

Meaninglessness, Alienation, and Loss of Culture/Identity (MAL) as Determinants of Chronic Disease

John Stevens, Andrew Binns, Bob Morgan, Garry Egger

He who has a why to live, can bear almost any how.

Friedrich Nietzsche

INTRODUCTION: "MAL" IN A LIFESTYLE CONTEXT

In considering a structure for Lifestyle Medicine in Chapter 4, the acronym NASTIE MAL ODOURS was used to define a comprehensive list of lifestyle and environmental determinants of chronic disease. Most of these are physical and apparent to anyone dealing with lifestyle-related problems. Most have been covered elsewhere in the book and the evidence for a unifying link through metaflammation presented. The MAL component, however, is more psychological and less apparent. This includes the effects of a loss of **M**eaning in life, personal **A**lienation at the individual and/or community level, and a **L**oss of culture or identity. Although there is accumulating evidence in some of the cultural literature of a link between each of these and disease, and causal associations in the epidemiological literature, there is to date only limited empirical evidence to be found in the medical literature. There is even less evidence of a direct link between metaflammation and these factors. In the current chapter we consider the available evidence on each of these determinants of chronic disease and propose suggestions for dealing with these within a Lifestyle Medicine context.

THE MEANING OF MEANINGLESSNESS

While it has probably always been intuitively recognized as important for health, Austrian psychiatrist Victor Frankl brought the concept of meaning to the fore in health and well-being in his seminal 1959 book *Man's Search for*

Lifestyle Medicine. http://dx.doi.org/10.1016/B978-0-12-810401-9.00020-6

Meaning (Frankl, 1959). Imprisoned in a WWII German concentration camp, Frankl found that the survival of fellow Jewish prisoners was related to the extent individuals had meaning or purpose in their life. This could be, but was not necessarily associated with, spiritual or religious beliefs. It could also spring from attachments to family, culture, occupation, interests, etc. In Frankl's case, his survival depended on the meaning he attained through developing a psychotherapeutic theory around the significance of this concept called *logotherapy* (from the Greek "logos" or meaning).

A growing body of literature now supports Frankl's view associating meaning with health (Roepke et al., 2014). The ability to derive meaning or purpose in life is associated with better physical health, reduced risk of suicide (Kleiman and Beaver, 2013), stroke (Kim et al., 2013), and myocardial infarction (Kim et al., 2012), as well as reduced overall mortality (Hill and Turiano, 2014) and psychological well-being when faced with chronic pain (Dezutter et al., 2015). Stigma and purpose in life are associated with depression, which can act as a mediator for other health problems (Davis et al., 2010). Further work has linked purpose to physiological changes (Kim et al., 2014), including reduced inflammatory markers (Morozink et al., 2010). Associations of metaflammation with quality of life (Nowakowski, 2014) and other biomarkers (Shankar et al., 2011) also suggest this would be the case.

In one study, people with more purpose were found to have better patterns of health care utilization, hence possibly explaining better health outcomes (Kim et al., 2014). Lack of meaning is linked to behavioral factors, such as smoking, stress, inactivity, poor nutrition, etc., that impact risk of poor physical outcomes. According to Roepke et al. (2014), meaning can mediate the relationship between other variables and physical health as well as be mediated by other factors in its relationship to physical health. Presence of a sense of meaning has also been found to be a robust measure of healthy lifestyle behaviors in European adolescents (Brassai et al., 2015).

Importantly, meaning, along with other components of psychological well-being, has become the focus of several intervention studies (Ruini and Fava, 2012) including classical logotherapy (Robatmili et al., 2015), designed to improve a person's life outlook, thus providing a potential point of intervention for improving health outcomes in a Lifestyle Medicine context. Interestingly, those at risk who escape Meaninglessness show themselves to have a social cohesion and resilience. They can make sense of what is happening to them and can respond.

ALIENATION AND ESTRANGEMENT

Alienation is a form of withdrawal or estrangement that can result from many different experiences—discrimination (Beatty et al., 2014), social isolation (Lacey et al., 2014), or rejection from friends, parents, family, peers, or society.

At a less obvious level it can result from the alienation from society that many feel, although may not recognize. In the words of Scottish activist Jimmy Reid:

> *It is the cry of men who feel themselves the victims of blind economic forces beyond their control. It's the frustration of ordinary people excluded from the processes of decision making. The feeling of despair and hopelessness that pervades people who feel with justification that they have no real say in shaping or determining their own destinies.*
>
> Reid (1972)

Mediating factors such as psychosocial stress (Ketterer et al., 2011; Johnson et al., 2013), loneliness (Hackett et al., 2012), and loss of control (Hamer and Chida, 2011; Hostinar et al., 2014) may play a part in the link shown in these studies between alienation, inflammation, and long-term chronic disease outcomes. Experimental trials have even shown a neural link with metaflammatory processes caused by social rejection (Slavich et al., 2010).

Of particular interest is the growing evidence relating to the effects of adverse childhood experiences (ACEs) on long-term alienation, metaflammation, and chronic disease. A groundbreaking survey in the United States of over 17,000 predominantly middle-class American adults by the Centers for Disease Control and health care provider Kaiser Permanent (CDC, 2014; Felitti and Anda, 2009) showed that more than 1 in 8 (12.6%) had experienced four or more ACEs on a 10 point ACEs scale (see Professional Resources), with a close relationship between ACE scores and health outcomes 50 years later. More importantly, the relationship was strikingly dose dependent. Chronic obstructive pulmonary disease was 2.5 times higher in those with an ACE score of >4; depression 4 times; heart disease 3 times, and suicide 12 times. Population-attributable risk figures (the proportion of a problem in a population that can be attributed to specific risk factors) showed that 54% of current depression and 58% of suicide attempts in women in the United States sample could be ascribed to ACEs.

The original ACE study came from observations made during a big weight loss study in the United States. The study had a high dropout rate, particularly among obese patients who had *successfully* lost weight. Exploring the reasons for this, the researchers found that obesity is often used as a shield against sexual abuse, and hence is often caused by unconscious or occasionally conscious behaviors like overeating and inactivity that were put in place as solutions to negative early experiences.

Part of the outcome effects of ACEs is undoubtedly caused by apparently innocuous lifestyle factors like poor nutrition and inactivity, but also by more self-harming but self-medicating behaviors like smoking, alcohol and drug abuse, or sexual promiscuity that result from alienation and psychosocial stress. These are dangerous enough in themselves, but there are also physiological links with altered brain structure in a highly plastic developing brain, and effects on the hypothalamic–pituitary–adrenal axis, which can affect later health. Metaflammatory responses are more common in later adulthood among children experiencing adverse early experiences (Lacey et al., 2013), possibly because of alterations to the immune system as a result of trauma in early life (Beaumeister et al., 2015). At least one study has shown that this can be reversed in low socioeconomic status (SES) youth through family-oriented psychosocial intervention (Miller et al., 2006).

What Can Be Done in a Standard Consultation?

- ☐ Actively listen to your patient's story.
- ☐ Be prepared to sensitively ask them about their childhood.
- ☐ It is good practice to ask "what happened to you?"
- ☐ Feedback to the patient that you understand, believe them, and show empathy for their anguish.
- ☐ Ask if the patient would be interested in completing the ACE questionnaire to get an ACE score of severity.
- ☐ Refer to a trauma-informed psychologist or counselor experienced in this field.
- ☐ Consider recommending other therapies such as art therapy or activities that may help them find meaning in their lives.
- ☐ Consider using antidepressants in select cases while being aware of the limitations of this approach.

LOSS OF CULTURE AND IDENTITY

Loss of culture and/or identity arise from dispossession, displacement, conflict, climatic events, or natural disasters and result in loss of purpose or alienation and the effects of these discussed earlier.

The effects of loss of culture are seen most evidently in indigenous First Nation cultures such as in Australia (Browne-Yung et al., 2013; Morgan, 2006), the United States, or Canada (Brown et al., 2012) where dispossession and cultural disruption is most marked. Close and complex spiritual links with the land mean that environmental dispossession has negative consequences for health, which are only just beginning to be understood (Richmond and Ross, 2009).

Displacement of communities, often through conflict (Doocy et al., 2013; Daoud et al., 2012) but also climatic events (Shehab et al., 2008) or natural disasters (Parker et al., 2015), also typically result in adverse health outcomes often associated with disruption of culture and reduced social, cultural, and economic capital (Greene et al., 2011). At least one study has shown that loss of self-control in these situations can lead to increased metaflammatory processes associated with an increased prevalence of chronic disease (Hostinar et al., 2014).

Despite the paucity of data on this in the health literature, there are indications of connections with chronic disease outcomes in several population groups (Doocy et al., 2013; Shehab et al., 2008).

SUMMARY

This chapter has briefly touched the surface of some of the deep psychological determinants of chronic disease. These have been recognized for some time as having adverse effects, but have only recently been linked with chronic disease outcomes. Referring back to Table 4.2, their medial determinants may be through psychosocial stress. However, their more proximal effects may be through other determinants of the NASTIE MAL ODOURS acronym: poor nutrition, less activity, more risk taking, unsafe work places, less sleep, more cigarettes and alcohol consumption, and fewer self-preservation skills, which in turn appear to lead to increased inflammation and morbidity.

THE CLINICIAN'S ROLE

While the clinical options for dealing with these psychological determinants of disease seem limited, they should not be ignored. Experiences with logotherapy in those with lack of purpose, art therapy, parenting education in lower SES children, developed resilience of those at risk from traumatic events, and simply recognizing and sympathetically discussing the issue with ACE-affected patients, offer a possible way forward.

New procedures in Lifestyle Medicine such as shared medical appointments (SMAs) as discussed in Chapter 4, and shown to be popular particularly in underserved groups (Stevens et al., 2016), may help ease the situation. Most importantly, the process of Lifestyle Medicine should teach us that attention should be diverted from treating the disease to treating the individual, in some cases capitalizing on existing cultural and community strengths that have otherwise been lost.

Practice Tips

Meaningless, Alienation, and Loss of Culture

	Medical Practitioner	Practice Nurse/Allied Health Professional
Assess	• Early experiences (if appropriate) • The influence of early trauma • Discuss childhood health status	• By screening • Collect background information • Early experiences
Assist	• By active listening • Discovery of psychological "blocks" • In finding appropriate support groups	• By listening nonjudgmentally • By providing emotional support • Using internet support
Arrange	• Referral to remedial programs if possible • Art therapy, family therapy, family-oriented psychosocial interventions, etc.	• Ongoing support • Connection with interest/sport/art, etc. groups

Key Points

Clinical Management

- Develop a rapport with the individual to allow a discussion of psychological factors associated with MAL.
- Increase awareness of the importance of MAL factors as independent risks for chronic disease.
- If appropriate and acceptable to the patient, get him/her to complete the 10-point ACE scale (see Professional Resources).
- Refer to appropriate treatments such as logotherapy, art therapy, etc., where possible and appropriate.
- Consider seeing clients in SMAs (see Chapter 4) for indigenous groups, individuals from displaced cultures, depression, or others alienated by aspects of society.

REFERENCES

Beatty, D.L., Matthews, K.A., Bromberger, J.T., Brown, C., 2014. Everyday discrimination prospectively predicts inflammation across 7-Years in racially diverse midlife women: study of women's health across the nation. J. Soc. Issues 70 (2), 298–314.

Beaumeister, D., Akter, R., Ciufolini, S., Pariante, C.M., Mondelli, V., June 2, 2015. Childhood trauma and adulthood inflammation: a meta-analysis of peripheral C-reactive protein, interleukin-6 and tumour necrosis factor-α. Mol. Pyschiatry.

Brassai, L., Piko, B.F., Steger, M.F., 2015. A reason to stay healthy: the role of meaning in life in relation to physical activity and healthy eating among adolescents. J. Health Psychol. 20 (5), 473–482.

Brown, H.J., McPherson, G., Peterson, R., Newman, V., Cranmer, B., 2012. Our land, our language: connecting dispossession and health equity in an indigenous context. Can. J. Nurs. Res. 44 (2), 44–63.

Browne-Yung, K., Ziersch, A., baum, F., Gallagher, G., 2013. Aboriginal Australians' experience of social capital and its relevance to health and wellbeing in urban settings. Soc. Sci. Med. 2013 (97), 20–28.

CDC (Centers for Disease Control and Prevention), 2014. Injury prevention and control: division of violence prevention. http://www.cdc.gov/violenceprevention/acestudy/.

Daoud, N., Shankardass, K., O'Campo, P., Anderson, K., Agbaria, A.K., 2012. Internal displacement and health among the Palestinian minority in Israel. Soc. Sci. Med. 74 (8), 1163–1171.

Davis, M., Ventura, J.L., Wieners, M., et al., 2010. The psychosocial transition associated with spontaneous 46,XX primary ovarian insufficiency: illness uncertainty, stigma, goal flexibility, and purpose in life as factors in emotional health. Fertil. Steril. 93 (7), 2321–2329.

Dezutter, J., Luyckx, K., Washholt, A., 2015. Meaning in life in chronic pain patients over time: associations with pain experience and psychological well-being. J. Behav. Med. 38 (2), 384–396.

Doocy, S., Sirios, A., Tileva, m, Story, J.D., Burnham, G., 2013. Chronic disease and disability among Iraqi populations displaced in Jordan and Syria. Int. J. Health Plann. Manage. 28 (1), e1–e12.

Felitti, V.J., Anda, R.F., 2009. The relationship of adverse childhood experiences to adult medical disease, psychiatric disorders, and sexual behavior: implications for healthcare. In: Lanius, R., Vermetten, E. (Eds.), The Hidden Epidemic: The Impact of Early Life Trauma on Health and Disease. Cambridge University Press, London.

Frankl, V., 1959. Man's Search for Meaning: an Introduction to Logotherapy. Touchstone. NY.

Greene, D., Tehranifar, P., Hernandez-Cordero, L.J., Fullilove, M.T., 2011. I used to cry every day: a model of the family process of managing displacement. J. Urban Health 88 (3), 403–416.

Hackett, R., Hamer, M., Endrighi, R., Brydon, L., Steptoe, A., 2012. Loneliness and stress-related inflammatory and neuroendocrine responses in older men and women. Psychoneuroendocrinology 37 (11), 1801–1809.

Hamer, M., Chida, Y., 2011. Life satisfaction and inflammatory biomarkers: the 2008 Scottish Health Survey. Jap. Psychol. Res. 53 (2), 133–139.

Hill, P.L., Turiano, N.A., 2014. Purpose in life as a predictor of mortality across adulthood. Psychol. Sci. 25 (7), 1482–1486.

Hostinar, C.E., Ross, K.M., Chen, E., Miller, G.E., August 11, 2014. Modeling the association between life course socioeconomic disadvantage and systemic inflammation in healthy adults: the role of self-control. Health Psychol.

Johnson, T.V., Abbasi, A., Master, V.A., 2013. Systematic review of the evidence of a relationship between chronic psychosocial stress and C-reactive protein. Mol. Diagn Ther. 17 (3), 147–164.

Ketterer, M., Rose, B., Knysz, W., Farha, A., Deveshwar, S., Schairer, J., Keteyian, S.J., 2011. Is social isolation/alienation confounded with, and non-independent of, emotional distress in its association with early onset of coronary artery disease? Psychol. Health Med. 16 (2), 238–247.

Kim, E.S., Strecher, V.J., Ryff, C.D., 2014. Purpose in life and use of preventive health care services. Proc. Natl. Acad. Sci. U. S. A. 14. http://dx.doi.org/10.1073/pnas.1414826111.

Kim, E.S., Sun, J.S., Park, N., Peterson, C., 2013. Purpose in life and reduced incidence of stroke in older adults: 'The Health and Retirement Study'. J. Psychosom. Res. 74 (5), 427–432.

Kim, E.S., Sun, J.S., Park, N., Peterson, C., Kubzansky, L.D., 2012. Purpose in life and reduced risk of myocardial infarction among older U.S. adults with coronary heart disease: a two-year follow-up. J. Behav. Med. http://dx.doi.org/10.1007/s10865-012-9406-4.

Kleiman, E.M., Beaver, J.K., 2013. A meaningful life is worth living: meaning in life as a suicide resiliency factor. Psychiatry Res. 210 (3), 934–939.

Lacey, R.E., Kumari, M., Bartley, M., August 22, 2014. Social isolation in childhood and adult inflammation: evidence from the national Child Development Study. Psychoneuroendocrinology.

Lacey, R., Kumari, M., McMunn, A., 2013. Parental separation in childhood and adult inflammation: the importance of material and psychosocial pathways. Psychoneuroendocrinology 38 (11), 2476–2484.

Miller, G.E., Brody, G.H., Yu, T., Chen, E., 2006. A family-oriented psychosocial intervention reduces inflammation in low-SES African American youth. Proc. Natl. Acad. Sci. U. S. A. 2014, 21.

Morgan, B., 2006. Toward a Redefinition of the Principle of Aboriginal Self Determination (Unpublished PhD thesis). University of Sydney.

Morozink, J.A., Friedman, E.M., Coe, C.L., Ryff, C.D., 2010. Socioeconomic and psychosocial predictors of interleukin-6 in the MIDUS national sample. Health Psychol. 29 (6), 626–635.

Nowakowski, A., 2014. Chronic inflammation and quality of life in older adults: a cross-sectional study using biomarkers to predict emotional and relational outcomes. Health Qual. Life Outcomes 12, 141.

Parker, G., Lie, D., Siskind, D.J., Martin-Khan, M., Raphael, B., Crompton, D., Kisely, S., 2015. Mental health implications for older adults after natural disasters – a systematic review and meta-analysis. Int. Psychogeriatr. 27, 1–10.

Reid, J., 1972. Rectorial Address. Glascow University, Sottish Left Review.

Richmond, C.A., Ross, N.A., 2009. The determinants of First Nation and Inuit health: a critical population health approach. Health Place 15 (2), 403–411.

Robatmili, S., Sohrabi, F., Shahrak, M.A., Talepasand, S., Nokani, M., Hasani, M., 2015. The effect of group logotherapy on meaning in life and depression levels of Iranian students. J. Adv. Couns. 37 (1), 54–62.

Roepke, A.M., Jayawickreme, E., Riffle, O.M., 2014. Meaning and health: a systematic review. Appl. Res. Qual. Life 9 (4), 1055–1079.

Ruini, C., Fava, G.A., 2012. Role of well-being therapy in achieving a balanced and individualized path to optimal functioning. Clin. Psychol. Psychother. 19 (4), 291–304.

Shankar, A., McMunn, A., Banks, J., Steptoe, A., 2011. Loneliness, social isolation, and behavioural and biological health indicators in older adults. Health Psychol. 30 (4), 377–385.

Shehab, N., Anastario, M.P., Lawry, L., 2008. Access to care among displaced Mississippi residents in FEMA travel trailer parks two years after Katrina. Health Aff. (Millwood) 27 (5), 16–29.

Slavich, G.M., Way, B.M., Eisenberger, N.I., Taylor, S.E., 2010. Neural sensitivity to social rejection is associated with inflammatory responses to social stress. Proc. Natl. Acad. U. S. A. 107 (33), 1417–1422.

Stevens, J., Dixon, J., Binns, A., Morgan, B., Richardson, J., Egger, G., 2016. Exploring the use of Shared Medical Appointments as a culturally appropriate, safe and effective health care model with Indigenous men. Aust. Fam. Phys. in press.

Professional Resources

Assesșing Adverse Childhood Experiences

Adverse childhood experiences questionnaire

Before your 18th birthday, had any of the following things happened to you?

If you have suffered major trauma in your life that might be disturbed by answering any of the following questions, please finish the questionnaire here and leave the following questions unanswered.

- ☐ I'd prefer not to answer the following questions.
- ☐ A parent or other adult in the household often or very often swore at you, insulted you, put you down, or humiliated you, or acted in a way that made you afraid that you might be physically hurt.
- ☐ A parent or other adult in the household often or very often pushed, grabbed, slapped, or threw something at you or ever hit you so hard that you had marks or were injured.
- ☐ An adult or person at least five years older than you touched or fondled you or had you touch their body in a sexual way or attempt or actually have oral, anal, or vaginal intercourse with you.
- ☐ You often or very often felt that no one in your family loved you or thought you were important or special, or your family didn't look out for each other, feel close to each other, or support each other.
- ☐ You often or very often felt that you didn't have enough to eat, had to wear dirty clothes, and had no one to protect you. Or your parents were too drunk or high to take care of you or take you to the doctor if you needed it.
- ☐ A biological parent was lost to you through divorce, abandonment, or other reason.

Professional Resources—cont'd

- ☐ Your mother or stepmother, often or very often, was pushed, grabbed, slapped, or had something thrown at her, or sometimes, often, or very often kicked, bitten, hit with a fist, or hit with something hard, or ever repeatedly hit over at least a few minutes or threatened with a gun or knife.
- ☐ You lived with someone who was a problem drinker or alcoholic, or who used street drugs.
- ☐ A household member was depressed or mentally ill, or attempted suicide.
- ☐ A household member went to prison.

(Scores of >4 indicate significant risk from ACEs)

From US Centers for Disease Control. Also see (http://www.npr.org/blogs/health/2015/03/02/387007941/take-the-ace-quiz-and-learn-what-it-does-and-doesnt-mean.)

Chapter 21

Preventing and Managing Injury at the Clinical Level

Kevin Wolfenden, Garry Egger

Accidents aren't accidents.

William Haddon (1974)

INTRODUCTION: INJURY PREVENTION AS PUBLIC HEALTH

When William Haddon, an engineer from the New York Health Department, proposed in 1974 that there is no such a thing as an "accident" but that injuries are in epidemic proportions, he redefined an area of health. As an epidemic, Haddon claimed that injury could be treated just like any other epidemic, by dealing with a host, a vector, and an environmental cause, all of which need to be considered in a global approach to the problem. In this way motor vehicle injuries became dissected not just in terms of treatment of the victim but modifying vectors (e.g., speed) and changing environments (e.g., dangerous roads). This brought injury into the realm of preventable health-related problems (Haddon, 1974).

THE EPIDEMIC OF INJURY

Worldwide, injury continues to be an important component of the global burden of disease, even though there has been a decline in its relative contribution over recent years (Haagsma, 2015). This is also the case for Western countries such as Australia where injuries currently account for 10% of the total burden of disease (AIHW, 2014a).

Transport injuries, self-harm (including suicide), and falls are the major contributors to the global injury burden (Haagsma, 2015). Similarly, these three types of injuries are the major causes of both injury deaths and hospitalizations in Western countries. In Australia for example, falls account for 32% of injury deaths and 40% of injury hospitalizations (AIHW, 2014a). Intentional injuries are the second most common cause of injury death (20%) and transport

Lifestyle Medicine. http://dx.doi.org/10.1016/B978-0-12-810401-9.00021-8

injuries account for 14% of injury deaths and 12% of injury hospitalizations (AIHW, 2014a).

Injury patterns change with the type of injury, age, sex, location, and severity.

Injuries can be categorized into unintentional (e.g., falls) or intentional (arising from self-harm or harm to others).

This chapter focuses on injuries that commonly present to general practitioners and that can be managed within a lifestyle medicine model. Because these differ significantly between countries and cultures, we focus here in particular on injuries in Australia and on:

- Injuries such as those that occur from falls (in the young and the elderly), transportation, drowning, poisoning, and the intentional injury of suicide;
- Developing a framework within which to understand the causes of these injuries, which can be applied to the understanding of, and intervention for, injury at the clinical level;
- A framework to guide practitioners in addressing injury issues within the context of lifestyle medicine.

UNDERSTANDING INJURY

Injury differs from the common term "accident" in that the latter implies something that can neither be understood nor altered. "Accident" fails to distinguish between the outcome (damage) and how it arose. Injury, on the other hand, is both understandable in cause and preventable.

The most common framework for understanding the causes and causal sequence leading to injury is the matrix developed by Haddon (1980), referred to earlier. This provides a model including the factors associated with an injury event (e.g., those related to the person injured, or the "host"; the means by which the physical damage was inflicted, or the "vector"; the physical and social environment leading to the injury); the time sequence of these factors (before the injury, during the injury, after the injury event). This is shown in tabular form as it may apply to falls in the elderly in Table 21.1.

The matrix provides a means of understanding the causal factors leading to injury, as well as factors that might be modified to prevent or mitigate against injury.

The matrix also provides a framework for practitioners to:

- ***Assess*** a patient who may be injured and the injury process over time, by identifying factors that may have played a role in a patient's injury;
- ***Assist*** in treating, preventing, or minimizing injury by changing causal factors at the preevent, event, or postevent phases;
- ***Arrange*** for patients to access those activities that counteract causal factors or mitigate damage.

TABLE 21.1 An Example Haddon Matrix for Falls in the Elderly

	Preevent	Event	Postevent
Host (human)	Musculoskeletal integrity Intoxication Medication Personal fitness Judgment Fatigue	Vision Flexibility Bone integrity Imbalance Eyesight	Age Physical condition Preexisting illness
Vector	Surface condition (i.e., slipperiness)/slope/ moisture, etc.	Speed of action Hardness of surface	Contaminants on the surface (e.g., infections)
Physical environment	Home/institutional environment Step distance Support (i.e., rails etc.) Safety design	Guard rails Fall surface	Recovery opportunities (rails) Communications Emergency services Rehabilitation services
Sociocultural environment	Alcohol attitudes Family arrangements Attitudes to safety	Warning system Family/ institutional support	Trauma support Personnel training

The role of the clinician in anything but the treatment of injury may seem limited. However, in this chapter we consider aspects of a number of injuries including falls, transportation, drowning, avoidable sports injuries, poisoning, and suicide, where lifestyle interventions may help reduce the injury rate.

What Can Be Done in a Standard Consultation?

- Consider individual patient risks for injury in older adults and children
- Discuss potential for falls and falls prevention in older adults and children
- Discuss with parents the importance of preventing poisoning and drowning in children
- Offer advice on sports injury prevention (e.g., use of mouth guard) in children
- Empathetically weigh up risks, and consider risk factors for suicide

INJURIES FROM FALLS

Fall injuries are a particular problem for the elderly and the young.

Fall deaths and hospitalization rates are high in the elderly (rising rapidly over 65 years of age), with elderly males having a higher rate of deaths than females, but the reverse being the case in regard to hospitalizations of the elderly (Henley and Harrison, 2015; AIHW, 2013).

Falls are also the most common cause of injuries leading to childhood admission to hospital, where male children have almost twice the level of fall injuries (Harris and Pointer, 2012).

Not surprisingly the pattern of injuries differs between the young and elderly. For those elderly admitted to hospital, falls commonly involve slipping, tripping, and stumbling, with a considerable number suffering falls from stairs, steps, ladders, and beds (AIHW, 2013). A frequent type of injury for the elderly is injury to the hip and thigh, often a fracture of femur, particularly for females (AIHW, 2013).

For children admitted to hospital, falls involving playground equipment (particularly climbing apparatus and trampolines) are the most common, particularly in the 5–9 ages (Pointer, 2013). Here common equipment involved in falls also included ice/roller skates and skateboards. Fractures to the forearm are also common in this age group.

The location of injury resulting in hospitalization also varies with age. Overall the home is the most common location for all age injuries. Young children (0–10 years) most commonly have fall injuries sustained at home or school; whereas for young men (15–24 years) injuries most commonly in sport areas (Berry and Harrison, 2006).

There are a variety of activities that have been shown to reduce falls in the elderly and the young, with best results being obtained when a range of initiatives are undertaken in combination (Norton, 1999).

For the elderly there are both intrinsic and external factors leading to falls that can be modified. Intrinsic factors often relate to the persons' general health status including deteriorated mobility and gait, poor muscle tone and strength, decreased sensation in lower limbs, reduced bone density, decreased vision, poor balance, or dizziness (often from the effects of medication).

The clinician may modify these factors through:

- Regular monitoring and treatment of preexisting conditions (such as cardiovascular disease, depression, arthritis, cataracts/vision problems);
- Encouraging resistance and other appropriate exercise training;
- Reducing osteoporosis/enhance bone density;
- Hormone replacement therapy for the postmenopausal;
- Vitamin D and calcium supplementation;
- Medication review including use of antidepressants, antipsychotics, antihypertensives, and diuretics, and in particular polypharmacy.

External factors identified that pose an injury risk to the elderly include slippery floors, loose carpets, rugs, slippery shoe soles, and poor lighting. As a result, checklists for identifying hazards in households inhabited by older people have been developed such as the "Stay on Your Feet" checklist that provides a comprehensive checklist to identify hazards in the home and contains suggested actions (see Injury Prevention, falls prevention in Health Topics www.health.nsw.gov.au). Clinicians can assist in modifying these factors through encouraging the use of these checklists, reminding patients of the importance of appropriate footwear, consideration of the need for use of hip protectors in high-risk fall patients, and encouraging the use of personal alarms and alerts to avoid the "long lie" experience of the fallen.

Falls in children usually result from heights and usually require the need for parental supervision, the use of protective equipment (e.g., pads), and awareness of hazards in the home and play environment. Clinicians can assist by encouraging parents to obtain and apply the checklist of fall hazards in the home and play areas that are available through Kidsafe or Farmsafe (see Other Resources section).

TRANSPORTATION INJURIES

Transportation injuries cover a range of transportation modes, the most common of which for deaths and hospitalizations are motor vehicles (accounting for about two-thirds of serious transport injuries), followed by cyclists and pedestrians. Motor cyclists have the highest rate of transport injury per vehicle, being about 10 times that for cars (Henley and Harrison, 2015; Henley, 2012).

Serious transport injury rates for males are twice those of females, with young adults (15–24 years) having the highest injury rates (Henley, 2012).

Over the last 20 years there has been a significant reduction in overall transport injury rates (Henley and Harrison, 2015). This reduction can be attributed to a range of initiatives developed to address factors identified in the matrix above, most notably the introduction of seatbelt legislation and enforcement, introduction of drink-driver legislation and enforcement, improved car safety design, improved road design, speed reduction initiatives, cycle helmet legislation and enforcement, and driver education, restriction, and reinforcement (e.g., driver license point systems).

Whilst medical practitioners have been at the forefront advocating for these broad social policy changes, individual practitioners can play a leading role on an individual and local community level by testing drivers' eyesight, coordination abilities, general health (e.g., epilepsy, coronary risk) at medical examination for license, and encouraging healthy attitudes to drink driving, children's use of bicycle helmets and protective gear for skating/skate boards, and use of authorized baby capsules.

DROWNING

Drowning and near drowning, whilst accounting for a small percentage of all injury deaths and hospitalizations, is the major injury cause of death for children under 5 years of age. The highest rates of drowning are in the very young and very old (Kreisfeld and Henley, 2008).

The place of drowning differs with age, with the private swimming pool (and bathtubs for the very young) the most common site for the young; whilst for adults the most common setting is natural water. For all-age unintentional drowning, men have over three times the drowning rate of women (Kreisfeld and Henley, 2008).

Attention can be given to this at the clinical level by checking access to water and swimming skills of toddlers with parents and referring to swim schools, increasing parents' awareness of risk, encouraging parents to learn resuscitation skills, asking about appropriate pool fencing, and advising about bathing practices in infants.

POISONINGS

Unintentional poisoning death rates are somewhat difficult to identify because of coding complexities and difficulty of differentiating the unintentional from drug dependence and intentional poisoning. However, for unintentional poisoning deaths by drugs, the overall rates vary with age, the young adult to middle age (20–50 years) having the highest rates. Furthermore, males account for two-thirds of the deaths (Kreisfeld et al., 2004).

The pattern of hospitalizations for unintentional poisoning by drugs differs from the death patterns. For hospital admissions, one in five were children 0–4 years. Rates are low in older children (5–14 years) but rise for young people (15–24 years) before gradually declining with age (Tovell et al., 2012).

Interestingly there are more females than males involved in overall poisonings. The drugs most commonly involved are benzodiazepines and 4-aminophenol derivatives such as paracetamol—the home being by far the most common location of poisoning (Berry and Harrison, 2006). Many medicinal poisonings for very young children arise from incorrect administration from parents, increasing the importance of correct advice from clinicians to parents.

The incident pattern for hospital admissions for unintentional poisoning with substances other than medicines, although much less frequent, shows similar patterns to medicinal poisonings with highest rates in the 0–4 years age group (particularly toddlers), tapering with age after young adulthood. However, more males are hospitalized than females. The most frequent agents of injury are toxic food, alcohol, and pesticides—the most frequent location being the home (Berry and Harrison, 2006).

Reducing the accessibility of the poisoning agents has been shown to be a key strategy for reducing poisoning, particularly childhood poisoning. Clinicians can also help by advising parents of the accessibility risks (potential dangers of household cleaning agents, correct labeling of containers), providing clear information on child medication use, and providing literature on the immediate management of poisoning.

SPORTS INJURIES

While the encouragement of sporting involvement is essential for good physical health, and most sports injuries are not severe, they can be prevented or the damage limited by early treatment.

Most hospitalizations occur in the young, with overall sports injury rates doubling after age 12 and increasing six- to sevenfold after age 16 (Egger, 1990). Three-quarters of those hospitalized are male. Most commonly sporting injuries come from contact sports (football codes, netball, basketball, etc.) and occur in the lower limb joints, namely the knees and ankles. Sports involving wheeled vehicles have the highest hospitalization and life-threatening rates (Kreisfeld et al., 2014).

These injuries can only partially be avoided by appropriate strengthening and, where evidence based, the use of protective clothing such as ankle strapping (Australian Rules), knee protection (netball), and shin pads (soccer). Compulsory use (such as with shin pads in soccer), reduce the clinician's need to be involved here. However, questioning about the use of mouth guards, headgear (cricket, rugby), padding, and other protective gear (e.g., in motor sports) can help prevent some of the major causes of injury.

SUICIDE

More Australians die from suicide each year than die from motor vehicle accidents.

Suicide is most common amongst 15–34 year olds and is much more common amongst males (AIHW, 2014b).

For every completed suicide there are estimated to be many times that number of attempts made.

Female rates for hospitalization for intentional self-harm (which includes attempted suicide) are 40% higher than for males (AIHW, 2014b).

Most people seek help prior to a suicide attempt, and the majority of those who die have consulted a primary health care professional in the weeks prior to their death (Pirkis and Burgess, 1998). It is here that the astute clinician may be able to have an effect.

Suicide attempts are influenced by a range of risk factors including the person's history (e.g., abuse victim, previous attempts, mental illness) and current mental state (e.g., depression), current external stresses (e.g., recent divorce, financial problems), access to the means of self-harm (e.g., firearms, medications), and the individual's access to protective factors such as family or peer support.

Risk factors and warning signs are shown in Table 21.2.

Whilst the overall suicide rate in Australia has trended downward in recent years (Henley and Harrison, 2015), the prevention of suicide at the individual

TABLE 21.2 Risk Factors and Warning Signs for Suicide

Risk Factors	Warning Signs
Change in illness status	Depression, despair, feeling of uselessness
History of mental illness,	Impulsivity or agitation
History of abuse (sexual, physical, neglect)	Anger, feelings of rejection
Family disruption	Psychosis or psychotic thoughts
Violence in family	Putting oneself down
Change in dwelling location	Feeling unsupported
Change in finances	Being distant/remote
Loss of employment	Talk about suicide
Change in relationships	Giving away possessions
Abuse at home/school	Being on edge
Access to suicide implements	Excessive drug/alcohol use

level is complex. However, the NSW Department of Health has provided some useful guidelines for the assessment of suicide risk and its management, available on the NSW Department of Health website (see Other Resources section). This document highlights the importance of the clinician in doing the following:

- engaging the client
- assessing the risk factors
- conducting a preliminary risk assessment (history, presentation, psychiatric assessment)
- taking immediate action to ensure patient's safety, if at significant risk
- making a referral if appropriate

Whilst there is no current rating scale that has proven predictive value, Table 21.3 is provided as a guide to assessing the level of current risk. This is adapted from the NSW Department of Health (2004) Suicide Risk Assessment Guide.

The level of risk will assist the clinician in determining the patient's management. Those patients at high risk (e.g., high risk of changeability, or presence of severe level of "At Risk Mental State") may need to be provided with actions to ensure their safety and reassessed within 24 h.

TABLE 21.3 Suicide Risk Assessment Guide

Issue	High Risk	Medium Risk	Low Risk
At-risk mental state • depressed • psychotic • hopelessness • guilt, shame • anger, agitation • impulsivity	Severe level exists (particularly hopelessness)	Moderate level exists (particularly hopelessness)	Low level or absent
History of suicide attempt	Any previous attempt or gesture	Possible attempt or family/friend attempt	No history of attempt
Suicidal thought • extent • intention • lethality	Thoughts of suicide persistent Clear plan and lethal means	Frequent thoughts of suicide Threats with low lethality	Nil or vague thought of suicide No plans
Substance disorder • current alcohol or drug misuse	Current intoxication or dependence	Risk of intoxication or abuse	Nil or infrequent use
Means • knowledge of and access to means	Access to lethal means (e.g., firearms)	Sought knowledge of lethal means	No evidence of seeking access to lethal means
Strengths and supports • help seeking • support of others	Patient refusing help Lack of supportive relationships	Patient ambivalent Limited supportive relationships	Patient accepting help Insightful
Assessment Confidence • information quality • changeability of the assessment	Clinician not confident/ unable to verify information from patient Patient condition dynamic	Some areas of limited confidence in the information from patient Patient condition stable	Good patient engagement Information verified from other sources Low changeability

Note that this is a guide only and does not replace clinical decision-making.
Adapted from NSW Department of Health Suicide Risk Assessment Guideline, 2004.

Understanding and utilizing listening consulting skills by a clinician is valuable, along with recognition of warning signs and careful control of medications. Referral to specialist services where necessary is obviously also vital.

Practice Tips

Reducing Injury

	Medical Practitioner	Practice Nurse
Assess	• Screen for risk factors in the elderly as in RACGP "red book" • Opportunistically check eyesight, bone density, and functional coordination and musculoskeletal integrity of older people • Review prescribed medications with regard to potential for role in all injury • Check use of correct baby capsules • Check use of swimming pool fencing • check swimming skills of toddlers • Assess for risk of suicide	• Provide an in-home fall checklist • Perform home medication review in the elderly where feasible • Encourage use of appropriate footwear • Encourage parents to conduct a hazard check in the home and play areas • Check for signs of suicide risk
Assist	• Encourage parents to take preventive action with young children, i.e., scalds, poisoning, drowning prevention • Encourage change of attitudes to "accidents" as nonpreventable • Encourage use of protective gear in play and sport, e.g., helmet strapping, mouth guards, etc.	• Encourage parents to take specific preventive action, e.g., turning hot water down to <50°C; safe storage of poisoning agents • Reinforce use of protective gear (helmets, etc.) and strapping, mouth guards, etc. in sport • Provide literature on prevention and immediate management of fall injuries/water safety, etc. • Provide information (i.e., from a consumer association) on safety products, e.g., baby capsules
Arrange	• Appropriate exercise programs for the elderly for falls protection • If necessary suggest vitamin D/sunlight treatment for bone density • Occupational therapist for home environment check of the aged referral to mental health specialist for those at risk of suicide	• Exercise specialist/ physiotherapist for exercise program for the elderly • Refer to accredited swim schools for toddlers • Refer to kidsafe/farmsafe/ bikesafe organizations, etc. where appropriate

SUMMARY

Injury is a leading cause of death and hospitalization worldwide and in Western countries like Australia. Those that can be impacted the most in the clinical setting are those in older adults and children. Morbidity from injury becomes more important with aging, however, with fall injuries being one of the major causes of hospitalization and health costs in older people. Transportation injuries, drownings, poisonings, and sports injuries are all preventable causes of injury in children. Suicide and violence (not covered here) are the two main forms of injury influencing older adolescents and the middle ages. Contrary to popular opinion, injuries are not "accidents" and, as such, are largely preventable using an epidemiological approach to their management. Clinicians can aid in this by adopting a few basic lifestyle-oriented principles and adopting a modern approach to injury prevention.

Key Points

Clinical Management

- Work toward changing patients' perception of injury as preventable and not synonymous with "accident."
- Regard injury management as within the ambit of "lifestyle medicine."
- Undertake regular medication and mobility review in elderly patients in regard to injury risk.
- Be alert to the risk factors for suicide in patients experiencing stress or with mental illness.
- Use opportunistic counseling to assist in injury awareness and prevention.
- Encourage parents to identify hazards in the home, travel, and play environments.
- Have available a range of information pamphlets and resource directories on injury in the surgery.

REFERENCES

AIHW (Australian Institute of Health and Welfare), 2013. Hospitalizations Due to Falls by Older People, Australia 2009–10. Injury research and statistics series no 70. Canberra.

AIHW (Australian Institute of Health and Welfare), 2014a. Australia's Health 2014. Australia's health seriers no 14. Canberra.

AIHW (Australian Institute of Health and Welfare), 2014b. Suicide and Hospitalised Self Harm in Australia: Trends and Analysis. Injury Research and Analysis. Injury research and statistics series no 93. Canberra.

Berry, J., Harrison, J., 2006. Hospital Separations Due to Injury and Poisoning, Australia 2001–02. Injury research and statistics series no 26. Australian Institute of Health and Welfare, Canberra.

Egger, G., 1990. Sports injuries in Australia: causes costs and prevention. Health Prom. J. Aust. 1 (2), 28–33.

Haagsma, J., December 2015. The global burden of injury: incidence, mortality, disability adjusted life years and trends from the Global Burden of Disease Study 2013. Inj. Prev.

Haddon, W., 1974. Strategies in preventive medicine: passive v active approaches to reduce human wastage. J. Trauma 4, 353–354.

Haddon, W., 1980. Advances in the epidemiology of injuries as a basis for public policy. Public Health Rep. 95 (5), 411–420.

Harris, C.E., Pointer, S.C., 2012. Serious Childhood Community Injury in New South Wales 2009–10. Injury research and statistics series no 76. Australian Institute of Health and Welfare, Canberra.

Henley, G., 2012. Serious Injury Due to Land Transport Accidents, Australia 2007–08. Injury research and statistics series no 59. Australian Institute of Health and Welfare, Canberra.

Henley, G., Harrison, J.E., 2015. Trends in Injury Deaths, Australia: 1990–01 to 2009–10. Australian Institute of Health and Welfare, Canberra. Injury research and statistics series no 74.

Kreifeld, R., Henley, G., 2008. Deaths and Hospitalizations Due to Drowning, Australia 1990 to 2003–04. Injury research and statistics series no 39. Australian Institute of Health and Welfare, Adelaide.

Kreisfeld, R., Newson, R., Harrison, J., 2004. Injury Deaths, Australia 2002. Injury research and statistics series no 22. Australian Institute of Health and Welfare, Adelaide.

Kreisfeld, R., Harrison, J.E., Pointer, S., 2014. Australian Sport Injury Hospitalizations 2011–12. Injury research and statistics series no 92. Australian Institute of Health and Welfare, Canberra.

Norton, R., 1999. Preventing falls and fall related injuries among older Australians. Australas. J. Ageing 18, 4–10.

NSW Department of Health, 2004. A Framework for Suicide Risk Assessment and Management for NSW Health Staff. NSW Department of Health, Sydney.

Pirkis, J., Burgess, P., 1998. Suicide and recency of health care contacts. A systematic review. Br. J. Psychiatry 173 (12), 462–474.

Pointer, S., 2013. Hospitalization Injury in Children and Young People 2011–12. Injury research and statistics series no 91. Australian Institute of Health and Welfare, Canberra.

Tovell, A., Mckenna, K., Bradley, C., Pointer, S., 2012. Hospital Separations Due to Injury and Poisoning. Australia 2009–10. Injury research and statistics series no 69. Australian Institute of Health and Welfare, Canberra.

Chapter 22

Rethinking Chronic Pain in a Lifestyle Medicine Context

Chris Hayes, Caroline West, Garry Egger

Neuroplasticity has the power to produce more flexible but also more rigid behaviors – A phenomenon I call 'the plastic paradox'

Norman Doidge, The Brain That Changes Itself

INTRODUCTION: A CASE STUDY

Brian (age 44) developed low back pain following a lifting injury at work. At the time of referral to a pain clinic some 6 months later, he was fearful of ongoing tissue damage in his back and was guarded in both his thinking and actions. He had developed abnormal patterns of movement and posture, and his overall level of activity had reduced. At times he would "push though pain" resulting in flare-ups. At other times he would avoid activity altogether. He felt anxious and depressed. He was angry about the circumstances of the injury, there were problems at home, and his diet was poorly balanced. The fact that his pain persisted beyond the usual time for healing and without major structural change on imaging suggested that brain interpretation and nervous system sensitization were key contributors. Brian was helped to develop a pain recovery plan, which included deprescribing medications, improving his diet with increased intake of fresh vegetables, and a program of mindful physical activity. Brian was encouraged to become more aware of his emotions and let go of his anger and unnecessary fear. This broad-based approach led to gradual improvement of his pain.

Brian's story in this case study shows that pain, like other lifestyle-related health problems, can no longer be looked at in a simple, unimodal fashion. While acute pain from tissue or nerve damage may respond to unimodal biomedicine, the management of chronic pain usually requires a more holistic approach, taking account of the patient's life story, patterns of thinking, actions, relationships, and lifestyle as well as any biological contributors (National Pain Strategy, 2010).

This shift in thinking about chronic pain has been driven by growing prevalence and the absence of consistent benefit from traditional treatments. It also comes from the realization that much chronic pain correlates poorly with

Lifestyle Medicine. http://dx.doi.org/10.1016/B978-0-12-810401-9.00022-X

biological pathology, and that some pathology (e.g., disc rupture) can exist in the absence of pain (Brinjikji et al., 2015). In fact, unnecessary imaging for low back pain can often make the problem worse, with more reported pain and trips to the doctor (Kendrick et al., 2001). Since the famous work of Dr. Henry Beecher with soldiers during World War II, the importance of context has been recognized. Clearly there is more involved in pain than meets the "pain" receptor.

TYPES AND EXTENT OF PAIN

Pain has been defined as "an unpleasant sensory and emotional experience associated with actual or potential tissue damage, or described in terms of such damage." Chronic pain is usually defined as lasting 3 months since onset, or as pain that extends beyond the expected period of healing (Turk and Okifuji, 2001).

In a typical advanced industrialized country like Australia, around one in five people of working age and one in three older adults suffers chronic pain, and up to 80% of these are missing out on treatment that could improve their health and quality of life (National Pain Strategy, 2010). The cost to the community was put at $Aus34 billion per annum in 2007, which makes it the third most costly health problem after cancers and heart disease. Five different types of pain are commonly recognized:

1. Acute pain, which is a normal and time-limited response to trauma or other noxious experiences;
2. Subacute pain, progressing toward chronic pain if not prevented;
3. Recurrent pain (e.g., migraine headaches);
4. Chronic, continuous (persistent) noncancer pain;
5. Cancer-related pain.

The source of pain has traditionally been conceptualized as relating to tissue (nociceptive pain) or nerve (neuropathic pain) injury. However, the recognition of sensitization in the nervous system and the roles of psychoneuroimmunology (Kiecolt-Glaser and Glaser, 1995) and brain plasticity (Doidge, 2008) have expanded thinking about the multifaceted contributors to chronic pain. Chronic pain can exist in the absence of nociception or neuropathy.

While in theoretical terms the categorization of pain may seem relatively simple, the IASP Classification of Chronic Pain (Merskey and Bogduk, 1994) reminds us that at the clinical coalface this is not so, with over 200 identified subtypes of pain. The key to diagnosis is predominantly in the history (McIntosh and Elson, 2008):

- the nature of the pain and words used to describe it
- its location
- radiation—does the pain travel?
- duration—how did it start, how long has it been present?
- intensity

- effects on life, i.e., mood, sleep, exercise, eating, relationships
- history of other traumas
- previous treatments

PAIN: HISTORICAL CONTEXT

Acute pain has always existed in humans as a warning of danger or injury. However, its continuance after the initiating stimulus poses genuine problems. In some ancient cultures, this type of pain was attributed to mystical sources and the displeasure of the gods. Hence shamans and witch doctors were used to exorcise the pain, with no thought toward biological contributors.

The first suggestions involving the brain and nervous system in the mediation of pain came from the early Greeks and Romans. In the mid-17th century, philosophers developed an interest in pain and Descartes' "specificity theory" continues to be influential today. Descartes proposed that peripheral activation of the nervous system led to a brain response, just as the pulling of a rope rings a bell Fig. 22.1

In the 19th century, opiates and salicylates were enlisted in the battle against pain, and surgical anesthesia was developed. These dramatic medical discoveries were supplemented by a myriad of dubious treatments based on other contemporary discoveries such as electricity (Fig. 22.2).

In the 1960s, Melzack and Wall (1965) introduced the gate control theory of pain, which proposed that the extent of central response to peripheral stimulation was determined by processing at the level of the spinal cord. Engel described the biopsychosocial model of illness in 1977, and this was

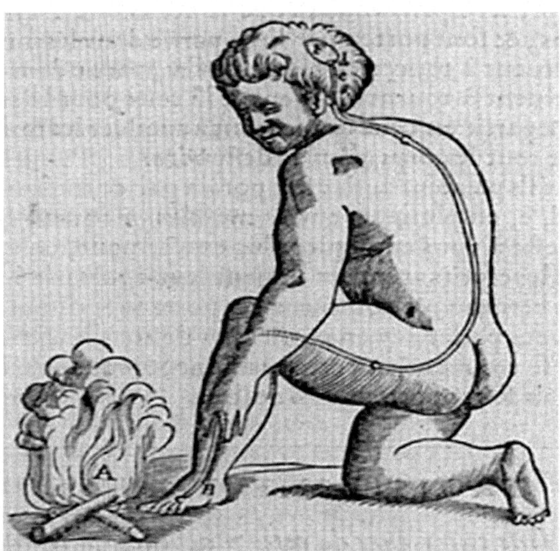

FIGURE 22.1 Rene Descartes' "hard-wired" specificity theory (1664).

FIGURE 22.2 Using new technology—electricity and pain circa the 19th century. *Phillips, P., 1984. Kill or Cure: Lotions, Potions and Quacks of Early Australia. Greenhouse Publisher, Richmond Victoria.*

later adapted to pain medicine (Loeser, 1980). However, Loeser's well-known "onion skin" model favored a view of chronic pain as biological in origin, with psychosocial and plasticity factors having only a secondary modulating influence. Sufferers were generally treated medically (often with pharmacotherapy) in the first instance. If those treatments proved unsuccessful, the psychosocial layers were addressed. However, the highest expectation of such psychosocial management was that patients would simply learn to accept and live with their pain. This "traditional" approach is now giving way to a more holistic and integrated model that allows for the possibility of learning to live without pain (Moseley and Butler, 2015).

RETHINKING CHRONIC PAIN: THE NEED FOR A NEW PARADIGM

Chronic pain has, no doubt, been with humans through the ages. Yet as the resources of modern medicine have been brought to bear on this ancient problem, only limited treatment gains have been achieved. Problems of tolerance, opioid-induced hyperalgesia (Ballantyne and Shin, 2008) and addiction (although this is usually observed in those with a history of substance abuse (Porter and Jick, 1980) have major implications for pharmacotherapy. In fact, opioid deprescribing is now commonly recommended for people on maintenance opioids for chronic noncancer pain (National Prescribing Service; NSW TAG). Implanted devices such as intrathecal pumps and spinal cord stimulators have limited efficacy (Hayes et al., 2012) and high complication rates. Other procedural interventions are also limited in degree and duration of benefit. Furthermore, any unimodal use of biomedical strategies runs the risk of distracting the recipient from active management. The application of cognitive behavioral therapy (CBT) to chronic pain has broadened the therapeutic approach but the extent of benefit remains unclear (Eccleston et al., 2009).

Most experts agree that the prevalence of chronic pain has increased over the last two to three decades. Why should this be so? There are a number of possible reasons including the aging population, greater early involvement in high-intensity contact sports, changes in occupational activities (e.g., sitting for long periods in office work), and even a change in community tolerance and reporting of pain. A more ubiquitous potential driver, however, is the modern Western lifestyle. Not coincidentally, chronic pain is strongly associated with multiple other long-term comorbidities impacting both physical and mental health (Caughey et al., 2010; Lima et al., 2009).

The recent finding that chronic, low-grade, systemic inflammation or "metaflammation" is associated with lifestyle-related "inducers" (see Chapters 3 and 4) may provide an intriguing explanation of this convergence of morbidities (Egger and Dixon, 2010). Metaflammation is known to affect the vascular endothelium, in part mediated by activated macrophages, and is thus thought to play an underpinning role in atherosclerotic disease. Recent information, however, suggests it also affects a range of other organs and systems (Libby, 2010). A parallel process may occur in the central nervous system (CNS), for example, with activated glial cells contributing to inflammation and the facilitation of pain related pathways (Watkins et al., 2001; Marchand et al., 2005). Two types of glial cells have a potential role. Firstly, microglia are known to be activated by signals from damaged brain tissue, infection, and, interestingly, opioid therapy (possibly contributing to tolerance and opioid-induced hyperalgesia). Once activated, microglia release inflammatory cytokines that can sensitize the brain and spinal cord (Guo and Schluesener, 2007; Milligan and Watkins, 2009).

Secondly, astrocytes provide the structural support matrix for the CNS and constitute the blood–brain barrier or neurovascular junction. There is some evidence to suggest that astrocyte activation and the resultant inflammation can also contribute to the facilitation of pain related pathways (Cao and Zhang, 2008). The question yet to be answered is whether or not the lifestyle inducers of metaflammation in the vascular endothelium can also activate glial cells and contribute to chronic pain? It is reasonable to propose that obesity, poor diet, inactivity, inadequate sleep, smoking, and a range of environmental factors that are known to cause vascular inflammation (Egger and Dixon, 2010) could also have neurovascular effects. This could also help explain the increased prevalence of chronic pain in patients with these lifestyle risks.

"PLASTICITY" AND CHRONIC PAIN

An expanded model of care in pain management requires the recognition of "plasticity" in terms of both the nervous system, bodily tissues, and

TABLE 22.1 Models in Chronic Pain Management

Traditional (Dualistic) Model	Emerging (Holistic) Model
• medical *or* psychological focus	• +social *and* environmental focus
• more clinician centered	• more patient centered
• limited benefits for limited time	• significant, long-term benefits
• individual treatment approach	• individual+group treatments
• patient as recipient of treatment	• patient as partner in treatment
• potential dependency/complications	• limited dependency/complications
• distracts from "active" management	• involves "active" management
• "siloed" health system approach	• integrated health system approach
• neural plasticity disregarded	• neural plasticity vital for treatment
• individual health perspective only	• individual+population perspective
• ongoing/discontinued biomedical treatment	• time-limited biomedical treatment

psychological processes of the individual, and health systems design, from community and primary care through to the tertiary sector (Davies et al., 2011a). Examples of the differences between traditional and emerging approaches is shown in Table 22.1.

PLASTICITY AND PERSON-CENTERED CARE

The notion of brain "plasticity" is fundamental to new developments in pain management. There is recognition that the same neural changeability that contributes to the persistence of pain could potentially allow its resolution. Evidence suggests that "focused attention" can increase neural plasticity and hence be used to positively reprogram brain pathways (Doidge, 2008, 2015). "Person-centered care," shared decision-making, and seeing the person as a whole rather than as fragmented parts are also fundamental in any new approach (Pruitt and Epping-Jordan, 2005). Research shows that patient-centered care (see Chapter 6) can reduce the use of diagnostic tests and referrals by 50% and also improve illness recovery (Stewart et al., 2000).

The "whole-person" approach recognizes the importance of the mindbody connection in causation and management (White et al., 2016). Thus the emotional response to trauma and other life events becomes an important determinant of the healing process (Broom, 2007; Sarno, 2006). A whole-person view also incorporates the potential for increased gain from biomedical intervention. A new generation of pharmacotherapies are currently in development targeting the processes of neural sensitization and reorganization. These agents may ultimately play a role in reducing risk of chronic pain after injury.

For many patients, the goal may not be abolition of pain but adaptation to the pain. Accepting pain rather than expecting it to be medically cured or banished can often lead to improving quality of life, the focus shifting to managing the pain and living with it. Programs that include mindfulness meditation

(Zeidan et al., 2010), relaxation training, and CBT may help with this process. Nevertheless, the cutting edge of pain research acknowledges the potential for abolition of chronic pain and the "higher" goal of learning to live without pain.

PLASTICITY OF SYSTEMS

At a systems level, partnership and integration are key aspects of the emerging paradigm (Pruitt and Epping-Jordan, 2005). Various managed care approaches have been developed for chronic diseases with perhaps the best known coming from the Kaiser Permanente organization in the United States (Slaughter, 2000). Effective partnership between primary and tertiary care requires a system of stratification with the targeting of more intensive input to those with more complex problems.

Population health is another perspective to be considered at a systems level (Pruitt and Epping-Jordan, 2005). An Australian study of particular interest in this regard showed that a television-based community education campaign had a significant impact on beliefs about back pain and reduced work-related back pain presentations for up to three years after the intervention (Buchbinder and Jolley, 2005).

Contemporary examples of Australian systems plasticity related to chronic pain include the implementation of preassessment education and orientation groups (Davies et al., 2011b; Hayes and Hodson, 2011) and the use of a group rather than individual patient assessment format (Smith et al., 2016). Benefits include early neuroscience education, wait list reduction at pain services, and the fostering of an active self-management approach. Additional examples of systems plasticity with a focus on the broader community are the use of social media to publicize key messages about chronic pain and its treatment (White et al., 2016; YouTube: Brainman stops his opioids and Brainman chooses 2014) and the development of educational websites (http://www.aci.health.nsw.gov.au/chronic-pain and http://www.painhealth.csse.uwa.edu.au/).

Lifestyle emerges as a central dimension in pain management at the level of both the individual and society. Poorly balanced nutrition, overuse of recreational drugs, inadequate sleep, reduced physical activity, and stress all play a role in chronic pain as they do in other chronic disease processes.

PRACTICAL THERAPEUTICS: IMPLEMENTING THE PRINCIPLES OF PLASTICITY

The whole person management approach illustrated in Fig. 22.3 has been developed as a practical tool for use in guiding patients in the design of a pain management plan. There is an emphasis on therapeutic balance and the person in pain choosing which aspects of life to work on. The aim is to reprogram unhelpful patterns in mind and body (Fig. 22.3).

Overlapping components of the whole person model include:

Biomedical: This includes medical treatments such as medication, nerve blocks, and surgery, as well as "hands-on" therapies such as physiotherapy, chiropractic, osteopathy, and acupuncture. There is an overlap with mindbody techniques such as

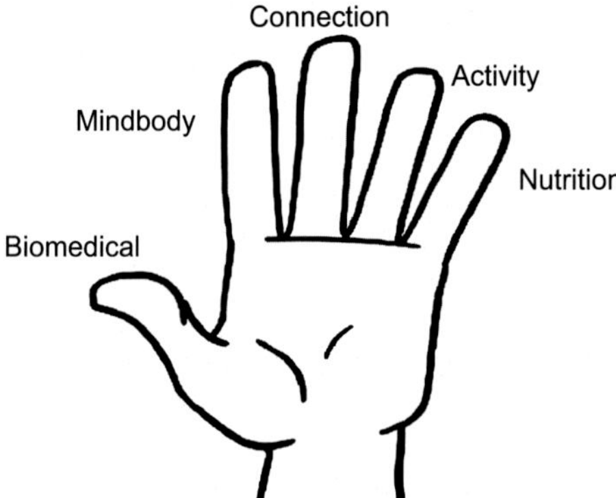

FIGURE 22.3 A whole person model. *Hunter Integrated Pain Service (HIPS) From White, R., Hayes, C., White, S., Hodson, F.J., 2016. Using social media to challenge unwarranted clinical variation in the treatment of chronic noncancer pain: the "Brainman" story. J. Pain Res. 9, 701–709.*

craniosacral and Feldenkrais therapy. Biomedical treatments typically work well for acute pain but are often less effective in chronic pain. The risks, such as medication side effects, and benefits need to be carefully considered. In the emerging paradigm, biomedical treatments can be used for a time-limited period in the first instance, creating a window of opportunity in which to develop active management strategies.

Mindbody: Thoughts and emotions have an immediate impact on the body. Unhelpful thought patterns (beliefs and expectations) and associated emotions (anxiety, for example) can contribute to physical health problems via the nervous, immune, and endocrine systems. In the reverse direction, physical health problems can produce changes in thoughts and emotions. The exercise of charting a timeline is one way of looking for important links between stressful periods of life and the onset of health problems such as chronic pain. Learning to be more aware of mind and body and the links between the two is a key aspect of treating pain. Cognitive behavioral therapy can help address issues around depression (which commonly coexists with pain), feelings of helplessness, anger, frustration, change in job status, and relationship concerns. Recent adaptions within cognitive behavioral therapy that show promise for chronic pain include acceptance commitment therapy and mindfulness (McCracken and Vowles, 2014). The use of any of these therapeutic approaches can help to wind down a sensitized nervous system with the potential to reduce pain over time.

Connection: Many people with chronic pain have a sense of disconnection or isolation relating to people (social), place (environment), or purpose. Recent research shows that social rejection causes a similar brain response to "physically" triggered pain (Eisenberger et al., 2003; Eisenberger, 2012). Thus one component of treating pain involves reestablishing lost connections. For some this is about spending more time in nature, for others volunteering or joining a group. In whatever form it takes, reconnecting can help to reduce nervous system sensitization and pain.

Activity: Our actions, like our thoughts and emotions, can easily become stuck in unhelpful patterns. Learning to "reprogram" activity is an important part of the overall brain retraining strategy. "Pacing" means finding the right balance and avoiding doing too little or too much. Gradually building activity helps to overcome the fear that there may be something dangerous and structurally wrong with the body. A comfortable daily walk is a commonly used treatment strategy in this area. The thought and the resultant action are clearly closely linked and can often be considered together in a process of "mindful" movement.

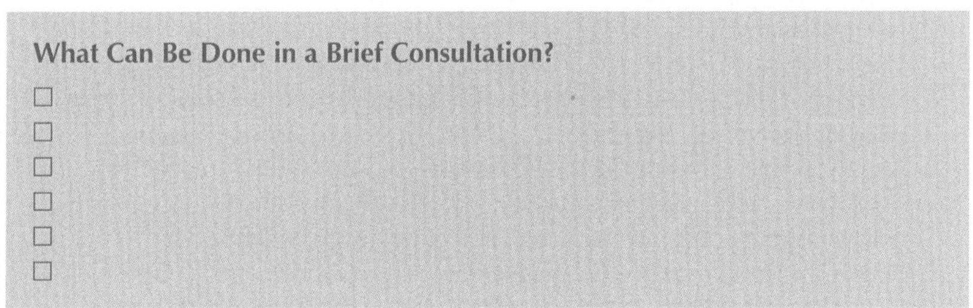

Nutrition: Nutrition can be considered to include all that we ingest (what we breathe, drink, and eat). Balancing the ratio of dietary protein to carbohydrate, avoiding excess refined carbohydrates, and optimizing intake of vegetables, fruits, antioxidants, and omega-3 essential fatty acids can bring benefit in multiple ways including reduction of chronic inflammatory processes.

SUMMARY

An expanded approach to treating chronic pain invites us to consider changes in the way we approach individuals with pain and also how we design health systems. At the level of the individual, a "whole person" approach brings greater therapeutic optimism. Nutrition and lifestyle approaches are balanced with traditional management elements in attempting to engage mindbody plasticity. The aim is both to improve coping strategies and to reduce pain intensity. In the context of multifaceted causation, the question arises of an individual's contribution to the persistence of his or her pain. As in other chronic conditions, this issue needs to be handled delicately in an environment that focuses on invitation and empowerment rather than blame.

At a systems level, plasticity is required, with the aim of developing a truly integrated approach. The prevalence and high cost of chronic pain require interprofessional collaboration and the balancing of resources across the spectrum from community to tertiary care. In improving systems integration, the challenge lies in developing a stratified approach that includes appropriate education at the community level, coordination of management in primary care, and improving access and therapeutic approaches at the tertiary level.

Key Points

Clinical Management

- Develop a system to "triage" patients according to complexity of pain management needs;
- Consider the "whole person" rather than a fragmented approach to pain management;
- Recognize that neural "plasticity" may perpetuate pain and paradoxically may also help to reduce it;
- Consider selective use of pharmacological treatment in chronic pain for a time-limited period to provide a "window of opportunity" for developing active management strategies;
- Use a team approach to pain management potentially including "hands-on" therapists (time limited), exercise specialists, physiotherapists, psychologists, nutritionists, nurses, and doctors;
- Aim to change unhealthy lifestyle behaviors (i.e., poor nutrition, inactivity, inadequate sleep, stress, and anxiety), with the anticipation that this may contribute to gradual pain reduction and other health benefits.
- Do not underestimate psychological, cultural, and work issues as mediating factors in chronic pain.

Practice Tips

Chronic Pain

	Medical Practitioner	Practice Nurse or Allied Health Professional
Assess	• type, extent, duration, location of pain • immediate medication need • need for further investigation • level of assistance required	• potential allied health care involvement • level of impairment • potential for self-management • suitability of allied health providers
Assist	• medication prescription • validation of holistic approaches • workplace negotiation	• "pacing" of activities • setting of goals • nutrition • relaxation • development of pain management plan
Arrange	• team care arrangement • triaging of care • specialist referral if required	• allied health team • group self-management if available • multidisciplinary team meetings/reporting

REFERENCES

Ballantyne, J.C., Shin, N.S., 2008. Efficacy of opioids for chronic pain: a review of the evidence. Clin. J. Pain 24 (6), 469–478. http://dx.doi.org/10.1097/AJP.0b013e31816b2f26.

Brinjikji, W., Luetmer, P.H., Comstock, B., Bresnahan, B.W., Chen, L.E., et al., 2015. Systematic literature review of imaging features of spinal degeneration in asymptomatic populations. Am. J. Neuroradiol. 36 (4), 811–816.

Broom, B., 2007. Meaning-Full Disease. Karnac Books, London.

Buchbinder, R., Jolley, D., 2005. Effects of a media campaign on back beliefs is sustained 3 years after its cessation. Spine 30 (11), 1323–1330.

Cao, H., Zhang, Y.Q., 2008. Spinal glial activation contributes to pathological pain states. Neurosci. Biobehav. Rev. 32 (5), 972–983.

Caughey, G.E., Roughead, E.E., Vitry, A.I., et al., 2010. Comorbidity in the elderly with diabetes: identification of areas of potential treatment conflicts. Diabetes Res. Clin. Pract. 87 (3), 385–393.

Davies, S.J., Hayes, C., Quintner, J.L., 2011a. System plasticity and integrated care: informed consumers guide clinical reorientation and system reorganisation. Pain Med. 12 (1), 4–8.

Davies, S., Quinter, J., Parsons, R., Parkitny, L., Knight, P., Forrester, E., et al., 2011b. Preclinic group education sessions reduce waiting times and costs at public pain medicine units. Pain Med. (12), 59–71.

Doidge, N., 2008. The Brain that Changes Itself. Viking Penguin, NY.

Doidge, N., 2015. The Brain's Way of Healing. Scribe Publications, London.

Eccleston, C., Williams, A.C., Morley, S., 2009. Psychological therapies for the management of chronic pain (excluding headache) in adults. Cochrane Database Syst. Rev. 15 (2), CD007407.

Egger, G., Dixon, J., 2010. Inflammatory effects of nutritional stimuli: further support for the need for a big picture approach to tackling obesity and chronic disease. Obes. Rev. 11 (2), 137–145.

Eisenberger, N.I., 2012. The neural bases of social pain: evidence for shared representations with physical pain. Psychosom. Med. 74 (2), 126–135.

Eisenberger, N.I., Lieberman, M.D., Williams, K.D., 2003. Does rejection hurt? An FMRI study of social exclusion. Science 302, 290–292.

Engel, G.L., 1977. The need for a new medical model: a challenge for biomedicine. Science 196 (4286), 129–136.

Lima, M.G., Barros, M.B., César, C.L., et al., 2009. Impact of chronic disease on quality of life among the elderly in the state of São Paulo, Brazil: a population-based study. Rev. Panam. Salud Publica 25 (4), 314–321.

Guo, L.H., Schluesener, H.J., 2007. The innate immunity of the central nervous system in chronic pain: the role of Toll-like receptors. Cell. Mol. Life Sci. 64, 1128–1136.

Hayes, C., Hodson, F.J., 2011. A whole-person model for care for persistent pain: from conceptual framework to practical application. Pain Med. 12, 1738–1749.

Hayes, C., Jordan, M.S., Hodson, F.J., Ritchard, L., 2012. Ceasing intrathecal therapy in chronic non-cancer pain: an invitation to shift from biomedical focus to active management. PLoS One 7 (11), e49124. http://dx.doi.org/10.1371/journal.pone.0049124.

Kendrick, D., Fielding, K., Bentley, E., Kerslake, R., Miller, P., Pringle, M., 2001. Radiography of the lumbar spine in primary care patients with low back pain: randomised controlled trial. BMJ 2001 (322), 400–405.

Kiecolt-Glaser, J.K., Glaser, R., 1995. Psychoneuroimmunology and health consequences: data and shared mechanisms. Psychosom. Med. 57 (3), 269–274.

Libby, P., 2010. How our growing understanding of inflammation has reshaped the way we think of disease and drug development. Clin. Pharmacol. Ther. 87 (4), 389–391.

Loeser, J.D., 1980. Perspectives on pain. In: Turner, P. (Ed.), Proceedings of the First World Congress on Clinical Pharmacology and Therapeutics. McMillan, London, pp. 313–316.

Marchand, F., Perretti, M., McMahon, S.B., 2005. Role of the immune system in chronic pain. Nat. Rev. Neurosci. 6, 521–532.

McCracken, L.M., Vowles, K.E., 2014. Acceptance and commitment therapy and mindfulness for chronic pain: model, process, and progress. Am. Psych. 69 (2), 178–187.

Melzack, R., Wall, P.D., 1965. Pain mechanisms: a new theory. Science 150 (3699), 971–979.

Merskey, H., Bogduk, N., 1994. Classification of Chronic Pain. International Association for the Study of Pain.

Milligan, E.D., Watkins, L.R., 2009. Pathological and protective roles of glia in chronic pain. Nat. Rev. Neurosci. 10, 23–36.

Moseley, G.L., Butler, D.S., 2015. 15Years of explaining pain: The past, present and future. J. Pain 16 (9), 807–813.

National Pain Strategy, March 2010. Developed by the National Pain Sumit. Canberra. www.painsummit.org.au (accessed April, 2010).

National Prescribing Service. http://www.nps.org.au/conditions/nervous-system-problems/pain/for-individuals/pain-conditions/chronic-pain/for-health-professionals/opioid-medicines/discontinuing-opioids.

NSW TAG http://www.ciap.health.nsw.gov.au/nswtag/documents/publications/guidelines/pain-guidance-july-2015.pdf.

Porter, J., Jick, H., 1980. Addiction rate in patients treated with narcotics. NEJM 302, 123.

Pruitt, S.D., Epping-Jordan, J.E., 2005. Preparing the 21st century global healthcare workforce. BMJ 330, 637–639.

Sarno, J.E., 2006. The Divided Mind: The Epidemic of Mindbody Disorders. Harper Collins, NY.

Slaughter, J., 2000. Kaiser Permanente: integrating around a care delivery model. J. Ambul. Care Manage. 23 (3), 39–47.

Smith, N., Jordan, M., White, R., Bowman, J., Hayes, C., 2016. Assessment of Adults Experiencing Chronic Non-cancer Pain: A Randomized Trial of Group Versus Individual Format at an Australian Tertiary Pain Service. http://dx.doi.org/10.1093/pm/pnv048 pnv048. First published online: 6 January 2016.

Stewart, M., Brown, J.B., Donner, A., McWhinney, I.R., Oates, J., et al., 2000. The impact of patient-centered care on outcomes. J. Fam. Pract. 49, 796–804.

Turk, D.C., Okifuji, 2001. Pain terms and taxonomies of pain. In: Loeser, J.D., Bonica, J.J. (Eds.), Bonica's Management of Pain, third ed. Lippincott Williams & Wilkins, Philadelphia. ISBN: 0683304623.

Watkins, L.R., Milligan, E.D., Maier, S.F., 2001. Glial activation: a driving force for pathological pain. Trend Neurosci. 24 (8), 450–455.

White, R., Hayes, C., White, S., Hodson, F.J., 2016. Using social media to challenge unwarranted clinical variation in the treatment of chronic noncancer pain: the "Brainman" story. J.Pain Res. 9, 701–709.

Zeidan, F., Gordon, N.S., Merchant, J., Goolkasian, P., 2010. The effects of brief mindfulness meditation training on experimentally induced pain. J. Pain 11 (3), 199–209.

Professional Resources

Measuring Pain

Brief Pain Inventory

Reproduced with acknowledgment of the Pain Research Group
The University of Texas MD Anderson Cancer Center, United States

Date: ______________________

Name: ______________________

1. On the diagram, shade in the areas where you feel pain. Put an X on the area that hurts most.

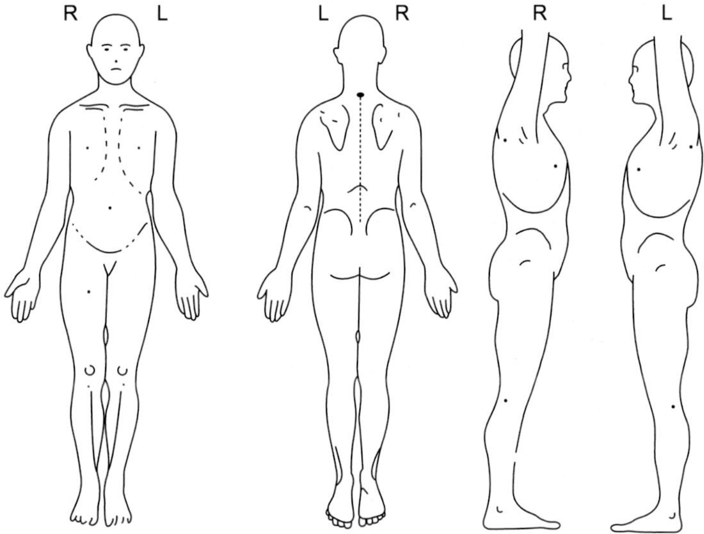

2. Please rate your pain by circling the one number that best describes your pain at its worst in the last 24 hours.

0	1	2	3	4	5	6	7	8	9	10

No pain — Pain as bad as you can imagine

3. Please rate your pain by circling the one number that best describes your pain at its least in the last 24 hours.

0	1	2	3	4	5	6	7	8	9	10

No pain — Pain as bad as you can imagine

4. Please rate your pain by circling the one number that best describes your pain on average.

0	1	2	3	4	5	6	7	8	9	10

No pain — Pain as bad as you can imagine

Continued

Professional Resources—cont'd

5. Please rate your pain by circling the one number that tells how much pain you have right now.

0	1	2	3	4	5	6	7	8	9	10

No pain — Pain as bad as you can imagine

6. What treatments or medications are you receiving for your pain?
7. In the last 24 hours, how much relief have pain treatments or medications provided? Please circle the one percentage that best shows how much relief you have received.

0%	10%	20%	30%	40%	50%	60%	70%	80%	90%	100%

No relief — Complete relief

8. Circle the one number that describes how, during the past 24 hours, pain has interfered with your:

   a. General activity

0	1	2	3	4	5	6	7	8	9	10

Does not interfere — Completely interferes

   b. Mood

0	1	2	3	4	5	6	7	8	9	10

Does not interfere — Completely interferes

   c. Walking ability

0	1	2	3	4	5	6	7	8	9	10

Does not interfere — Completely interferes

   d. Normal work (includes both outside the home and housework)

0	1	2	3	4	5	6	7	8	9	10

Does not interfere — Completely interferes

   e. Relations with other people

0	1	2	3	4	5	6	7	8	9	10

Does not interfere — Completely interferes

   f. Sleep

0	1	2	3	4	5	6	7	8	9	10

Does not interfere — Completely interferes

   g. Enjoyment of life

0	1	2	3	4	5	6	7	8	9	10

Does not interfere — Completely interferes

Professional Resources—cont'd

Brief Pain Inventory Scoring Instructions

1. Pain Severity Score
   This is calculated by adding the scores for questions 2, 3, 4, and 5 and then dividing by 4. This gives a severity score out of 10.
2. Pain Interference Score
   This is calculated by adding the scores for questions 8a, b, c, d, e, f, and g and then dividing by 7. This gives an interference score out of 10.

Chapter 23

Understanding Addictions: Tackling Smoking and Hazardous Drinking

John Litt, Caroline West

Addiction must be viewed at the outset as a personal problem rather than a chemical problem. To discuss the alcohol problem, the heroin problem or the marijuana problem makes as much sense as discussing suicide as the hanging problem, the wrist slashing problem and the bridge jumping problem.

INTRODUCTION: THE DIFFERENT FORMS OF ADDICTION

Smoking

Amongst preventable diseases worldwide, smoking offers perhaps the greatest scope for prevention. Daily smoking rates have declined to just over one in eight of the population (AIHW, 2014a,b). One in two smokers will die from a smoking-related illness, and more than half of these deaths will occur in middle age (25–54 years). On average, smoking causes a 12-year loss of life expectancy and consumes around 7% of health care expenditures (National Center for Chronic Disease, 2014). Cessation before the age of 40 years reduces the risk of death associated with continued smoking by about 90% (National Center for Chronic Disease, 2014). Yet, despite the enormous adverse impact of smoking on health, many people throughout the world (and particularly now the developing world) continue to smoke daily. Interestingly, up to two-thirds of smokers in advanced Western countries are thinking about quitting in the next six months and nearly half have made an attempt to quit in the last 12 months (Scollo and Winstanley, 2015). However, there are mixed views on the percentage achieving permanent abstinence, from around 5% (Baillie et al., 1995; Edwards et al., 2014) to two-thirds (Smith et al., 2015).

Lifestyle Medicine. http://dx.doi.org/10.1016/B978-0-12-810401-9.00023-1

Alcohol

Hazardous alcohol consumption is associated with considerable morbidity and mortality (Rehm et al., 2015b; National Health and Medical Research Council, 2009). The leading causes of death are liver disease followed by road crash injury, cancer, and suicide. More people (particularly those amongst 15- to 19-year-olds) die from the acute effects of alcohol, while more of those 45 years and over died from chronic effects.

One in five Australians drinks at levels that puts them at risk from lifetime harm (AIHW, 2014a,b). Alcohol is a major contributor to the global burden of disease. Up to 10% of cancers, 20% of intentional injuries, and 7% of all deaths and 3.9% of disability-adjusted life years can be attributed to excessive alcohol use. Heavy drinking also increases the risk of high blood pressure, stroke, unintentional injuries, and cancer. It is implicated in a significant number of hospital admissions and imposes a significant cost burden on society. Heavier consumption has been consistently associated with poorer quality of life and increased mortality (Rehm et al., 2015b; National Health and Medical Research Council, 2009). The effects of high-risk drinking on the body are shown in Fig. 23.1.

Lifetime risky drinkers (1 in 5 in the population) are defined as people who consume more than two standard drinks per day (on average over a 12-month period). Single-occasion risky drinkers (1 in 4 in the population) are defined as people consuming five or more standard drinks on a single drinking occasion (AIHW, 2014a,b).

WHY IS ADDRESSING SMOKING AND DRINKING SO CHALLENGING?

To understand this question, we need to answer another question: Why do people drink and smoke? The common reasons for both smoking and drinking include: relaxation purposes, it increases confidence, better sleep, reducing anxiety, dealing with stress/panic attacks, depression, dependence, or to counteract withdrawal symptoms.

At the clinical level there are a number of reasons why proposing cutting down (Heather, 2012; Yoast et al., 2008; Litt, 2007) or stopping these behaviors is difficult (Twyman et al., 2014; Litt, 2007). These include the following.

CLIENT FACTORS

Both smoking and drinking are social activities that provide both enjoyment and a way for many people to reduce stress and put them at ease. These are potent influences on maintaining both habits. For drinking, there are strong social norms supporting a culture of high alcohol use.

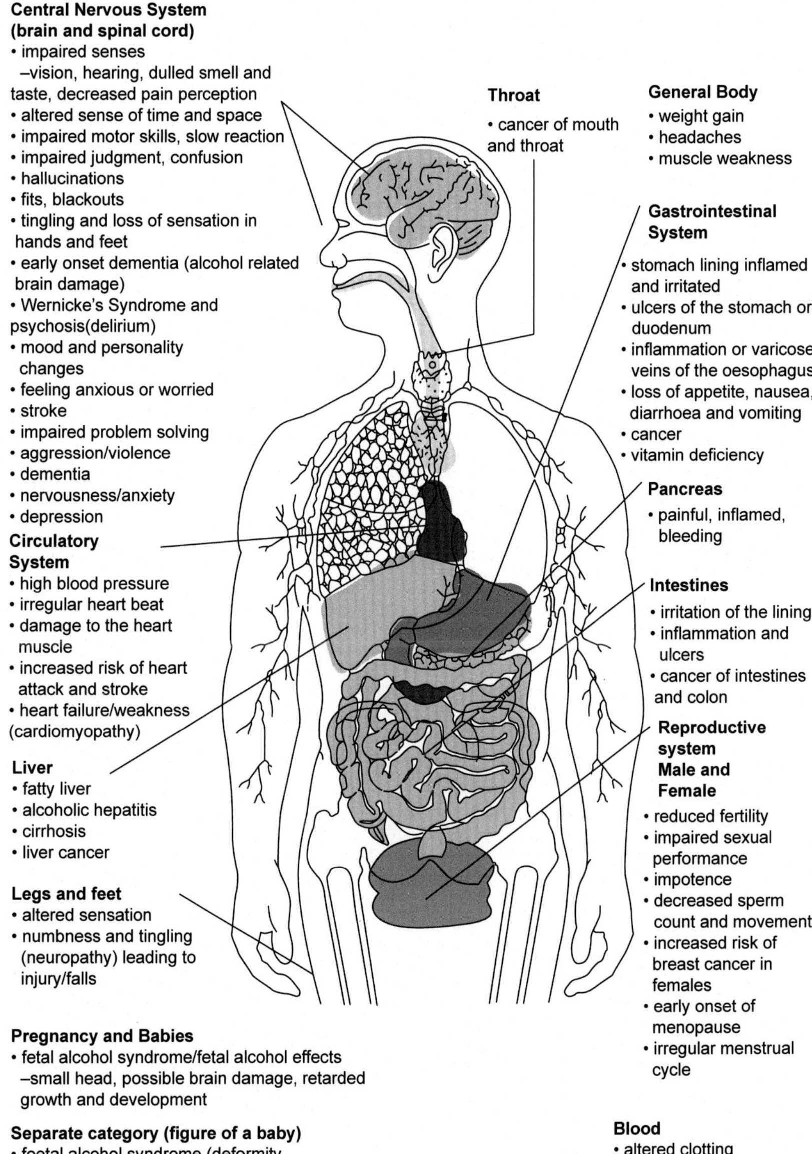

FIGURE 23.1 Effects of long-term use and misuse of alcohol.

Lack of Knowledge of Health Effects and Benefits of Treatment

Most studies report that smokers are aware of the health risks associated with smoking. Nevertheless, smokers, when contrasted with ex-smokers, are more likely to:

- have less health-related knowledge about the effects of smoking
- express less concern about the health effects of smoking
- downplay the health-related risks of smoking
- report greater potential for harm from nicotine medications over the nicotine in cigarettes
- be less aware of effective treatment strategies for quitting
- be overly optimistic about their ability to quit at any time

Heavy drinkers who perceive that their drinking has more risk associated with it are more likely to consider cutting down. For men drinking, on average, six to eight standard drinks per day, one in five is likely to develop an alcohol use disorder (National Health and Medical Research Council, 2009; Rehm et al., 2015a). Lack of awareness of the effectiveness of a number of treatments is also a barrier for both smokers and hazardous drinkers.

Lack of Concern

Many heavy drinkers view their consumption as moderate and not harmful. This can contribute to the mixed perceptions that individuals have on both the role and effectiveness of health care workers in assisting them to cut down their drinking. Prevention activities or health-related behaviors are also influenced by the individual's perception of his or her susceptibility or risk of contracting a disease that is related to the lifestyle behavior and whether he or she believes that the behavior is harmful to health. Less than 1 in 10 with evidence of alcohol problems or dependence seeks help or treatment (Oleski et al., 2010; Teesson et al., 2006; Wu and Ringwalt, 2004).

Sensitivity of the Topic

Many individuals find discussion of their smoking or drinking habits a sensitive issue. The negative stereotyping associated with a health-related behavior or illness can impact on the likelihood that a person seeks help or assistance.

Underestimation of the Difficulty Associated With Changing Behavior

For smokers and hazardous drinkers, a significant barrier to making changes is the perception that they may be able to sort the problem out themselves without seeking help, if they really wish to.

Many health-related behaviors are difficult to change. For smokers, there are a large number of barriers that make it difficult to quit and that also influence the smoker's preparedness to seek help. These include:

- addiction to nicotine
- perceived difficulty with quitting
- fear of weight gain
- concerns about dealing with stress
- lack of support
- presence of other smokers in the household

Reluctance to Seek Help

Less than one-third of smokers seek help from their general practitioner (GP) to assist them with quitting, and <10% of smokers in Australia have called a QUITline. A number of smokers report a sense of fatalism or that it is too late to do something. Nevertheless, most smokers are interested in quitting and more than half had made at least one attempt to quit. Hazardous drinkers, similarly, find it difficult to cut back. The barriers to doing this include:

- physical dependence on alcohol
- lack of support
- difficulty coping with stress and using alcohol as a coping strategy
- stigmatization associated with the label of problem drinker
- reluctance to seek help, especially from medical sources
- lack of self-efficacy

CLINICIAN FACTORS

Willingness to Enquire About Smoking and Drinking

Health care workers are ambivalent in asking about some aspects of lifestyle. Some feel they cross the boundary of privacy when they ask about health-related behaviors that are seen by their clients as personal choices. Such enquiry may be justifiable only when the health-related behavior can be linked to the client's presenting complaint. Thus, health care workers may be reluctant to intervene if their clients do not see their drinking as a problem. As most clients who are drinking at hazardous levels do not seek care, it is not surprising that most of this group are not being offered assistance or even treatment.

In the general practice setting, less than one in three patients is asked about their drinking and two-thirds are asked about their smoking. Men are more likely to be asked about their drinking than women.

Knowledge

The knowledge of health care workers about the guidelines for low-risk levels of drinking is variable. Similarly, there is a wide spectrum of awareness and/or confidence of the effectiveness of behavioral interventions for unhealthy lifestyles.

Attitudes and Beliefs

Many health care workers are pessimistic about their effectiveness in facilitating behavior change in patients who smoke, or who drink at hazardous levels, despite good evidence to the contrary.

Skills and Practices

Health care workers underutilize effective strategies with both hazardous drinkers and smokers, including motivational interviewing techniques and brief behavioral interventions. Referral of either smokers to the QUITline or hazardous drinkers to appropriate treatment services is also uncommon. Health care workers' own health practices can also influence their likelihood of providing lifestyle counseling. Health care workers who smoke, do little exercise, or eat a less than healthy diet, are less likely to counsel about the related lifestyle area.

PRACTICE SETTING

Lack of Time

Lack of time is a major barrier to the effective delivery of prevention. Most lifestyle-related activities, including counseling, is provided opportunistically. The disposable time available to tackle issues on the clinician's agenda, after dealing with the reasons that the patient has come to the doctor, is generally limited. This influences the nature and extent of the counseling that can be provided.

Lack of Organizational Infrastructure

The most successful lifestyle implementation strategies use the skills of the entire health care team and the practice infrastructure to provide timely, pertinent information during a medical visit and subsequently (Keurhorst et al., 2015).

In the GP setting, a supportive organizational infrastructure for lifestyle intervention includes:

- Practice policies, e.g., standing orders, protocols and procedures;
- Clarification of roles and responsibilities;
- Accessible, evidence-based guidelines;
- Availability and use of appropriate resources, e.g., practice nurses;
- Multifaceted delivery options, e.g., delegation to a practice nurse, use of multidisciplinary clinics and group sessions, appropriate referral groups or agencies;
- Information management and information technology systems, e.g., practice registers (age/sex/disease), recall and reminder systems, health summaries;
- Waiting room materials, e.g., notice board, patient education, and information material;
- Screening and information gathering materials and strategies, e.g., patient questionnaire, inventory of practice prevention, clinical audit;

- Consultation materials, e.g., decision aids, computer decision support systems, and patient education and information material.

The barriers discussed herein contribute to the reluctance of health care workers to be involved in both screening for alcohol consumption and associated risks and problems *and* providing brief intervention.

WHAT WORKS?

Smoking Cessation Guidelines

Smoking status should be assessed for every patient over 10 years of age. Patients who smoke, regardless of the amount they smoke, should be offered regular brief advice to stop smoking (Zwar et al., 2014).

Alcohol Guidelines

All patients should be asked about the quantity and frequency of alcohol intake. Those drinking, on average, more than two standard drinks a day or four standard drinks per occasion should be offered brief advice to reduce their intake. The Australian guidelines to reduce health risks from drinking alcohol are shown in Table 23.1.

TABLE 23.1 Australian NHMRC Alcohol Guidelines 2009 (NH & MRC, 2009)

Guideline I
Reducing the Risk of Alcohol-Related Harm Over a Lifetime
The lifetime risk of harm from drinking alcohol increases with the amount consumed.
For healthy men and women, drinking no more than two standard drinks on any day reduces the lifetime risk of harm from alcohol-related disease or injury.
Guideline 2
Reducing the Risk of Injury on a Single Occasion of Drinking
On a single occasion of drinking, the risk of alcohol-related injury increases with the amount consumed.
For healthy men and women, drinking no more than four standard drinks on a single occasion reduces the risk of alcohol-related injury arising from that occasion.
Guideline 3
Children and Young People Under 18 Years of Age
For children and young people under 18 years of age, not drinking alcohol is the safest option. 1. Parents and carers should be advised that children under 15 years of age are at the greatest risk of harm from drinking and that for this age group, not drinking alcohol is especially important. 2. For young people aged 15–17 years, the safest option is to delay the initiation of drinking for as long as possible.

AN ACRONYM FOR TREATMENT: THE 5 A'S

GPs are a credible, authoritative, effective, and cost-effective source of advice on how to stop smoking or cut down on hazardous drinking.

The number of smokers needed to treat to prevent one death is 18, assuming optimal (best practice) treatment. If minimal (brief advice is provided), the number needed to treat is 52 (Hughes, 2010). For subjects drinking at hazardous levels, 1 in 10 needs to be counseled to produce one subject who will reduce drinking to low risk levels (Ballesteros et al., 2004).

The key components of brief interventions for smoking cessation and assisting hazardous drinkers are the 5 A's—Ask, Assess, Advise, Assist, and Arrange—reduced in this instance for better recall to the three A's used throughout this book: Assess, Assist, and Arrange. The first step is to routinely identify patients who drink hazardously or who smoke. Using a patient survey in the waiting room can help to identify these groups Table 23.2 outlines the key components of the 5 A's related to alcohol.

RAISING THE ISSUE

The next step is to raise the issue of patients' smoking or drinking in a sensitive manner. Just asking about drinking may have an impact (McCambridge

TABLE 23.2 Brief 5A's Approach to Assessing and Assisting Those Whose Drinking May Be Affecting Their Health

Ask	"How do you feel about your drinking at the moment?" Quantify both frequency and amount.
Assess	Level of drinking (using Q1-3 of AUDIT-C); if high risk, consider more intensive intervention. Interest and confidence in cutting or stopping (e.g., ask, "Are you willing to try cutting your alcohol consumption?"). Ask them to rate their interest on a scale of 0–10 where 0 = no interest and 10 = very interested. Repeat the process for assessing confidence barriers to change (e.g., "What would be the hardest thing about cutting down?"). Record alcohol patterns in patient's record.
Advise	Provide brief, nonjudgmental advice about positive benefits of cutting down.
Assist	Offer resources and support, including alcohol information/ resources. Consider referral and encourage social support.
Arrange	Follow-up review. A separate consultation about alcohol.

Modified from the revised Lifescripts alcohol guidelines and the RACGP SNAP guide, 2nd edition, 2015 (RACGP, 2015).

and Kypri, 2011). Practical tips to minimize the likelihood of reactivity and aggravation include:

- Normalize the enquiry (ask all your patients and provide a context for asking).
- Avoid making a judgment prematurely. Premature advice or attempting to "close the loop" too soon is likely to be met with resistance.
- Provide the patient with time and space to work through the consequences of the information. Seeing a connection between a behavior and health consequences is a necessary first step in the change process. Just because the doctor regards it as a problem doesn't mean the patient does. Acceptance and preparedness to change takes time.
- Understand the patient's perspective. Ask how he or she feels about their behavior. What do they get out of it? What are the benefits? Acknowledge that changing behavior is difficult.
- Maintain rapport/therapeutic alliance by working with the patient. Ensuring agreement on what the problem is and respecting the patient's autonomy both facilitates the communication process and improves the likelihood of success.
- Leave the door open. Patients see their GP 5 or 6 times a year. The next consultation could provide an opportunity to follow up where you left off.

ASSESSMENT

The next step is to use motivational interviewing techniques to assess the patient's interest and confidence in changing by quitting smoking or cutting down on drinking. Explore the patient's perceptions of any costs and the perceived benefits of cutting down on drinking or quitting smoking. Patient sensitivity to enquiry is lower and responsiveness is higher when information is collected indirectly through tools such AUDIT-C (see Professional resources), rather than face-to-face. Self-report screening questionnaires are more sensitive and accurate than the biological markers of heavy alcohol use. A careful systematic enquiry is the most valid measure of alcohol intake. Alcohol-related harms increase with increasing AUDIT-C scores (Rubinsky et al., 2013).

Dependence on nicotine or alcohol should also be assessed. Time to first cigarette (less than 1 h), smoking more than 10 cigarettes a day, and evidence of withdrawal with previous quit attempts are associated with nicotine dependence. Features of alcohol dependence include:

- tolerance
- inability to control use despite evident harms
- withdrawal symptoms (e.g., tremor, sweating, anxiety, nausea and vomiting, agitation, and insomnia)

Assess also whether the patient's alcohol intake is affecting their health and relationships (Haber et al., 2009). This can be done using the acronym

LIMPED. It is useful to remember that hazardous and harmful drinking often go undetected.

L	**Legal** problems (e.g., driving under the influence of alcohol, assaults, etc.)
I	**Injury** (e.g., any trauma or motor vehicle accidents). Alcohol contributes to1 in 5 of all injuries, 1 in 3 falls and drownings, 30% of motor vehicle accidents, and 1 in 8 suicides.
M	**Medical problems**. Patients drinking more than the NHMRC recommended levels long term are more likely to have a range of medical problems. These include a range of cancers, including lips, mouth, throat and esophagus, stomach, pancreas, and liver; fatty change (hepatic steatosis) and inflammation (steatohepatitis) of the liver, cirrhosis of the liver; brain insult/injury including cognitive problems and dementia and Wernicke–Korsakoff syndrome (thiamine deficiency) resulting in brain hemorrhages; increased risk of cardiovascular diseases (e.g., hypertension, hemorrhagic stroke, ischemic stroke, heart failure, and cardiomyopathy); peripheral neuropathy (limb muscle weakness); and male sexual impotence.
P	**Psychological and psychiatric problems** (e.g., anxiety, depression, sleeping difficulties). Heavy drinking can also aggravate symptoms in people with milder degrees of anxiety and depression. While alcohol consumption may bring some short-term relief from anxiety or stress, it tends to worsen mood in the longer term, especially at higher levels of consumption.
E	**Employment problems** (e.g., fall off in work performance, accidents, etc.). Alcohol contributes to 10% of industrial accidents and 13% of all employees on sick leave have alcohol problems.
D	**Domestic problems** (e.g., domestic, other, violence, financial worries). Forty to 50 percent of all incidents involving aggression occur while under the influence of alcohol.

If the patient acknowledges that any of these areas may be relevant, ask about whether they (or others) have concerns about this. It is also useful to assess the barriers to change and high-risk situations. Many of these were listed earlier.

MAKING CHANGES

Providing clear actionable advice, with an explanation of how the person's symptoms can be related to their smoking or drinking (the mechanisms of the cause) can help, for example:

For drinking: The level of alcohol you are consuming is probably contributing to your stomach pain. Alcohol increases the production of acid in the stomach, which delays stomach emptying and helps to weaken the protective layer of mucus lining the stomach. Both these effects are likely to be contributing to your gastritis. Cutting down on your drinking will help resolve the inflammation.

For smoking: When you smoke around your children, the tobacco smoke tends to weaken the cilia in the respiratory tract. The cilia are like little vacuum cleaners that help to mop up dust and infective agents (bacteria and viruses) and remove them from the respiratory tract. Children exposed to environmental tobacco smoke are more prone to respiratory infections, including meningococcal meningitis (4–6 times more likely to get it), and have more exacerbations of their asthma.

The goals for change should be both incremental and realistic. This is more challenging for smoking, where quitting is the only long-term option. Cutting down can be an interim strategy but it is often associated with a change in the patient's inhalation pattern to redress the declining nicotine levels. For many hazardous drinkers, cutting down is a reasonable end point. Abstinence is more appropriate if there is evidence of alcohol dependence or other physical or mental impairment from alcohol use.

Behavioral self-management training can be used to foster self-management strategies in the patient who is drinking heavily. Strategies include: self-monitoring, e.g., using a drink diary; coping skills, training, setting drinking limits; controlling rates of drinking; identifying high-risk drinking situations; and identifying rewards for limiting drinking.

PHARMACOTHERAPY

Pharmacotherapy is useful for the patient who is dependent upon either nicotine or alcohol. Nicotine replacement therapies (NRTs) are very effective when there is nicotine dependence. Most of the NRTs contain less nicotine than the patient generally smokes. Occasionally the patient may need two forms of NRT, e.g., the nicotine patch for basal levels of nicotine, supplemented by using the nicotine lozenge in the morning to deal with early morning cravings.

For patients dependent on alcohol (typically drinking 6–10 standard drinks every day or most days) together with evidence of tolerance and associated harms, acamprosate (which blocks glutamate receptors and activates gamma-aminobutyric acid receptors) and naltrexone (an opioid receptor antagonist) are useful agents. Both can be used to reduce cravings and increase the likelihood of the patient remaining abstinent. Acamprosate also reduces the relapse rates, has no significant drug interactions, and has fewer side effects than naltrexone. Naltrexone has the advantage of once-a-day dosing and a fairly rapid onset. Naltrexone reduces cravings for alcohol, the number of drinking days, and the total amount consumed in about 40–60% of patients. The disadvantages are the blocking of opioid analgesia and the numerous drug interactions. Unfortunately, patient adherence with both medications is relatively low.

All patients who receive some assistance with quitting or cutting down should be followed up. Follow-up significantly increases the likelihood of a patient quitting or cutting down on their drinking. For smokers, referral to the QUIT program is particularly useful. An active QUIT (phone) line support program in the first three months of quitting can significantly boost long-term quit rates to 20–30%. It is also useful to routinely offer follow-up appointments.

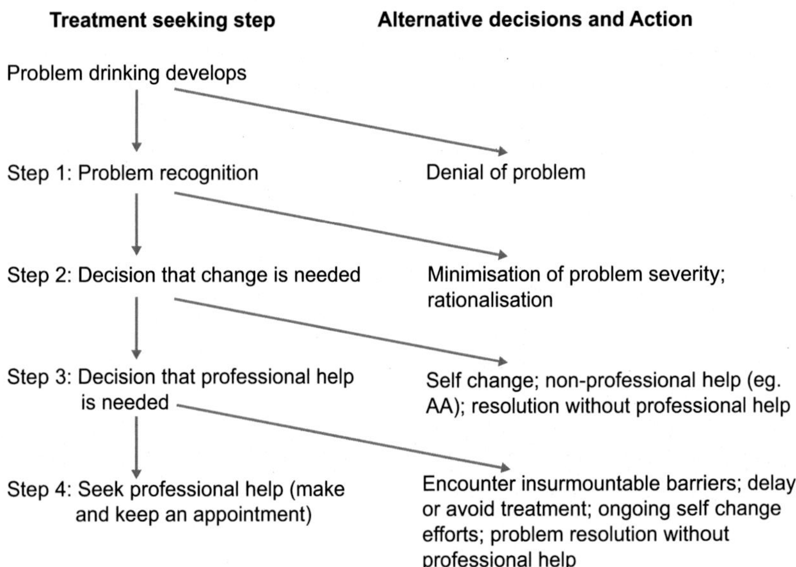

FIGURE 23.2 Steps in the process of seeking help for alcohol consumption (Saunders et al., 2006).

WHAT STIMULATES CHANGE?

The most commonly given reasons for change by hazardous drinkers or those with alcohol-related problems are related to health and concerns with money and relationships. For smokers, health-related concerns and financial costs top the list of factors that stimulate the need to quit.

While patient motivation is a strong predictor of treatment outcomes in patients who drink at hazardous levels, many hazardous drinkers express little interest in cutting down and some are only prepared to cut down when they hit "rock bottom."

It is also useful to understand the factors that underpin a person's decision *not* to change. Smokers who have never tried to quit, commonly:

- do not want to quit
- are in good health (so don't see any impact of their smoking)
- have friends and family who are not trying to get them to quit
- report that they have had no advice or any helpful advice to quit from a health professional (Sharma and Szatkowski, 2014)

For subjects with problem drinking, it can be helpful to recognize the steps in the process of seeking help as the barriers may differ (see Fig. 23.2).

Alcohol Dependence

For patients with moderate to severe alcohol use disorders (i.e., those meeting five or more DSM-IV-TR criteria for alcohol abuse or dependence (American Psychiatric Association, 2000)), it is best to recommend abstinence from alcohol. See Table 23.3.

TABLE 23.3 Diagnostic Criteria for Alcohol Abuse and Dependence

Abuse: Patient must meet one or more criteria without meeting criteria for dependence
Determine whether, in the past 12 months, the patient's drinking has repeatedly caused or contributed to the following: Risk of bodily harm (drinking and driving, operating machinery, swimming) Relationship trouble (family or friends) Role failure (interference with home, work, or school obligations) Run-ins with the law (arrests or other legal problems)
Dependence: Patient must meet three or more criteria
Determine whether, in the past 12 months, the patient has experienced the following: Not able to stick to drinking limits (repeatedly exceeded them) Not able to cut down or stop drinking (repeated failed attempts) Spent a significant amount of time drinking (or anticipating or recovering from drinking) Continued drinking despite problems (recurrent physical or psychological problems) Spent less time on other matters (activities that had been important or pleasurable) Shown an increase in tolerance (needed to drink more to produce the same effect) Shown signs of withdrawal (tremors, sweating, or insomnia when trying to quit or cut down)

The threshold criterion for any alcohol use disorder is a dysfunctional pattern of substance use causing clinically significant impairment or distress.

Practice Tips

Quitting Smoking, Decreasing Alcohol Consumption

	Medical Practitioner	Practice Nurse
Assess	• question about smoking or drinking in a sensitive manner • check for possible health effects from smoking/drinking • assess alcohol dependence • check whether alcohol intake is risky (i.e., use AUDIT-C and, if necessary, LIMPED if suspect moderate-heavy use and/or consequences relating to alcohol use)	• administerAUDIT-C questionnaires

	Medical Practitioner	Practice Nurse
Assist	• provide clear, nonjudgmental advice • provide feedback on the impact of a patient's drinking or smoking and explain the mechanisms • set realistic and incremental goals • offer pharmacotherapy if alcohol dependent	• provide written information about drinking or smoking
Arrange	• offer or refer for behavioral self-management training • ensure follow-up • identify other health professionals for a care plan (i.e., psychologist) • refer to existing programs, e.g., QUIT, alcohol treatment services, etc.	• prepare care plan • contact and coordinate other health professionals

Key Points

Clinical Management

- Most patients won't volunteer information about their smoking and drinking. You need to routinely use opportunities to ask about smoking and drinking issues.
- Use demonstrated tools to assist assessment, e.g., AUDIT-C (Babor et al., 2001).
- Develop a 1-min routine that you can use regularly with most patients.

REFERENCES

AIHW (Australian Institute of Health and Welfare), 2014a. National Drug Strategy Household Survey Detailed Report 2013. Drug statistics series no. 28. Cat. no. PHE 183. Australian Institute of Health and Welfare, Canberra.

AIHW (Australian Institute of Health and Welfare), 2014b. Australia's Health. Australia's health series no. 14. Cat. no. AUS 178. Australian Institute for Health and Welfare (AIHW), Canberra.

American Psychiatric Association, 2000. Diagnostic and Statistical Manual of Mental Disorders, fourth ed. American Psychiatric Association, Washington, DC.

Babor, T., Higgins-Biddle, J., Saunders, J., Monteiro, M., 2001. AUDIT: The Alcohol Use Disorders Identification Test, Guidelines for Use in Primary Care. World Health Organization, Geneva.

Baillie, A., Mattick, R., Hall, W., 1995. Quitting smoking: estimation by meta-analysis of the rate of unaided smoking cessation. Aust. J. Public Health 19, 129–131.

Ballesteros, J., Duffy, J., Querejeta, I., Arino, J., Gonzalez-Pinto, A., 2004. Efficacy of brief interventions for hazardous drinkers in primary care: a systematic review and meta-analysis. Alcohol. Clin. Exp. Res. 28, 608–618.

Bush, K., Kivlahan, D.R., McDonell, M.B., Fihn, S.D., Bradley, K.A., September 14, 1998. The AUDIT alcohol consumption questions (AUDIT-C): an effective brief screening test for problem drinking. Ambulatory Care Quality Improvement Project (ACQUIP). Alcohol Use Disorders Identification Test. Arch. Intern. Med. 158 (16), 1789–1795.

Edwards, S.A., Bondy, S.J., Callaghan, R.C., Mann, R.E., 2014. Prevalence of unassisted quit attempts in population-based studies: a systematic review of the literature. Addict. Behav. 39, 512–519.

Haber, P., Lintzeris, N., Proude, E., Lopatko, O., 2009. Quick Reference Guide to the Treatment of Alcohol Problems. Companion Document to the Guidelines for the Treatment of Alcohol Problems. Australian Government, Department of Health and Ageing, Canberra.

Heather, N., 2012. Can screening and brief intervention lead to population-level reductions in alcohol-related harm? Addict. Sci. Clin. Pract. 7, 15.

Hughes, J.R., 2010. A quantitative estimate of the clinical significance of treating tobacco dependence. Am. J. Prev. Med. 39, 285–286.

Keurhorst, M., van de Glind, I., Bitarello do Amaral-Sabadini, M., Anderson, P., Kaner, E., Newbury-Birch, D., Braspenning, J., Wensing, M., Heinen, M., Laurant, M., 2015. Implementation strategies to enhance management of heavy alcohol consumption in primary health care: a meta-analysis. Addiction.

Litt, J.C., 2007. Exploration of the Delivery of Prevention in the General Practice Setting (Ph.D.). Flinders.

McCambridge, J., Kypri, K., 2011. Can simply answering research questions change behaviour? Systematic review and meta-analyses of brief alcohol intervention trials. PLoS One 6, e23748.

National Center for Chronic Disease, 2014. The Health Consequences of Smoking-50Years of Progress: A Report of the Surgeon General. Reports of the Surgeon General. Centers for Disease Control and Prevention (US), Atlanta, GA.

National Health and Medical Research Council, 2009. Australian Guidelines to Reduce Health Risks from Drinking Alcohol. NHMRC, Canberra.

Oleski, J., Mota, N., Cox, J.B., Sareen, J., 2010. Perceived need for care, help seeking, and perceived barriers to care for alcohol use disorders in a national sample. Psychiatr. Serv. 61, 1223–1231.

RACGP, 2015. Smoking, Nutrition, Alcohol, Physical Activity (SNAP). RACGP, Melbourne.

Rehm, J., Anderson, P., Manthey, J., Shield, K.D., Struzzo, P., Wojnar, M., Gual, A., 2015a. Alcohol use disorders in primary health care: what do we know and where do we go? Alcohol Alcohol.

Rehm, J., Gmel, G., Probst, C., Shield, K., 2015b. Lifetime-risk of Alcohol-attributable Mortality Based on Different Levels of Alcohol Consumption in Seven European Countries. Implications for Low-risk Drinking Guidelines. Centre for Addiction and Mental Health, Toronto, Ontario.

Rubinsky, A.D., Dawson, D.A., Williams, E.C., Kivlahan, D.R., Bradley, K.A., 2013. AUDIT-C scores as a scaled marker of mean daily drinking, alcohol use disorder severity, and probability of alcohol dependence in a U.S. general population sample of drinkers. Alcohol. Clin. Exp. Res. 37, 1380–1390.

Saunders, S.M., Zygowicz, K.M., d'Angelo, B.R., 2006. Person-related and treatment-related barriers to alcohol treatment. J. Subst. Abuse Treat. 30, 261–270.

Scollo, M., Winstanley, M., 2015. Tobacco in Australia: Facts and Issues. Cancer Council Victoria, Melbourne.

Sharma, A., Szatkowski, l, 2014. Characteristics of smokers who have never tried to quit: evidence from the British Opinions and Lifestyle Survey. BMC Public Health 14, 346.

Smith, A.L., Chapman, S., Dunlop, S.M., 2015. What do we know about unassisted smoking cessation in Australia? A systematic review, 2005–2012. Tob. Control 24, 18–27.

Teesson, M., Baillie, A., Lynskey, M., Manor, B., Degenhardt, L., 2006. Substance use, dependence and treatment seeking in the United States and Australia: a cross-national comparison. Drug Alcohol Depend. 81, 149–155.

Twyman, L., Bonevski, B., Paul, C., Bryant, J., 2014. Perceived barriers to smoking cessation in selected vulnerable groups: a systematic review of the qualitative and quantitative literature. BMJ Open 4, e006414.

Wu, L., Ringwalt, C., 2004. Alcohol dependence and use of treatment services among women in the community. Am. J. Psychiatry 161, 1790–1797.

Yoast, R.A., Wilford, B.B., Hayashi, S.W., 2008. Encouraging physicians to screen for and intervene in substance use disorders: obstacles and strategies for change. J. Addict. Dis. 27, 77–97.

Zwar, N., Richmond, R., Borland, R., Peters, M., Litt, J., Bell, J., Caldwell, B., Ferretter, I., 2014. Supporting Smoking Cessation: A Guide for Health Professionals. RACGP, Melbourne.

Professional Resources

Measuring Addiction

The AUDIT-C Questionnaire for Measuring Alcohol Addiction

The Alcohol Use Disorders Identification Test (AUDIT-C) is an alcohol screen that can help identify patients who are hazardous drinkers or have active alcohol use disorders (including alcohol abuse or dependence).

AUDIT-C

The Alcohol Use Disorders Identification Test is a publication of the World Health Organization, @ 1990

Q1: How often did you have a drink containing alcohol in the past year?

Answer	Points
Never	0
Monthly or less	1
Two to four times a month	2
Two to three times a week	3
Four or more times a week	4

Q2: How many drinks did you have on a typical day when you were drinking in the past year?

Answer	Points
None, I do not drink	0
1 or 2	0
3 or 4	1
5 or 6	2
7 to 9	3
10 or more	4

Q3: How often did you have six or more drinks on one occasion in the past year?

Answer	Points
Never	0
Less than monthly	1
Monthly	2
Weekly	3
Daily or almost daily	4

Scoring: The AUDIT-C is scored on a scale of 0–12 (scores of 0 reflect no alcohol use). In men, a score of 4 or more is considered positive; in women, a score of 3 or more is considered positive. Generally, the higher the AUDIT-C score, the more likely it is that the patient's drinking is affecting his/her health and safety.

The Alcohol Use Disorders Identification Test (Bush et al., 1998)

Chapter 24

Medicines: The Good, The Not So Good, and The Sometimes Overused

Julian Henwood, Stephan Rössner, Andrew Binns

It is an art to know when to administer medication, but a much greater and more difficult one to know when to stop.

Professor Peter Pillans, Clinical Pharmacologist

INTRODUCTION: DRUGS—GOOD AND BAD

In 1996 *Time* magazine featured a cover story on the new wonder drug combination for weight loss, dexfenfluramine (Redux) with phentermine, or "fen-phen" as it was affectionately known. According to *Time* (and some significantly influenced academic proponents), the combination was a "miracle" cure set to significantly impact the obesity epidemic. The rest, as they say, is history. Less than two years after the *Time* feature, dexfenfluramine was taken off the market to short-circuit a number of potential lawsuits for the rise in primary pulmonary hypertension and mitral valve problems in users (Connolly et al., 1997).

Without doubt, medications, whether they are based on plant compounds (e.g., digitalis, atropine) or derived from laboratory synthesis (e.g., penicillin, statins), are amongst the greatest success stories in modern medicine. We only have to look to the development of the antibiotics and their huge impact on disease in the early part of the 20th century to see this. There is also no doubt that while the human body is well equipped to self-correct much disease through a highly sophisticated immune system, the addition of "medicines," both natural and synthetic, has significantly aided this process throughout human history. However, to reverse an old proverb, for every silver lining, there is often a cloud. It is easy to forget that, although drugs can be selective, they can be nonspecific with respect to effects on other target organs, causing side effects, or adverse drug reactions, which may also impact on lifestyle.

Lifestyle Medicine. http://dx.doi.org/10.1016/B978-0-12-810401-9.00024-3

Medications and Side Effects

Just about every drug that has a therapeutic effect also has a side effect. So while there is immense value in the use of appropriate medication, the benefits of each medication need to be weighed against the costs of their side effects. In some situations very serious side effects can be accepted, since they may cure potentially lethal diseases. For drugs treating cancer for example, even life-threatening side effects may be acceptable if they significantly increase the chances of survival. The more benign a condition is, the fewer the acceptable side effects.

At the same time, it is obvious that compliance to a medication schedule is dependent on the ability of a patient to adhere to the side effects of a drug. This is more likely where the side effects are outweighed by the benefit. Hypertension serves as an example to illustrate this point. Fifty years ago hypertension was treated with mutilating surgery, when nerves were cut on both sides along the spine with modest effects on blood pressure. Barbiturates were then prescribed, which calmed patients to the extent that they could not perform their work. Gradually drugs were developed that controlled blood pressure with fewer and fewer side effects. However, as late as the 1970s methyldopa was the most common drug used. It caused loss of mental alertness and drowsiness, and impaired autoimmune function and liver function. Despite these relatively common serious side effects, methyldopa was described as the drug that most physicians found to be the most effective and least objectionable drug for the majority of patients with moderate degrees of hypertension.

It is obvious that conditions related to lifestyle-based diseases (e.g., weight loss medication) requiring medication should have a profile with minimal side effects. In many cases this will result in a decision to continue with that medication (albeit cautiously), but in others an alternative approach might be sought. In some cases at least, such an alternative may simply be a change in lifestyle. Early type 2 diabetes, for example, might be treated with an insulin-sensitizing drug or a graded exercise program with dietary changes.

In this chapter, we look at potential problems that can arise from *only* going down the drug path with medications that impact on lifestyle and health. Note that this discussion in general excludes over-the-counter (OTC) and alternative/complementary therapies. It also conforms to the suggestion, contained in Table 1.1, that side effects of treatment need to be considered in lifestyle medicine as part of the outcome, and that these cannot always be justified by a treatment outcome.

COMMON SIDE EFFECTS

While almost all drugs cause side effects, there are a number of reported side effects that are characteristic of many different drugs. These include headache, nausea, reduction in libido, impotence, dryness of the mouth, weight gain (or loss), fatigue, muscular pain, and dizziness.

We have concentrated here only on those medications that (1) have small but significant side effects but are widely used in the population, or (2) have significant adverse effects in a small number of users, and hence may interfere with a lifestyle medicine orientation. As with the doctor who is taught to suspect pregnancy in every fertile woman, the underlying principle behind the discussion is to treat every medication as having potential side effects, and a decision to treat as a balance between benefits versus risk. We have not looked extensively at herbal and nonprescription medications because, unlike medications that have to go through extensive clinical trials, there is often no reliable evidence-base for these.

IATROGENESIS AND MEDICINES

The term *iatrogenesis* means literally, "doctor-induced disease." It was popularized by Ivan Illich in the 1970s to discuss the influences of modern health services.

In the current context, iatrogenesis is used to describe the less-than-positive outcomes of medication use under certain circumstances. This includes medications that may:

- cause weight gain
- cause other side effects that may be counterproductive to a lifestyle-based solution
- interact adversely with other medications/foods or substances

Medications That Can Cause Weight Gain

Weight gain is a common side effect of many medications (Ness-Abramof and Apovian, 2005). However, the precise mechanism of action that leads to this remains surprisingly unclear in most instances. According to Kopelman (2006) it is likely to involve a number of interlinked mechanisms that include altered mood, enhanced hunger, and altered metabolic processes (including thermogenesis). Such weight gain is likely to occur on a background of genetic predisposition, as shown in Fig. 24.1.

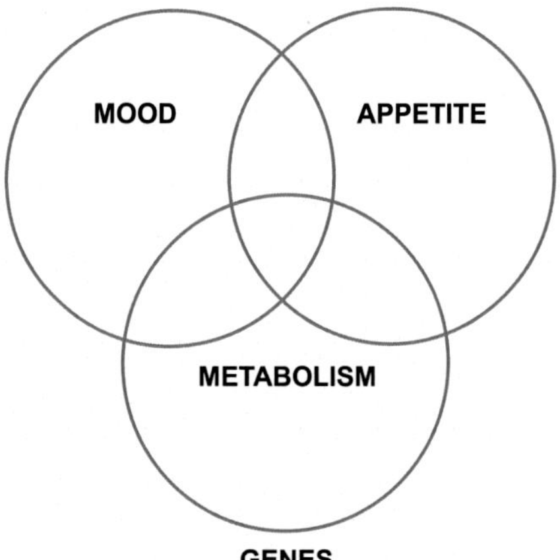

FIGURE 24.1 A schematic illustration of the linkages between mood, hunger regulation, and metabolic pathways in relation to drug-induced weight gain. *Modified from Kopelman P., September 2006. Iatrogenesis and Weight Gain. Paper Presented to the 10th International Congress on Obesity, Sydney.*

TABLE 24.1 Common Medications That Can Have Weight Gain as a Side Effect

Category	Use	Common Generic Names
Corticosteroids	Antiallergy; antiinflammatory	Dexamethasone Prednisone Triamcinolone
Antidiabetics	Reduce blood sugars/ insulin resistance	(a) Insulin (b) Sulfonylureas (c) Thiazolidinediones
Phenothiazines	Antipsychotics Atypical antipsychotics	Chlorpromazine Olanzapine Risperidone Quetiapine
Tricyclic antidepressants	Reduce depression	Amitriptyline Clomipramine Imipramine
SSRIs/SNRIs	Reduce depression	Paroxetine Mirtazapine
Depot contraceptive implant contraceptive IUCD	Contraception	Medroxyprogesterone Etonogestrel Levonorgestrel
Antihypertensives	Beta-blockers	Atenolol Metoprolol
Anticonvulsants		Carbamazepine Sodium valproate Phenytoin

A list of medication categories, and examples of generic and common brand names (in some countries), known to cause weight gain is shown in Table 24.1.

Prescription medications causing most weight gain are the antipsychotics (almost all), some antidiabetics, oral corticosteroids (all if taken long enough), and certain antidepressants (varies with medication and reaction to depression). Lesser effects come from the anxiolytics, some antihypertensives such as beta-blockers, and some anticonvulsants. Indeed, one anticonvulsant (topiramate) has been approved in combination with phentermine for weight loss in the United States.

While hormone replacement therapy (HRT) treatments have been widely touted as causing weight gain, there is clear evidence that this is not the case (Kongnyuy et al., 1999). Weight gain is common during menopause. Women on HRT actually have less weight gain when compared with women of the same age not using HRT, although there may still be some absolute weight gain. Restriction of HRT may be based on other concerns (i.e., breast cancer with long-term use), but weight gain should not be a consideration.

The clinician and the patient, in the knowledge that prescription of drugs that cause weight gain will provide clear clinical benefit, should anticipate the possibility of weight gain with these medications. Methods to avoid weight gain must be preplanned and implemented to minimize any detrimental effect of additional body weight at the time of prescribing the drug.

Medications That Cause Other Side Effects Counterproductive to Lifestyle Change

There are a number of categories of medication that need to be considered under this heading. These include medications for the following:

Weight Loss Medications

As well as weight gain from a variety of medications, there are potential side effects of medications prescribed specifically for weight loss. There are only a limited number of these, and all were discovered by serendipity—none of the current batch of weight loss medications were developed specifically for this purpose. Sibutramine, which began as an antidepressant, was used as an "appetite suppressant," with potential contraindications in uncontrolled hypertension (as it increases blood pressure and heart rate in some patients). It has since been withdrawn due to an association with increased cardiovascular events and stroke. Orlistat is a gastric lipase inhibitor, the side effects of which are the actual treatment (e.g., steatorrhea). In some countries, it is now available over the counter. Rimonabant, a cannabinoid receptor blocker that reduces food intake can cause depression, which could be counterproductive to weight loss in some individuals. This has also been withdrawn from the market due to potentially serious side effects of depression and suicidality. A caffeine/ephedrine mix (which is not available in many countries) has resulted in some cardiac deaths with the ephedrine or ephedra (ma huang) additive. Phentermine is one of the oldest used weight loss medications and may have fewer side effects than most, perhaps because it is usually only prescribed over short periods (3 months maximum). However, insomnia and tachycardia make it unsuitable for use in a small proportion of overweight patients, and it is contraindicated for those on selective serotonin reuptake inhibitors (SSRIs), serotonin–norepinephrine reuptake inhibitors (SNRIs), and tricyclics because of risk serotoninergic side effects. All weight loss medications still require a lifestyle change for a successful outcome and hence should be seen as adjuncts to treatment rather than as the main treatment.

Since several antiobesity drugs have been withdrawn because of side effects, it is natural that the pharmaceutical industry has looked for other strategies to develop effective products. One such strategy has been to combine low doses of existing weight-reduction drugs for which toxicology and safety have been established for other indications. Two such products are presently on some but not all markets, namely topiramate/phentermine and naltrexone/bupropion. Both show some although not impressive results. In spite of the low doses used in the combinations, long-term safety has been questioned, and on some markets these drug combinations are not available. Liraglutide, initially used in the treatment of type 2 diabetes, was found not only to improve carbohydrate metabolism but also reduce body weight in slightly higher doses than used for diabetes. The injectable drug, registered as Saxenda, is now available in several countries. After two years, 52% of patients on liraglutide 3 mg had maintained more than 5% weight loss and 26% more than 10%. The side effects are mainly gastrointestinal but in most cases transient and mild.

Cholesterol Lowering/Modifying

The discovery of statins for modifying heart disease risk levels of blood lipids has been equated in importance with the development of antibiotics in the early 20th century. Study after study has shown their value in reducing mortality and morbidity from heart disease and stroke and even in reducing risk of related problems such as type 2 diabetes (Jones, 2007).[1]

One significant side effect of the statins, however, and another group of lipid-lowering medications, the fibrates, is myalgia or muscle related pain or soreness, including cramping and increased muscle tears (Gaist et al., 2001; Baer and Wortmann, 2007). At the extremes, this may manifest as rhabdomyolysis (although the risk of this is very rare, less than three in 100,000 patient years; Mukhtar and Reckless, 2005; Law and Rudnicka, 2006). Of particular concern is the higher rate of myalgia and myositis in exercising individuals (Urso et al., 2005), such as athletes (Sinzinger and O'Grady, 2004), although this may have a genetic component (Vladutiu et al., 2006). As physical activity is a lifestyle recommendation for lipid lowering, this class of drug, particularly in heavily exercising individuals, may counteract this. One alternative for susceptible individuals may be a reversion to an older form of cholesterol-lowering medication, nicotinic acid (also known as niacin), which is often not tolerated because of flushing side effects (Capuzzi et al., 1998). An extended-release formulation of niacin reduced the side effect of flushing but increased the risk of liver damage (Bhardwaj and Chalasani, 2007), and the combination product with laropiprant (to reduce flushing) increased side effects when used with a statin (HPS2-THRIVE study); this has been withdrawn from the market. A slow-release form (inositol hexanicotinate) is available, which may have a lower incidence of flushing. Lowering the dosage or trying alternative statins may also

1. Although it should also be noted that the "beatification" of the statins has also been questioned (Mascitelli and Pezetta, 2007) because of the failure to often compare effects with nonatherogenic lifestyle change.

help. If this doesn't work, an assessment of benefit to risk will need to be made for each individual.

Antidepressants

While the new generation of SSRI and SSNI antidepressants appear to be effective with a milder side effect profile than previous medications of this type (tricyclics, monoamine oxidase inhibitors), they can still have side effects that may be counterproductive in some cases. Reduced libido and erectile dysfunction in many men, for example, may add to depression (Shabsigh et al., 1998), and sexual functioning is often decreased in those with depression (Seidman and Roose, 2001). Weight gain from some of this class of medications is also common. Alternatives include choosing SSRIs that don't cause weight gain (e.g., fluoxetine), or psychological approaches, at least in mild depression, or alternative OTC medications with an evidence base such as St. John's wort (Williams and Holsinger, 2005) or SAMe (Papakostas et al., 2003). Exercise as a nondrug alternative also has proven benefits.

Other Medications

Without elaborating on the wide range of medications available and minor side effects from these (and remember all drugs have side effects), it's worth pointing out some potential problems with interactions between lifestyle change and medication.

Modern exercisers, sports participants, and arthritis sufferers, for example, are often heavy users of painkillers such as NSAIDs (e.g., ibuprofen, naproxen) to reduce pain and, in the case of athletes, maintain performance in increasingly long and heavy contact playing seasons. Some studies have shown an increase in erectile dysfunction as a result of NSAID use in men (Shiri et al., 2006), elevated hypertension in men who use several forms of painkillers, and gut problems such as abdominal pain, nausea, etc. Reduced NSAID use with increased awareness of the problem is probably the best option for coping with this, as these are usually taken on a discretionary basis anyway. The COX-2 inhibitor rofecoxib was withdrawn because of the increased risk of heart attack and stroke. As an alternative, paracetamol 1 g, every 4–6 h, can suffice as an analgesic without these side effects.

Glucocorticoids can increase the risk of osteoporosis and bone fractures from activity if used for longer terms (Compston, 2003) and should be used cautiously. Mania (and paradoxically depression), psychosis, and diabetes risk can also be increased with excessive corticosteroid use. Short-term use of these is usually not a problem, but alternatives may need to be sought over the longer term.

While antihypertensive medications have improved significantly in their side effect profiles since the 1960 and 1970s, there are still potential problems with some diuretics and the risk of diabetes, and with some beta-blockers and the risk of fatigue and muscle pain, which would limit physical activity involvement.

Medications That Interact Adversely With Other Drugs/Foods etc.

As expected, many medications will interact with other medications, or indeed with chemicals in food to cause side effects or adverse reactions. Of these, the most relevant for our discussion here is the potential synergistic effect of prescribed and OTC antidepressant medications. The combination of St. John's wort and certain SSRIs or SNRIs may lead to the "serotonin syndrome" (Hu et al., 2005). Other combinations that may lead to the serotonin syndrome are shown in Table 24.2.

Grapefruit juice, which is normally considered a healthy addition to a nutritional diet, can be contraindicated with certain statins (e.g., simvastatin) as this causes an approximate doubling of the plasma levels (increased absorption from a decreased metabolism in the gut wall), resulting in an increased chance of myalgia and other side effects. Alcohol, which is recommended in moderation for a healthy heart, is known to have adverse effects with a list of medications too numerous to mention here.

In addition, to improve compliance, there is a trend toward combining medications in the one "polypill." While the combinations of these have usually been tested in separate prescriptions, the effects of combination will not be known until they are tested in large numbers.

What Can Be Done in a Standard Consultation?

- ☐ Attempt lifestyle change before initiating drug treatment where possible
- ☐ Check possible side effects of any medication and limitations to the patient
- ☐ Discuss any side effects that may limit lifestyle change
- ☐ Consider drug interaction possibilities, which may not be covered in the prescribing or product information sheet
- ☐ Consider/discuss effects of weight gain with medication

TABLE 24.2 Common Medications That May Lead to the Serotonin Syndrome

Drug Class	Examples
Antidepressants	Selective serotonin reuptake inhibitors (SSRIs), monoamine oxidase inhibitors (MAOIs), tricyclics, mirtazapine, venlafaxine
Antiparkinson drugs	Amantadine, bromocriptine, levodopa, selegiline, cabergoline, pergolide
Migraine therapy	Dihydroergotamine, naratriptan, sumatriptan, zolmitriptan
Illicit drugs	Cocaine, hallucinogenics – MDMA, LSD
Other agents	Tramadol, carbamazepine, lithium, reserpine, sibutramine, St. John's wort, bupropion, pethidine, morphine

Other Potential Problems

Because a low incidence of side effects is not picked up in a medication until it is used much more widely than in the clinical trials that allow it to be marketed, long-term problems are often not immediately obvious. This was the case with the fen-phen combination discussed at the start of the chapter. Although highly successful in weight loss, the severe side effects were ultimately seen as too risky for continued use.

Pharmaceutical business interests also play no small role in drug-related iatrogenesis. The industry (and its shareholders) has a vested interest in increased prescribing because of the profits at stake. Minor reductions in risk, in terms of mmol/L of cholesterol, etc., should be viewed in their proper context. For example, an absolute reduction of risk from 4% to 3% might be advertised as a 25% improvement (in relative risk), although the clinical significance of this may be relatively insignificant.

Small changes in the chemical formula of a medication for competitive gain may also lead to complications evidenced after the medication has been used in the population at large. This was seen with cerivastatin (Lipobay), which increased the risk of rhabdomyolysis beyond an acceptable level for a statin. One of the thiazolidinediones, the new class of antidiabetic insulin sensitizers used for diabetes treatment (troglitazone), was withdrawn because of the potential risks of liver damage, and cautions have been issued against another, rosiglitazone (suspended from the European market) because of a rise in apparent heart disease rates (Nissen and Wlski, 2007). As seen in Chapter 5, most antidiabetic medications cause weight gain, which may ultimately reduce glycemic control. The development of the thiazolidinediones raised hopes of a pharmaceutical solution for prevention of progression from prediabetes to diabetes. The Diabetes Reduction Assessment with Ramipril and Rosiglitazone Medication (DREAM) study in 2006 tended to support this suggestion (DREAM, 2006). However, this has been widely criticized on the basis of outcomes important to patients (Montori et al., 2007). Other drugs in this class are still widely used, e.g., pioglitazone. Although this class of drugs does cause weight gain, it is subcutaneous rather than visceral fat and therefore not such a metabolic risk. However, another side effect of the glitazones that may cause problems is fluid retention and resulting cardiac failure.

Finnish researchers have shown that a more effective solution to diabetes prevention is a greater emphasis on lifestyle change, which has now been shown to be effective over five years from a single dose, compared to medication that is required on an ongoing basis (Tuomilehto, 2007). Any side effects from this (i.e., increased exercise, improved nutrition) are only likely to be positive, and economic costs are less. Modeling outcomes, Dutch workers have shown that both community and individual approaches to prevention and early stage treatment are more cost-effective than intensive treatment options (Jacobs-van de Bruggen et al., 2007).

THE PROBLEM OF NONADHERENCE

On a different level to iatrogenesis is the problem of nonadherence to appropriately prescribed medication (Heidenreich, 2004). Indications are that less than half of all patients prescribed a lipid-lowering medication, for example, are still taking it six months later. This is of special concern considering that it may take from six months to a year for the beneficial effects of certain medications like statins to become apparent (Blackburn et al., 2005). Barriers to medication adherence include complex regimens, treatment of asymptomatic conditions (e.g., hypertension), and convenience factors. These barriers are particularly prevalent in the elderly, making them especially at risk for medication nonadherence. Where genuine efficacy is established, and the costs and benefits of treatment are evaluated, adherence through targeting motivation, as discussed in Chapter 3, is of particular relevance.

"DEPRESCRIPTION" AND POLYDRUG USE

Of the four stages of medication prescription (starting, changing, adding, stopping/reducing), the last is undoubtedly the least practiced. This is particularly the case with chronic diseases, where lifetime prescriptions targeting risk reduction have become standard practice in recent years. As a result, overuse of medications (polypharmacy), where more than five, and up to 10 or more, medications are taken regularly by an individual, has become commonplace. This is particularly so in the elderly. In Australia, for example, the average over-60-year-old is on an average of seven different medications (up to one in five of which is estimated to be inappropriate), with a significant number regularly taking 10 or more (Reeve et al., 2013). Adverse drug experiences (ADEs) are thought to hospitalize one in four community-living older people over any five-year period (Kalish et al., 2012).

Deprescription is therefore a procedural catchphrase for the future and none more so than for the ~60% patients suffering more than one chronic ailment. Deprescription not only reduces the risk of harm but entails stopping medicines unlikely to benefit, reduces drug burden, improves compliance, and reduces unnecessary expense for the taxpayer, according to Dr. Peter Pillans from the Princess Alexandria Hospital in Brisbane, Australia. It is not about denying effective treatment.

A patient-centered deprescription approach has five phases:

1. Compilation of a comprehensive medical history – with indications for each medication and possible ADEs;
2. Identification of potentially inappropriate medications – listing high-risk drugs and treatment goals;
3. Determination of whether medication can be ceased – with patient consent;

4. Planning and initiation of withdrawal – with tapering to reduce withdrawal reactions;
5. Monitoring support and documentation – with implementation and substitution of nonpharmacological therapies.

Best benefits are likely to be achieved if deprescription is part of a total lifestyle approach to care. Influences of nutrition, exercise, stress management, etc. may need to be considered to accomplish successful deprescription, but the latter needs to become part of the modern lifestyle prescription arsenal.

WHAT CAN THE CLINICIAN DO?

Lifestyle medicine involves the wise (and creative) use of medications as an adjunct to lifestyle interventions—instead of the other way around. Where medications present a potential problem, however, there are a number of options available to the clinician, including the following:

- If a patient can change his or her lifestyle, reduction or even elimination of some medications may be possible;
- Try to eliminate the need for medication in the first place by early intervention through a lifestyle change in prediabetes and early diabetes (e.g., exercise as an alternative to some early stage antidiabetics);
- Where available, substitute a medication with fewer side effects impacting on lifestyle change (e.g., SSRIs instead of MAOIs or tricyclics for depression; ACE inhibitors or angiotensin receptor blockers instead of beta-blockers).
- Lower dosage to a level that better balances side effects and benefits (e.g., reduced-statin dosage for myalgia);
- Use drugs in combinations that have a combined lower risk (e.g., a statin and ezetrol);
- If no option is available, inform the patient about the risk of side effects and the side effects–benefit ratio of this. If a drug must be used that causes weight gain, early preventative lifestyle measures should be recommended;
- In cases of polydrug use, consider a process of deprescription to improve pharmaceutical benefits and reduce complications.

SUMMARY

Pharmaceutical advances have been a boon to modern health. However, the rise in lifestyle-related diseases has meant a reevaluation of the benefits of some medications where these are used as a substitute for lifestyle change, causing side effects that may outweigh the benefits of the drug. Where possible, lifestyle change should be attempted before medication use, and drugs that have least impact on potential lifestyle change should be favored over those that make lifestyle change problematic.

Practice Tips

Preventing Misuse of Medications

	Medical Practitioner	Practice Nurse/Allied Health Professionals
Assess	• side effects of current medication usage • possible interactions of different drugs (including nonprescription therapies) and potential effects on lifestyle change • current literature and respected information sources for an update on side effects	• medication and complementary therapy use in conjunction with prescribing doctor • role of recommended lifestyle changes (e.g., exercise) on side effects of medication
Assist	• changing lifestyle instead of medication where possible • monitoring of potential risk factors resulting from change/decrease in medication use	• providing advice about changing lifestyle instead of medication where possible • monitoring of dosages used
Arrange	• use of lowest effective dose to minimize any side effects where possible • changing medication where this is likely to help in lifestyle change • Pharmacist involvement/advice where feasible	• follow-up visits • referrals to other allied health professionals

Key Points

Clinical Management

- In cases of lifestyle-related disease, consider medication as an adjunct to lifestyle change rather than vice versa;
- Where possible, look for a lifestyle alternative to medications that may have side effects that interfere with lifestyle change;
- Concentrate on patient's motivation and "readiness to change" (see Chapter 3) to get compliance to lifestyle change;
- Consider and discuss potential side effects of a medication with the patient and discuss alternatives;
- Monitor lifestyle-related side effects and their impact on lifestyle change (e.g., with statins, potential muscle soreness effects on exercise);
- If necessary, change medication or dose to enable better lifestyle change compliance.

REFERENCES

Baer, A.N., Wortmann, R.L., 2007. Myotoxicity associated with lipid-lowering drugs. Curr. Opin. Rheumatol. 19 (1), 67–73.

Bhardwaj, S.S., Chalasani, N., 2007. Lipid lowering agents that cause drug-induced hepatotoxicity. Clin. Liver Dis. 11, 597–613-vii.

Blackburn, D.F., Dobson, R.T., Blackburn, J.L., Wilson, T.W., 2005. Cardiovascular morbidity associated with nonadherence to statin therapy. Pharmacotherapy 25, 1035–1043.

Capuzzi, D.M., Guyton, J.R., Morgan, J.M., Goldberg, A.C., Kreisberg, R.A., Brusco, O.A., Brody, J., 1998. Effect and safety of an extended-release niacin (Niaspan): a long-term study. Am. J. Cardiol. 82, 74U–81U.

Compston, J., 2003. Glucocorticoid-induced osteoporosis. Horm. Res. 60 (Suppl. 3), 77–79.

Connolly, H.M., Crary, J.L., McGoon, M.D., Hensrud, D.D., Edwards, B.S., Edwards, W.D., Schaff, H.V., 1997. Valvular heart disease associated with fenfluramine-phentermine. N. Engl. J. Med. 337, 581–588.

DREAM (Diabetes reduction assessment with ramipril and rosiglitazone medication) Trial Investigation, 2006. Effect of rosiglitazone on the frequency of diabetes in patients with impaired glucose tolerance or impaired fasting glucose: a randomised control trial. Lancet 368, 1096–1105.

Gaist, D., Rodriguez, L.G., Huerta, C., Hallas, J., Sindrup, S.H., 2001. Lipid-lowering drugs and risk of myopathy: a population-based follow-up study. Epidemiology 12, 565–569.

Heidenreich, P.A., 2004. Patient adherence: the next frontier in quality improvement. Am. J. Med. 117, 130–132.

HPS2-THRIVE study. http://www.thrivestudy.org/docs_prof.htm.

Hu, Z., Yang, X., Ho, P.C., Chan, S.Y., Heng, P.W., Chan, E., Duan, W., Koh, H.L., Zhou, S., 2005. Herb- drug interactions: a literature review. Drugs 65, 1239–1282.

Jacobs-van der Bruggen, M.A.M., Vijgen, S.M., Hoogenveen, R.T., Boss, G., Baan, C.A., Bemelmans, W.J., 2007. Lifestyle interventions are cost effective in people with different levels of diabetes risk: results from a modeling study. Diab. Care 30, 128–134.

Jones, P.H., 2007. Clinical significance of recent lipid trials on reducing risk in patients with type 2 diabetes mellitus. Am. J. Cardiol. 99, 133B–140B.

Kalish, L.M., Caughey, G.E., Barratt, J.D., et al., 2012. Prevalence of preventable medication-related hospitalizations in Australia: an opportunity to reduce harm. Int. J. Qual. Care 24, 239–249.

Kongnyuy, E.J., Norman, R.J., Flight, I.H.K., Rees, M.C.P., 1999. Oestrogen and progestogen hormone replacement therapy for peri-menopausal and post-menopausal women: weight and body fat distribution. Cochrane Database Syst. Rev. (Issue 3):CD001018. http://dx.doi.org/10.1002/14651858.CD001018.

Kopelman, P., September 2006. Iatrogenesis and Weight Gain. Paper Presented to the 10th International Congress on Obesity. Sydney.

Law, M., Rudnicka, A.R., 2006. Statin safety: a systematic review. Am. J. Cardiol. 97 (Suppl.), 52C–60C.

Mascitelli, L., Pezetta, F., February 21, 2007. Questioning the 'beatification' of statins. Int. J. Cardiol. 123 (2).

Montori, V., Isley, W., Guyatt, G., 2007. Waking up from the DREAM of preventing diabetes with drugs. BMJ 334, 882–884.

Mukhtar, R.Y., Reckless, J.P., 2005. Statin induced myositis; a commonly encountered or rare side effect? Curr. Opin. Lipid 16, 640–647.

Ness-Abramof, R., Apovian, C.M., 2005. Drug-induced weight gain. Drugs Today (Barc.) 41, 547–555.

Nissen, S.E., Wlski, K., 2007. Effect of rosiglitazone on the risk of myocardial infarction and death from cardiovascular causes. NEJM 356, 1–15.

Papakostas, G.I., Alpert, J.E., Fava, M., 2003. S-adenosyl-methionine in depression: a comprehensive review of the literature. Curr. Psychiatry Rep. 5, 460–466.

Reeve, E., Shakib, S., Hendrix, I., Roberts, M.S., Wiese, M.D., 2013. The benefits and harms of deprescribing. Med. J. Aus. (7). http://dx.doi.org/10.5694/mja13.00200.

Seidman, S.N., Roose, S.P., 2001. Sexual dysfunction and depression. Curr. Psychiatry Rep. 3, 202–208.

Shabsigh, R., Klein, L.T., Seidman, S., Kaplan, S.A., Lehrhoff, B.J., Ritter, J.S., 1998. Increased incidence of depressive symptoms in men with erectile dysfunction. Urology 52, 848–852.

Shiri, R., Koskimaki, J., Hakkinen, J., Tammela, T.L., Auvinen, A., Hakama, M., 2006. Effect of nonsteroidal anti-inflammatory drug use on the incidence of erectile dysfunction. J. Urol. 175, 1812–1815.

Sinzinger, H., O'Grady, J., 2004. Professional athletes suffering from familial hypercholesterolaemia rarely tolerate statin treatment because of muscular problems. Br. J. Clin. Pharmacol. 57, 525–528.

Tuomilehto, J., 2007. Counterpoint: evidence-based prevention of type 2 diabetes: the power of lifestyle management. Diab. Care 30, 435–438.

Urso, M.L., Clarkson, P.M., Hittel, D., Hoffman, E.P., Thompson, P.D., 2005. Changes in ubiquitin proteasome pathway gene expression in skeletal muscle with exercise and statins. Arteriosc. Thromb. Vasc. Biol. 25, 2560–2568.

Vladutiu, G.D., Simmons, Z., Isackson, P.J., Tarnopolsky, M., Peltier, W.L., Barboi, A.C., Sripathi, N., Wortmann, R.L., Phillips, P.S., 2006. Genetic risk factors associated with lipid-lowering drug intensive myopathies. Muscle Nerve 34, 153–162.

Williams Jr., J.W., Holsinger, T., 2005. St John's for depression, worts and all. BMJ 330, E350–E351.

Chapter 25

Relationships, Social Inequity, and Distal Factors in Lifestyle Medicine: Tackling the Big Determinants

Garry Egger, Andrew Binns, Stephan Rössner, Maximillian De Courten

If civilisation is to survive, it must live on the interest of nature, not the capital.
Ronald Wright (A Short History of Progress, 2004).

INTRODUCTION: CAUSALITY REVISITED

In Chapter 2, levels of causality for the modern epidemic of lifestyle-related diseases were dissected step by step. This pointed to distal factors that initially seemed far removed from culpability, such as industrialization, population pressures, and economic growth. The paradox, which becomes apparent from this, is that the underlying system of growth and development that has allowed great advances in human living conditions has become a Damoclean sword, from which we need to find a safe and expedient escape. In the words of one writer, "…we behave like someone spending wildly on their credit card, in the belief that the bill will never come. But it's already in the mail" (Mass, 2006).

In the intervening chapters we discussed the management of lifestyle-related risk factors and proximal and medial determinants of these, with the emphasis being on interventions more immediate to the realm of a clinician. In this chapter we pull back again to look more at the broader medial and distal factors that need to be considered for a more global perspective on the problem. In particular, we'll look at the R (Relationships) and S (Social Inequity) of the NASTIE MAL ODOURS acronym and some of the more distal drivers of these. We will focus on the work by Dr. Michael Marmot on social and relationship issues (Marmot, 2004, 2015; Marmot et al., 2008) and that by Wilkinson and Picket (2010) on the effects of social inequality on health. Finally, we'll consider some other work on the effects of the greater distal drivers on changing health dynamics.

Lifestyle Medicine. http://dx.doi.org/10.1016/B978-0-12-810401-9.00025-5

Some of this may not appear immediately relevant to the clinician and hence will not be covered extensively here. Some of the clinical implications suggested will also be tentative and may sit uneasily with the standard clinician. However, it is hoped it might provide a deeper understanding of the future of lifestyle medicine and a better understanding of solutions to the problem at the public health, as well as the clinical, level.

RELATIONSHIPS

The quality of personal and social relationships, particularly in relation to the availability of perceived social support in the domestic, work, and home environments, is clearly linked to a range of chronic diseases, but specifically heart disease, stroke, and all-cause mortality (Uchino, 2006). The pathways for this aren't clear, and psychological mediators haven't been proven, but disease outcomes have been associated with poor relations such as relationship discordance, spousal ambivalence, and isolation that can stretch across the lifespan (Heinze et al., 2015).

People who have supportive close relationships have lower levels of a range of different chronic diseases generally and systemic inflammation (metaflammation) specifically, compared to people who have unsatisfactory relationships. This type of inflammation has been proposed by leading Canadian researcher J-P Despres (2012) as the possible link with heart disease.

Negative and competitive social interactions can increase proinflammatory cytokine activity on a daily level. In reverse, a Finnish study (Ovrum and Rickertsen, 2015) has shown that social support can alleviate the inflammation associated with childhood adversities. Improving individual awareness of the importance of social support and assisting in finding such support through programs such as "Act-Belong-Commit," developed in Western Australia (Chapter 16), should be integral to chronic disease management.

Poor relationships and social status can also have indirect effects on actual and perceived health through stress, anxiety, and/or depression, which are becoming increasingly accepted as medial determinants of heart disease. It's now clear that depression is an independent risk factor as illustrated by the National Heart Foundation's new guidelines on treating depression in heart disease. For many years the subjective quality of stress measurements led hard-nosed researchers to dismiss it as a potential factor in heart disease, but this was clearly countered in the 52-country Interheart study published in 2008, which identified it as not just involved but as a major determinant.

SOCIAL INEQUALITY

While not a "cause" of disease per se, inequality and social disadvantage have been shown to be robust predictors of ill health, as described by Marmot et al. (2008) and Marmot (2015). Disadvantage exists not just through socioeconomic status and income inequalities, but through economic stress and job insecurity, with metaflammation again a possible link (Marmot and Allen, 2014). According to the World Health Organization Commission on Social Determinants of Health, inequities—in power, money, and resources—are responsible for much of the inequalities in health within and between countries (WHO, 2008).

While the effects of socioeconomic status on health (and inflammation) are relatively clear, the effects of income disparities on health have been more controversial. English epidemiologists Wilkinson and Pickett (2010) in a descriptive analysis of ratios of rich to poor within and between Organization for Economic Co-operation and Development countries show a linear worsening of a number of health and social statistics (obesity, infant mortality, teenage pregnancies, etc.) in countries with greater income gradients.

Much has been made of the mechanisms underlying social disadvantage, socioeconomic (SE) status, and inequality. Marmot and his colleagues show that modifiable health behaviors (nutrition, inactivity, sleep, stress) and obesity could explain around 50% of the incidence in ailments like heart disease (Marmot, 2015). Poor relationships and social status can also have indirect effects on heart disease through stress, anxiety, and/or depression, which are becoming increasingly accepted as primary determinants of heart disease (Lim et al., 2012). It's now clear that depression is an independent risk factor as illustrated by the National Heart Foundation's new guidelines on treating depression in heart disease.

Interestingly, it appears that poverty per se, is not so much an issue in health as relative poverty and the width of the gap between the rich and poor, variously defined as the Gini Index or RP20 (gap between the highest and lowest 20% of income earners). Respective poverty has been linked with reduced longevity, and increased rates of morbidity and disability as if it were a "post code–related" etiology. Issues affecting the selection of post code (income, education, history, family background) are rarely decomposed to recognize health promotion targets.

Lifestyle and Inequality

More recently, the dominant determinant has come to be thought of as "lifestyle." Do those at the bottom end of the social scale simply engage in more of those lifestyles (smoking, drinking, poor diet) that are themselves determinants of ill health? With bigger inequality gaps, changes in behavior are brought about to protect such gaps. Walking on the streets, for example, is discouraged because of increased crime rates; smoking becomes an issue of class as much as satisfaction. Still, in a relatively equal per capita income society, there are variations in factors that can determine health outcomes—occupation and environment are crucial examples.

Norwegian researchers portioned out these determinants by comparing data on a range of socioeconomic factors and a range of disease-related lifestyle factors

(Ovrum and Rickertsen, 2015). Using a complicated statistical process, they used national bi-annual data to statistically calculate the links between lifestyles and SE status as a whole, and individual lifestyle behaviors and components of SE status independently.

They considered education, income, occupation, age, gender, marital status, psychological traits, and childhood circumstances as potential sources of inequality. They found that sources of inequality in health were not necessarily representative of sources of inequality in underlying lifestyles.

Education, for example, is generally an important source of inequality as measured by the Gini index, in both lifestyles and health. Income on the other hand, is unimportant in all of the lifestyle indicators examined, except physical activity. In other words, those with a high income are more likely to be physically active than those with lower incomes, but incomes surprisingly don't affect eating of fruits or vegetables. Gender is generally a better predictor of this type of eating behavior, with women faring better than men, irrespective of income or education, as is age, with the regular eating of fish.

If this type of information is used for health promotion, it would result in better targeting for information campaigns. Nutrition campaigns and information, for example, would be best targeted at men and the young, irrespective of class or education. As childhood obesity and dietary behavior is most influenced by maternal education, differences in childhood circumstances represent differences in opportunities that are beyond personal responsibility.

Activity campaigns on the other hand might be best targeted at those with less financial means for carrying these out.

"To the extent that our main public health policy goal is to improve unhealthy lifestyles, and health inequalities, health authorities should search for key sources of population differences in single, important production factors of health, including important lifestyles, and in turn design tailored policies for each of these factors," claimed the Norwegian report.

"Alternatively, if sources of inequality in health other than SE status are considered legitimate or unavoidable, one should be aware that the importance of education and income in explaining overall inequality may vary substantially across different lifestyle and health indicators… and policies should be targeted according."

SUMMARY

Relationships and inequality have a big effect on health. However, inequality does not always have the same impact—at least on lifestyle and health. Hence, there is a need to become better at selecting those factors that differentiate the different lifestyles between groups of individuals and not just concentrate on income or educational inequality (despite the fact that these also remain worthy of intervention). In thinking about chronic diseases and conditions, therefore, it is no longer enough to just consider the obvious lifestyle factors. Relationships and social factors might be difficult to deal with at the clinical level, but they do have to be seen as part of the problem to get a better perspective on managing the issues.

REFERENCES

Despres, J.-P., 2012. Abdominal obesity and cardiovascular disease: is inflammation the missing link? Can. J. Cardiol. 28 (6), 642–652.

Heinze, J.E., Kruger, D.J., Reischi, T.M., Zimmetman, M.A., 2015. Relationships among disease, social support, and perceived health: a lifespan report. Am. J. Community Psychol. 56 (3–4), 268–279.

Lim, S.S., Vos, T., Flaxman, A.D., Danaei, G., Shibuya, K., Adair-Rohani, H., et al., 2012. A comparative risk assessment of burden of disease and injury attributable to 67 risk factor clusters in 21 regions, 1990–2010: a systematic analysis for the Global Burden of Disease Study 2010. Lancet 380 (9859), 2224–2260.

Marmot, M., 2004. Status Syndrome: How Your Social Standing Directly Affects Your Health and Expectancy. Bloomsbury, London.

Marmot, M., 2015. The Health Gap: The Challenge of an Unequal World. Bloomsbury, London.

Marmot, M.G., Shipley, M.J., Hemingway, H., Head, J., Brunner, E.J., 2008. Biological and behavioural explanations of social inequalities in coronary heart disease: the Whitehall II study. Diabetologia 51 (11), 1980–1988.

Marmot, M., Allen, J.J., 2014. Social determinants and health equity. Am. J. Public Health 104 (Suppl. 4), S517–S519.

Mass, B., 2006. Quoted in 'Planet Earth: The Future' BBC Books. London, p. 172.

Ovrum, A., Rickertsen, K., 2015. Inequality in health versus inequality in lifestyles. Nordic J. Health Econ. 3 (1), 18–33. http://dx.doi.org/10.5617/njhe.972.

Uchino, B.N., 2006. Social support and health: a review of physiological processes potentially underlying links to disease outcomes. J. Behav. Med. 29 (4), 377–387.

Wilkinson, R., Pickett, K., 2010. The Spirit Level. Bloomsbury, London.

World Health Organisation (WHO), 2008. Commission on Social Determinants of Health. Closing the Gap in a Generation Final Report of the Commission on Social Determinants of Health. WHO, Geneva.

Section III

OTHER ISSUES FOR LIFESTYLE MEDICINE

Chapter 26

Sex and Lifestyle: Not Being Able to Get Enough of a Good Thing Because of a Lifetime of Getting Too Much of a Good Thing

Michael Gillman, Garry Egger

The phrase 'use it or lose it' is particularly appropriate for the genitalia.
Meuleman (2002)

INTRODUCTION

According to the sex therapist Dr. Rosie King, the prevalence of sexual dysfunction in men roughly corresponds to the decade of age in which this occurs. Hence, 40% of 40-year-old men will experience some problem, 50% of 50-year-olds, and 60% of 60-year-olds. It's this knowledge, she claims, that makes no man want to live to 100!

While this may reflect some medical license, it does indicate the importance of age in sexual functioning. Age is perhaps the most predictable determinant of sexual dysfunction (where dysfunction is defined as "....an issue of concern for a particular individual"), with problems typically increasing proportionately with age. Other factors are biology, illness, psychological health, past experience, and education. However, the next most common cause is lifestyle, and this can exacerbate or ameliorate the effects of aging.

Sex and lifestyle are linked in a number of ways, from sexually transmitted diseases to sexual performance malfunction. Because the former is more of a conventional health problem, which has an extensive literature of its own, it is only the latter with which we will concern ourselves here. This is extensive enough, covering a wide variety of conditions, including erectile dysfunction (ED), hypoactive sexual desire, and ejaculatory, orgasmic, sexual aversion, and hormonal disorders, to name but a few. As can be seen from the list, the biggest problems—at least in terms of performance—are likely to occur amongst males, hence, the greater concentration on male sexual performance and lifestyle in

Lifestyle Medicine. http://dx.doi.org/10.1016/B978-0-12-810401-9.00026-7

this chapter. This is not meant to ignore the performance or desire questions of females, and where there is clear evidence for these associated with lifestyle, they are considered. However, the bulk of the discussion will concern lifestyle and sexual performance in men.

PREVALENCE OF SEXUAL PROBLEMS

Because of the sensitivity of the topic, the prevalence of sexual problems in the community is difficult to accurately detect. Generally accepted figures suggest that around one in three men between the ages of 18 and 59 report dissatisfaction with some aspect of their sexual function (Laumann et al., 2008). This continues to rise over age 60, despite the fact that sexual desire may remain high. ED appears to be the most common problem, with surveys suggesting a prevalence of close to 20% of men over 20 years of age suffering from this (Selvin et al., 2007). Rates of ED amongst men with type 2 diabetes mellitus (T2DM) are estimated to be two to three times higher than those without (Selvin et al., 2007), but more critically, the changes in lifestyle and lifestyle-based etiology of many of these problems lead to a prediction of doubling in prevalence by the year 2025 (Fabbri et al., 2003).

Erectile dysfunction, sometimes called "impotence," is the repeated inability to obtain or maintain an erection firm enough for sexual intercourse. Erectile dysfunction can be *global*, in which case all erections are decreased, or *situational*, where strong erections may be obtained during rapid eye movement sleep but not with sexual stimulation. In older men, ED usually has a physical cause, such as disease, injury, or side effects of drugs. The most common cause in younger men is vascular disease with the same risk factors as for cardiovascular disease. Diseases such as diabetes, kidney disease, chronic alcoholism, multiple sclerosis, atherosclerosis, other vascular diseases, and neurologic disease account for about 70% of all ED cases.

As well as erectile dysfunction, sexual drive, ejaculatory function, sexual problem assessment, and sexual satisfaction appear to be the main areas where lifestyle can have an effect on male sexuality. Medical awareness of a patient's sexual health however is probably not high, and health professionals may often not render sufficient care to patients' sexual health due to their conservative attitude and lack of skill in addressing sexuality.

CAUSES: GENERAL

There is now clear evidence that lifestyle factors have a direct impact on the development of many sexual problems. Clearly stress, anxiety, and depression will impact on many of the conditions listed previously and addressing these will improve patient outcomes. Cardiovascular risk factors such as T2DM, hypertension, smoking, alcohol or drug abuse, a sedentary lifestyle and obesity, as well as chronic diseases such as arthritis, renal and hepatic failure, and pulmonary disease, have all been linked to the development of sexual problems in general, and ED, in particular. Many medications used to treat these problems have paradoxically been implicated in reducing sexual function (Chapter 24). Finally, sleep deprivation (Chapter 18) has also been suggested to inhibit nocturnal erections, thus decreasing the normal oxygenation of the corpora cavernosa necessary to maintain endothelial function (Kamp et al., 2005).

Whilst less commonly reported, lifestyle factors can also have an impact on female sexuality. As with men, aging is an inhibitory factor on sexual desire and ability, but according to Professor Beverly Whipple, one of the authors of *The Science of Orgasm*, the adage of "use it or lose it" is as relevant to women as it is for men, in this case for maintaining high estrogen levels and reducing vaginal dryness. Obesity, smoking, and inactivity are also lifestyle factors likely to have a dampening effect on female arousal. A small study of 30 women with ischemic heart disease suggested that sexual dysfunction was common (30%), and in fact 23% of these women experienced sexual problems before developing cardiac symptoms (Steinke, 2010). Larger studies are required before any definite links between cardiovascular disease and female sexual dysfunction can be made. Medications can also impact on female sexuality with a significant proportion of females having problems with arousal or orgasm due to certain medications (see following).

SEXUAL ORGAN HARASSMENT

CAUSES: SPECIFIC

Individual lifestyle factors that can cause sexual problems include the following:

Inactivity

Several studies have shown a higher risk of sexual problems, including ED, amongst those who are sedentary or inactive, with a dose–response relationship. One study has shown that a history of vigorous exercise is protective against ED across all ages (Leon et al., 2014). The reasons for this are not clear, but a consistent level of physical activity, as part of a natural lifestyle throughout evolution has probably resulted in a feed-forward mechanism for maintenance of sex hormone levels, which is disrupted with inactivity.

Smoking

Smoking is a risk factor for ED, in particular, because of the narrowing of the pudendal artery through atherogenesis. This reduces blood flow and velocity in smokers as well as the penile nitric oxide synthase activity required for an adequate erection. Data from the Massachusetts Male Aging Study show that cigarette smoking doubled the risk of developing ED in those men who were followed up (Huang et al., 2015).

Alcohol

Whist alcohol in moderation may have a stimulatory effect on sexual arousal, its excessive use, as summed up by Shakespeare in *Macbeth*, "provokes the desire but takes away the performance." Alcohol can cause orgasmic delay in women and can depress testosterone concentrations in men (Peught and Belenko, 2001). While short-term effects from social drinking may not be of concern, chronic drinking can lead to systemic disturbances such as liver and pancreatic problems, which can interfere with performance. Chronic alcoholism can also result in permanent impotence even after the cessation of drinking for many years.

Obesity

The scientific literature is unequivocal about sexual dysfunction and obesity (Abrahamian and Kautzky-Willer, 2016). In the Massachusetts Male Aging Study (Huang et al., 2015), having a body mass index (BMI) >28 doubled the incidence of ED at follow-up. An increase in BMI has been shown to reduce the quality of erections as demonstrated using the IIEF-5, a validated questionnaire-based method of assessing erectile dysfunction. Just by increasing the BMI by 1 kg per square meter, the IIEF-5 score can decrease by 0.141 independent of age (Derby et al., 2000). However, obesity, per se, may not be the issue, and as with metabolic health in general, remaining fit and active, even in the presence of obesity, may help reduce the problem. Obesity, probably through insulin resistance, is associated with polycystic ovaries in women (Orio et al., 2007) and consequent fertility problems. Obesity in couples has also been shown to result in subfecundity, with a dose–response relationship in time to pregnancy with obese male and female partners (Ramlau-Hansen et al., 2007). There is also an inverse relationship between bioavailable testosterone levels in men, sexual performance, and waist circumference (Kapoor et al., 2007).

Type 2 Diabetes

As type 2 diabetes results from several of the risk factors discussed before, it is not surprising that it is a major risk factor for sexual dysfunction in itself. As with obesity, there is a clear and unequivocal relationship between T2DM and sexual function, affecting ED and ejaculatory problems in men, but also influencing arousal, genital sensation, lubrication, and dyspareunia in up to 50% of female diabetes cases (Muniyappa et al., 2003). This can occur through the narrowing of peripheral arteries, as well as the development of insulin resistance and its effects on sex-based hormones. Hypogonadism affecting all aspects of sexual performance in men has been shown to be higher in diabetic than nondiabetic men (Kapoor et al., 2007). Screening for ED and other sexual problems in T2DM patients therefore may be warranted.

What Can Be Done in a Brief Consultation?

- ☐ Ask about sexual performance in patients with other metabolic risk factors.
- ☐ Determine whether this is a problem for the patient.
- ☐ Discuss specific aspects of the problem (ED? Ejaculation? etc.).
- ☐ Explain the lifestyle basis of many sexual performance problems.
- ☐ Consider changes in female desire, as well as changes in male performance.
- ☐ Explain possible adjunctive treatments.
- ☐ Order pathology tests, e.g., E/LFTs, serum testosterone, TSH etc.
- ☐ Refer or make another appointment to consider further.

Depression

Depression can be both a cause and consequence of sexual problems, and both can be comorbid with other diseases (Shabsigh et al., 2001). Although little has been published on the nondrug-induced effects of depression on sexual performance, some studies suggest that between 25% and 75% of both men and women with unipolar depression suffer either loss of libido, disorders of arousal, or erection or lubrication problems (Williams and Reynolds, 2006). One study has shown that the link in men at least may be through the mediation of decreased sexual activity in single, widowed, or divorced men as well as smoking, age, and physical inactivity (Nicolosi et al., 2004).

Medications

One of the most common side effects of medication is sexual dysfunction, either in desire or performance ability. Ironically, many of the medications used to treat diseases and risk factors that can also be risk factors for sexual problems such as heart disease, hypertension, hypercholesterolemia, diabetes, depression, etc., can cause or exacerbate sexual problems. Perhaps the most commonly used of these are the antidepressants, which can set up a vicious cycle where sexual inadequacy can lead to depression, leading to further sexual inadequacy (Clayton and Montejo, 2006). While some antidepressants (e.g., bupropion) appear to have less adverse effects on sexual performance, they are also generally less affective for depression treatment. Substitution of antidepressants, cognitive therapy for mild depression, and the use of sildenafil citrate as an antidote therapy (Rudkin et al., 2004) can help overcome these problems.

Medications, such as some antidepressants, also impact on female sexuality, with a significant proportion of females having problems with arousal or orgasm. Antihistamines can also cause vaginal dryness.

MANAGEMENT

In a study of around 600 men ages 30–70 years, midlife lifestyle modification appeared too late to reverse the adverse effects of nicotine, obesity, and alcohol consumption on sexual function, while physical activity appeared to be effective in reducing the risk of ED even if initiated in midlife (Esposito et al., 2004). A controlled study of obese men (BMI >30) revealed that an increase in physical activity from 48 to 195 min per week, and a decrease in BMI from 37 to 31 over 2 years, resulted in an increase of the IIEF-5 score from 13.9 to 17 (Salonia et al., 2003).

It does appear, however, that identification and correction of vascular ED and regular physical activity and normalizing BMI will help both in the prevention and the treatment of this disorder.

Health care professionals are seeing an increasing number of male patients wanting to know if they have low testosterone levels and if medication can correct this. True male hypogonadism, whereby there are unequivocal symptoms and signs of androgen deficiency and at least two separate early morning serum testosterone levels of less than 6 nmol/L is quite rare. This condition needs to be treated with testosterone replacement therapy to avoid longer-term consequences such as osteoporosis. However, it is unlikely to be the panacea for men's health problems (Yeap et al., 2012).

It is known that testosterone levels decrease with increasing age in men, with levels going down by around 0.8% per year over age 40 (Morales and Lunenfeld, 2002). Many terms have been coined for this condition such as "andropause," "male climacteric," androgen deficiency in the aging male (ADAM), and most recently, "late onset hypogonadism." Research is not clear on whether these men should be managed with hormone therapy.

Aging is also often associated with an increase in abdominal fat mass, a decrease in muscle mass, and a decrease in activity levels, and it is these that can independently affect testosterone levels. It is also known that testosterone disorders are implicated through metabolic derangements such as obesity, insulin resistance, and the metabolic syndrome. It has become apparent that visceral obesity leads to a decrease in insulin receptor numbers, which results in hyperinsulinemia. This hyperinsulinemia leads to a decrease in sex hormone-binding globulin levels and thus a decrease in available serum testosterone. Furthermore, it has been demonstrated that insulin receptor numbers can be restored to normal if insulin levels are decreased by weight loss.

Testosterone undergoes aromatization to estradiol particularly in adipose tissue. Therefore men with increased visceral fat will have higher levels of estradiol. This can lead to effects on the male breast tissue resulting in gynecomastia. Interestingly, physical activity appears to decrease this process of aromatization. Hence, a slightly reduced serum testosterone level may merely reflect an increased visceral fat mass, and the initial treatment should involve lifestyle behavioral modification aimed at decreasing the percentage of visceral fat before resorting to testosterone replacement therapy. Use of PDE5 inhibitors has been shown to be less effective in men (such as diabetics) with hypogonadal levels of testosterone. Testosterone replacement in these cases can convert sildenafil nonresponders into responders (Kapoor et al., 2007).

Other treatments might involve medication substitution or reduction, medications to enhance the sexual response such as the PDE5 inhibitors (e.g., sildenafil), cognitive and behavioral therapy, counseling, vacuum therapy treatments, injections, and, as a last resort, surgery. There are no approved medications for females for sexual problems, and lifestyle changes might be expected to have less effect here than with males.

The final word might go to Meuleman (2002), who suggests that "… given the fact that current symptomatic treatments are invasive, are costly, or have unproved long-term benefits, in daily practice, too little attention is paid to the modification of risk factors and secondary prevention. A more physically active lifestyle, quitting smoking, reduction of overweight, and a moderate alcohol intake may be viewed as integral components of a comprehensive risk reduction program."

SUMMARY

Lifestyle factors play a significant role in sexual performance. Key pervasive factors such as obesity, poor nutrition, inactivity, and smoking again stand out as the predominant influences, but other factors such as poor sleep, depression, and some forms of medication may also play a part. Sexual activity, desire, and performance, where they are seen as being a problem, should be considered in any overall lifestyle analysis.

Practice Tips

Managing Lifestyle-Related Sexual Problems

	General Practitioner	Practice Nurse
Assess	• Ask sensitively about sexual function in those with cardiovascular disease and metabolic risk • Check possible comorbidities • Check medication effects and possible alternatives • Assess for pharmacotherapy treatment	• Check smoking, drinking, activity levels, and obesity to see if these may impact on sexual function • Take blood pressure, height, weight, waist circumference
Assist	• Use first-line treatments such as lifestyle change, counseling, patient education, medication changes • Advise on "use it or lose it" principle • Propose PDE5 inhibitors, Vacuum constriction devices, intracavernosal injections, etc.	• Provide written materials for patient education • Assist with counseling if experienced in this area
Arrange	• Identify other health professionals for a care plan • Refer to mental health professional, endocrinologist, urologist, or sex therapist where appropriate	• Prepare care plan • Contact and coordinate other health professionals • Discuss options with other professionals

Key Points

Clinical Management

- Check sexual function as part of other major chronic disease symptoms;
- Consider problems such as ED as a multifactorial condition that may require a multidisciplinary approach to treatment (especially when depression is present);
- Consider adjuvant or drug substitution treatment approaches to depression where this is associated with sexual dysfunction;
- Discuss the range of treatment approaches available, e.g., PDE5 inhibitors, intracavernosal injections, vacuum constriction devises, etc.

REFERENCES

Abrahamian, H., Kautzky-Willer, A., 2016. Sexuality in overweight and obesity. Wien. Med. Wochenschr January 26. (ahead of print).

Clayton, A.H., Montejo, A.L., 2006. Major depressive disorder, antidepressants and sexual function. J. Clin. Psychiatry 67 (Suppl. 6), 33–37.

Derby, C.A., Mohr, B.A., Goldstein, I., et al., 2000. Modifiable risk factors and erectile dysfunction: can lifestyle changes modify risk? Urology 56 (2), 302–306.

Esposito, K., Giugliana, F., Di Palo, C., et al., 2004. Effect of lifestyle changes on erectile dysfunction in obese men. JAMA 291, 2978–2984.

Fabbri, A., Caprio, M., Aversa, A., 2003. Pathology or erection. J. Endocrinol. Invest. 26 (3 Suppl), 87–90.

Huang, Y.C., Vhin, C.C., Chen, C.S., Shindel, A.W., Ho, D.R., Lin, C.V.S., Shi, C.S., 2015. Chronic cigarette smoking impairs erectile function through increased oxidative stress and apoptosis, decreased nNOS, endothelial and smooth muscle contents in a rat model. PLoS One 10 (10), e0140728.

Kamp, S., Ott, R., Knoll, T., et al., 2005. REM-sleep deprivation leads to a decrease of nocturnal penile tumescence (NPT) in young health males: the significance of NPT-measurements. In: (611.13) XVII World Congress of Sexology, July 10–15, Montreal, Canada, 2005.

Kapoor, D., Aldred, H., Clark, S., Channer, K.S., High Jones, T., 2007. Clinical and biochemical assessment of hypogonadism in men with type 2 diabetes. Diabetes Care 30 (4), 911–917.

Laumann, E., Das, A., Waite, L., 2008. Sexual dysfunction among older adults: prevalence and risk factors from a nationally representative U.S. Probability sample of men and women 57–85 Years of age. J. Sex. Med. 5 (10), 2300–2311.

Leon, L.A.B., Eukushima, A.R., Rocha, L.Y., Madrion, L.B.M.M., Rodriguez, B., 2014. Physical activity on endothelial and erectile dysfunction: a literature review. Aging Male 17 (3), 125–130.

Meuleman, E.J.H., 2002. Prevalence of erectile dysfunction: need for treatment? Int. J. Impot. Res. 14 (Suppl. 1), S22–S28.

Morales, A., Lunenfeld, B., 2002. Investigation, treatment and monitoring of late onset hypogonadism in males. Official Recommendations of the International Society for the Study Aging Male. Aging Male 5, 74–86.

Muniyappa, R., Norton, M., Dunn, M.E., Banerji, M.A., 2003. Diabetes and female sexual dysfunction; moving beyond “benign neglect”. Curr. Diabetes Rep. 8 (2), 515–522.

Nicolosi, A., Moreira Jr., E.D., Villa, M., Glasser, D.B., 2004. A population study of the association between sexual function, sexual satisfaction and depressive symptoms in men. J. Affect. Disord. 82 (2), 235–243.

Orio, F., Falbo, A., Grieco, A., Russo, T., Oppedisano, R.M., Scchinelli, A., Giallauria, F., Santoro, T., Tafuri, D., Colao, A.M., Palomba, S., 2007. Polycystic ovary syndrome and obesity: non pharmacological options. Minerva Ginecol. 59 (1), 63–73.

Peught, J., Belenko, S., 2001. Alcohol drugs and sexual function: a review. J. Psychoact. Drugs 33 (3), 223–232.

Ramlau-Hansen, C.H., Thulstrup, A.M., Nohr, E.A., 2007. Subfecundity in overweight and obese couples. Hum. Reprod. Mar 7.

Rosen, R.C., 1999. Development and evaluation of an abridged, 5-item version of the International Index of Erectile Function (IIEF-5) as a diagnostic tool for erectile dysfunction. Int. J. Impot. Res. 11, 319–326.

Rudkin, l, Taylor, M.J., Hawton, K., 2004. Strategies for managing sexual dysfunction induced by antidepressant medication. Cochrane Database Syst. Rev. 18 (4), CD003382.

Salonia, A., Briganti, A., Dehò, F., et al., 2003. Pathophysiology of erectile dysfunction. Int. J. Androl. 26 (3), 129–136.

Seftel, A.D., 2006. Diagnosis of sexual dysfunction. Stand Pract. Sex Med. 62.

Selvin, E., Burnett, A.L., Platz, E.A., 2007. Prevalence and risk factors for erectile dysfunction in the US. Am. J. Med. 120 (2), 151–157.

Shabsigh, R., Zakaria, l, Anastasiadis, A.G., Seidman, A.S., 2001. Sexual dysfunction and depression: etiology, prevalence, and treatment. Curr. Urol. Rep. 2 (6), 463–467.

Steinke, E.E., 2010. Sexual dysfunction in women with cardiovascular disease: what do we know? J. Cardiovasc. Nurs. 25 (2), 151–158.

Williams, K., Reynolds, M.F., 2006. Sexual dysfunction in major depression. CNS Spectr. 11 (8 Suppl. 9), 19–23.

Yeap, B.B., Araujo, A.B., Witter, G.A., 2012. Do low testosterone levels contribute to ill-health during male ageing? Crit. Rev. Clin. Lab. Sci. 49 (5–6), 168–182.

Professional Resources

Measuring Sexual Function

Sexual Health Inventory for Men (SHIM)(Rosen, 1999)

Patient Instructions: Please answer the following questions, ticking the number that most applies, from 1 = least functional to 5 = most functional.

	1. Least functional	2	3	4	5. Most functional
How do you rate your confidence that you could get and keep an erection?					
When you had erections with sexual stimulation, how often were your erections hard enough for penetration?					
During sexual intercourse, how often were you able to maintain your erection after you had penetrated (entered) your partner?					
During sexual intercourse, how difficult was it to maintain your erection to the completion of intercourse?					
When you attempted sexual intercourse, how often was it satisfactory for you?					

Professional Instructions: Each question is rated on a Likert scale of 1 = least functional to 5 = most functional. Totals range from 5 to 25.

22–25 Normal erectile function
17–21 Mild erectile dysfunction
12–16 Mild to moderate erectile dysfunction
8–11 Moderate erectile dysfunction
<7 Severe erectile dysfunction

Brief points to cover in the physical examination of a male patient with erectile dysfunction (Seftel 2006)

- Check for gynecomastia
- Check for body hair distribution
- Check for penis size

Professional Resources—cont'd

- Check femoral pulses
- Check for inguinal hernia
- Check the foreskin and glans and urethra for sexually transmitted infections, other infections, or pathology
- Check the tunica for plaques
- Check the testes for size and consistency and the presence of two testes
- Check for varicoceles
- Check the rectum for anal tone, anal pathology, and prostate (after informed consent)

Chapter 27

Lifestyle and Oral Health

Bernadette Drummond, Garry Egger

Unlike most human organs, teeth need a rest.

INTRODUCTION: ORAL HEALTH

Imagine a serene afternoon on a Serengeti plain. All that's left of a big cat's meal is a steaming carcass and some blood dripping from its powerful jaws. As always, the remnants remain in the teeth for some time. Lions don't floss. Yet they usually maintain strong, healthy teeth throughout the duration of their lives and die with their choppers intact. Most humans, on the other hand, are lucky to get to three-score-and-ten with a full mouthful of teeth, and those that do, have often had significant "work" done.

So what does this tell us about human oral health? Basically, that many of the dental problems suffered by humans result from the lifestyle we lead and the way we care for our teeth. Obviously, these are not all to do with the modern way of life, as images of toothless serfs from earlier days testify to the great strides made in dentistry with modernity. However, as with many aspects of health canvassed in this book, there comes a time when positive advances are often overwhelmed by environmental changes that, while providing greater material prosperity, can have a hidden cost in human well-being.

Oral health is a large vista, with several professions dedicated to its management. We do not intend here to trample the whole field, but merely to sample some issues associated specifically with our way of life that may cause oral health problems. The two main forms of oral disease considered are dental decay/erosion and periodontal (gum) disease. As we will see, the main culprits already have been introduced: inadequate diet, sweetened beverages, inactivity, obesity, smoking, and diabetes. The link becomes even more obvious when we open our mouths.

LIFESTYLE, DENTAL CARIES, AND TOOTH EROSION

Dental caries is a contagious bacterial disease, with children generally gaining the infection from their mothers (but also possible is infection traded between

Lifestyle Medicine. http://dx.doi.org/10.1016/B978-0-12-810401-9.00027-9

teenagers, testing the associations that rapacious young people form). Having said this, the bacteria that cause the problem (mutans streptococci and lactobacilli in particular) exist in all humans and don't necessarily cause a problem unless a number of conditions prevail. They require fermentable carbohydrate as a nutrient (not just simple sugar, as is the general public perception) and produce acids that demineralize (dissolve) teeth on a particular area of a tooth covered by dental plaque. Coupled with acidic foods and beverages with a pH below 5.5, these can cause dental caries as well as erosion of tooth enamel (Bartlett, 2005). When erosion occurs, the whole surface of the tooth dissolves, leaving the enamel very thin and the tooth extremely sensitive to cold and hot foods.

If there is no food or bacteria, decay cannot occur, but of course, this is totally unrealistic. Hence protection is afforded in the natural world by saliva, which acts as a buffer, with calcium, fluoride, and other minerals, to protect against bacteria and acidosis, as well as help remineralize teeth and provide protection through antibodies and antibacterial action (Hara et al., 2006). Several factors in the modern lifestyle mitigate against the protective effects of saliva, such as medications, which decrease the salivary flow, or eating habits that do not allow saliva time to bathe the teeth and remineralize the enamel. Hence, protective actions need to be consciously taken or decay will occur.

EROSIVE FACTORS IN THE DIET

While fermentable carbohydrate is the key nutritive factor for dental bacteria, this is only relevant if there is enough for bacterial metabolism. Hence total food intake, as well as the quality of that intake can be relevant. Eating a whole chocolate bar can thus have the same effect on mouth bacteria as just eating one piece, as long as it is eaten all at one time. However, if it is eaten in several bursts then the bacteria will make acid for longer and this will have a worse effect on the tooth. Carbohydrates that "stick" to the teeth, such as toffee, will have a greater effect on decay production, not necessarily because of its sugar content, but because this is likely to be around between the teeth longer, and being between the teeth is not easily shifted through the use of a toothbrush. Flossing thus becomes important.

A growing issue is the carbohydrate-acid mix provided by some beverages, such as soft drinks, sports drinks, flavored sparkling water drinks, packet mixes, and fruit juices. Contrary to popular opinion, cola and other soft drinks are sometimes less acidic than fruit juices, so it is important to understand that all drinks containing acids can potentially erode teeth if they are used inappropriately. It has been shown that the sugared version of both cola and noncola drinks are more corrosive than their "diet" counterparts (Jain et al., 2007), but these are still not risk free. Removing the sugar will remove the risk of decay, but it will not remove the risk of erosion. The childhood trick of placing a tooth in a glass of cola to see the enamel eroded overnight works equally well for most fruit juices, suggesting (1) that the food or drink product itself is important for erosion and decay, and also (2) that this is most corrosive when the teeth are left to "bathe" in such solutions. Sipping drinks over an extended period, such as long-distance cyclists using "sports" drinks, or babies left to sip on juice while in the cot, can have the most damaging effects (Zero and Lussi, 2006). Wine tasters and judges are particularly at risk because of the regular contact of wine with teeth (Lussi and Jaeggi, 2006).

Tips for Good Teeth

- Take at least 2 h between eating or drinking if at all possible.
- Eat (or drink) sweet foods with meals, not between.
- Avoid "sticky" sweet foods and carbohydrates before bedtime.
- Chew (unsugared) gum (for at least 20 min) after a main meal.
- Don't sip soft drinks, fruit juices, flavored water, or sports drinks over time.
- Floss as well as brush to make sure plaque is removed between the teeth.
- Eat a variety of foods, including the recommended intake of dairy protein.
- Don't smoke, get regular exercise, and reduce obesity.
- Drink fluoridated water or ask your dentist for advice about the best way to use fluoride to improve your teeth.
- Wear a mouth guard if playing a contact sport.

PROTECTIVE FACTORS

Recovery against a carbohydrate onslaught takes time for the teeth. Hence, breaks in eating and drinking of at least 2 h are recommended, a process that is often uncommon in young children in the modern environment. The belief that a constant intake of liquid is needed to avoid dehydration is discussed in Chapter 9. It is certainly not a good practice when trying to protect the teeth from damage.

The timing of eating foods is also important. Carbohydrate-rich foods (particularly those that are sweet and/or sticky) should be avoided before bedtime when the salivary flow is at its lowest and therefore will not provide the best tooth protection. Sweet drinks, such as soft drinks, sports drinks, wine, and fruit juices, if drunk at all, should be taken with meals or other foods. A variety of foods, as encouraged in a healthy diet, is also best for teeth, preferably avoiding frequent use of sticky carbohydrate-rich foods.

As well as cleaning (and preferably flossing) teeth, chewing unsugared gum after a meal or snack can help stimulate saliva with antiacidic and antibacterial protective benefits. However, this appears to require at least 20 min of chewing. Some foods in the diet, like dairy products, also have a protective effect, possibly through the qualities of proteins such as casein phosphopeptide, which can coat the teeth, providing a rich source of calcium and phosphate to decrease demineralization and enhance remineralization (see following section). Even sweetened, flavored milk does not cause the same increase in acidity when compared with other sweetened drinks. Again, this supports the benefits of a higher protein diet in weight control, as discussed in Chapter 4. Fluoride, as added in extra amounts to many water supplies, also acts to slow down the demineralization process and enhance the remineralization process.

PERIODONTAL DISEASE AND DIABETES

A key feature of insulin resistance and type 2 diabetes is endothelial dysfunction and systemic inflammation, particularly in the small vessels of the body. This leads to possible infection and reduced healthy blood flow to the peripheral vessels that becomes characteristic of peripheral neuropathy, nephropathy, retinopathy, and other problems associated with diabetes. Less commonly known is the inflammation and endothelial dysfunction of small vessels of the gums and tissues surrounding the teeth, and the effects of this on the periodontal structure, often resulting in loss of teeth, and tooth, gum, and mouth disease (Amar et al., 2003). Periodontal problems are thus linked with systemic health problems including heart disease, chronic infections, respiratory disease, and diabetes (Joshipura et al., 2000), although the link is possibly bidirectional (Mealey and Rethman, 2003). In particular, the link between periodontal disease and diabetes is becoming increasingly clear (Mealey and Oates, 2006), and hence it's not unexpected that such problems would also be associated with diabetes risk factors such as obesity (Pischon et al., 2007). Inactivity, which itself is a cause of type 2 diabetes, has been shown to be linked to tooth loss in older people (Tada et al., 2003), presumably resulting from periodontal problems. Unhealthy diet is also an issue, with those aspects of diet mentioned regarding dental decay also being involved (Lussi et al., 2006). Last, but definitely not least, cigarette smoking is thought to be the main risk factor associated with chronic destructive periodontal disease, the risk being 5–20-fold higher in a smoker than a nonsmoker (Bergstrom, 2004).

What Can Be Done in a Brief Consultation?

- ☐ Check for potential periodontal problems in obese, inactive, or smoking patients.
- ☐ Ask about pattern of use of drinks and food (i.e., sipping, snacking).
- ☐ With parents, check on pattern of fluid intake of children as well as use of teeth care including flossing.
- ☐ Check to make sure there is some use of fluoride either in the drinking water or in other vehicles such as toothpaste or mouth rinse.
- ☐ Explain 5–20-fold risk of periodontal disease and tooth loss to smokers.

SPORTS AND DENTAL HEALTH

While an active sporting involvement is to be recommended for a healthy lifestyle, engagement in sports comes with injury risks. Mouth guards to protect against orofacial injury were first introduced in boxing in the United States in the 1920s. Since then, they have been made compulsory by sporting bodies in a number of sports. Modern materials and design have made a range of products effective in reducing the number of fractured teeth and providing some protection against jaw fracture as well. Studies comparing mouth guard users with nonusers in a range of sports have shown a relative risk of injury of 1.6–1.9 when a mouth guard is not worn. Hence, these are recommended for all sports with a potential for impact or fall injuries (Knapik et al., 2007).

SUMMARY

The lifestyle factors that cause poor oral health are generally the same culprits responsible for other metabolic diseases discussed in this book: poor nutrition, excessive sugary foods and drinks, smoking, obesity, and inactivity. The links between poor oral health and type 2 diabetes are also becoming clearer with new research, but whether this is causal or an outcome is yet to be determined. Good oral health requires attention to diet, overall body weight, and activity level. However, brushing, flossing, chewing sugarless gum, giving the teeth a rest between meals, and avoiding sticky carbohydrate foods before retiring for the night all add up to good oral health.

Practice Tips

Improving Oral Health

	Medical Practitioner	Practice Nurse
Assess	• check teeth in association with lifestyle information • periodontal damage in patients with diabetes risk • oral health in obese/overweight patients	• check oral health practices • check use of food and drinks in relation to teeth • check oral health practices in the obese/overweight • need for mouth guard in sports participants, particularly children
Assist	• explain about lifestyle factors and association with dental health • explain that poor dental health can affect general health • explain that parents' poor dental health can put children at risk	• provide written materials for oral health • inform where dental care can be obtained • explain links between oral health and lifestyle practices leading to metabolic problems
Arrange	• referral to dentist or dental specialist	• line up referral

Key Points

Clinical Management

- Consider oral health as a significant aspect of lifestyle medicine;
- Consider the dental as well as calorific implications of food and fluid intake;
- Associate dental decay and periodontal disease with other forms of metabolic problems (i.e., hypoglycemia);
- Become proactive on dental health with overweight children and their parents;
- Bring dental health into nutritional advice and obesity and weight control management.

REFERENCES

Amar, S., Gokce, N., Morgan, S., Loukideli, M., Van Dyke, T.E., Vita, J.A., 2003. Periodontal disease is associated with brachial artery endothelial dysfunction and systemic inflammation. Arterioscler. Thromb. Vasc. Biol. 23 (7), 1245–1249 .

Bartlett, D.W., 2005. The erosion in tooth wear: aetiology, prevention and management. Int. Dent. J. 55 (Suppl. 1), 277–284.

Bergstrom, J., 2004. Tobacco smoking and chronic destructive periodontal disease. Odontology 92 (1), 1–8.

Hara, A.T., Lussi, A., Zero, D.T., 2006. Biological factors. Mongr. Oral. Sci. 20, 88–99.

Jain, P., Nihill, P., Sobkowski, J., Agustin, M.Z., 2007. Commercial soft drinks: pH and in vitro dissolution of enamel. Gen. Dent. 55 (2), 150–154.

Joshipura, K., Ritchie, C., Douglass, C., 2000. Strength of evidence linking oral conditions and systemic disease. Compend. Contin. Educ. Dent. Suppl. (30), 12–23.

Knapik, J.J., Marshall, S.W., Lee, R.B., Darakjy, S.S., Jones, S.B., Mitchener, T.A., delaCruz, G.G., Jones, B.H., 2007. Mouthguards in sport activities: history, physical properties and injury prevention effectiveness. Sports Med. 37 (2), 117–144.

Lussi, A., Jaeggi, T., 2006. Occupation and sports. Mongr. Oral. Sci. 20, 106–111.

Lussi, A., Hellwig, E., Zero, D., Jaeggi, T., 2006. Erosive tooth wear: diagnosis, risk factors and prevention. Am. J. Dent. 19 (6), 319–325.

Mealey, B.L., Oates, T.W., 2006. Diabetes mellitus and periodontal disease. J. Periodontol. 77 (8), 1289–1303.

Mealey, B.L., Rethman, M.P., 2003. Periodontal disease and diabetes mellitus. Bidirectional relationship. Dent. Today 22 (4), 107–113.

Pischon, N., Heng, N., Bernimoulin, J.P., Kleber, B.M., Willich, S.N., Pischon, T., 2007. Obesity, inflammation, and periodontal disease. J. Dent. Res. 86 (5), 400–409.

Tada, A., Watanabe, T., Yokoe, H., Hanada, N., Tanzawa, H., 2003. Relationship between the number of remaining teeth and physical activity in community-dwelling elderly. Arch. Gerontol. Geriatr. 37 (2), 109–117.

Zero, D.T., Lussi, A., 2006. Behavioral factors. Mongr. Oral. Sci. 20, 100–105.

Chapter 28

Lifestyle and Environmental Influences on Skin

Hugh Molloy, Garry Egger

It's skin that keeps us in.

INTRODUCTION: THE SKIN AS AN ORGAN

Not so long ago, it would have been appropriate to start this chapter by stating that the skin is the body's biggest organ. Laid out on the ground it would cover an area of roughly 4×3 m in the average-sized human. As a structural unit responsible for different functions in the body (the definition of an "organ"), it would seem there are none bigger. It was only when adipose tissue (fat) was classified as an endocrine organ in the 1990s, and with the development of the obesity epidemic, that skin was relegated to number two in the size department (Scherer, 2006). Still, this doesn't detract from the often overlooked fact that skin is an organ, and it covers a large area of the human body. Its roles include keeping us contained, exerting an even pressure on all the tissues enclosed within the deep fascia (body stocking), protecting us against the rigors of the outside world, and assisting awareness of what is happening around us (Balognia and Schaffer, 2014).

WHY SKIN IN LIFESTYLE MEDICINE?

This chapter is not concerned with infectious skin diseases, or those with internal origins. Instead, it is aimed at providing a new way of looking at the many common skin problems related to the modern environment, and highlighting what can be done about these in the context of lifestyle medicine. It is our contention (spelled out in more detail in Molloy and Egger, 2007), that three factors (overheating, excessive dryness, and an overobsession with cleanliness) alone and/or in combination, are responsible for a large proportion of modern skin problems, ranging from persistent acne to certain forms of eczema and psoriasis. Yet they have all but been ignored in skin care education. It should be stressed that not all the approaches discussed herein have a strong evidence base as yet. Hence, much of what is discussed is based on logic and clinical experience.

Lifestyle Medicine. http://dx.doi.org/10.1016/B978-0-12-810401-9.00028-0

TABLE 28.1 A "Skindex" of Modern Skin Problems With a Possible Lifestyle Cause (Molloy and Egger, 2007)

1. dry, rough, or scaly skin that may or may not be itchy
2. redness, greasiness, rashes, persistent acne with no obvious cause
3. unexplained morning sneezing, runny nose, and/or dry throat
4. limp, lifeless, stringy, or greasy hair
5. dark, baggy rings around the eyes ("doona eyes")
6. comedones ("whiteheads") on the forehead, cheeks, and chin
7. redness and itchiness between the breasts
8. recurrent tinea
9. disturbed sleep
10. feeling tired/washed-out on waking up

The premise is that much (although obviously not all), of what happens to the skin in modern times results from our way of life rather than any mysterious or unknown microbes. Lifestyle medicine is a way of helping to manage and deal with skin problems such as those shown in Table 28.1.

"Dermatitis": The Biggest Lifestyle-Based Skin Problem

Dermatitis, meaning literally "an inflammatory condition (*itis*) of the skin (*derma*)" is a general term given to a range of noninfectious skin disorders. These are subdivided in many ways, based on appearance, location, underlying causes, etc. The approach we use here (not necessarily accepted by all dermatologists) is to consider most types of dermatitis to be reactions to skin "insult" of some form, either external or internal, psychological, or physical. The skin then reacts to such insult in a variety of ways. It may show up as eczema, acne, psoriasis, urticaria, rashes, pimples, or in any number of other forms, but most can be considered "patterns of reaction" to insult. No one pattern is necessarily discrete or as clear-cut as it may sound by the name that it is given.

Causes of Dermatitis and Other Modern Skin Problems

Although causes of skin problems can be many and varied, we are only concerned here with those associated with lifestyle factors that can be easily recognized and modified with clinical advice and assistance, without recourse to expensive treatment options. We have not included the role of medications here as these are beyond the scope of the lifestyle medicine approach considered here.

Environmental Factors

It's now thought that the effects of heat and drying on skin, which is often inherently dry anyway, are amongst the most unrecognized causes of skin damage seen today (Yokota et al., 2014). There are a number of environmental factors, both natural and unnatural, that can contribute to this.

The Outdoor Environment

Sunlight: While sunlight is necessary for an optimum level of vitamin D, excessive sunlight can cause skin problems (Mason and Reichrath, 2013). Skin cancers, which result from excessive years of 'baking" in the sun, are more common in fair-skinned than dark-skinned people, but they are still possible in those with totally black skin (Gloster and Neal, 2006). Good quality sunglasses with 100% UV protection (price is no indication) and a broad-brimmed hat are recommended as the best lines of defense. Sun damage to skin can range from relatively benign solar keratoses (commonly known as "sunspots") to highly malignant melanomas.

Cold: While winter reduces the potential for skin cancers through sun exposure, cold, dry winter air can dry the skin, leading to xerosis, which in turn can cause "winter itch." This is most common in older people with good, or even excessive hygiene, but who, through aging, have increasingly dry skin. However, it can also exist in younger and more active people (Adams, 2002). Dryness from cold can also be the cause of a range of other skin insults (Yokota et al., 2014). Regular use of an inexpensive moisturizer can help overcome this.

The Indoor Environment

Bedding: Modern bedding, and the overheating resulting from this is an often unrecognized potential causes of skin problems in the indoor environment (Okamoto-Mizuno et al., 1999). The continental quilt (duvet, doona), which was developed for the European climate, but has been taken up widely in warmer climates like Australia, is equivalent in insulation to five to six blankets, but can't be discarded in sections, like blankets can, during the night. Because heat can't escape from under a doona, body heat during sleep rises to where it can be released (i.e., from the nipple line to the head) as sweat into the open air. This occurs during periods of light sleep and often causes rubbing of the face and head scratching. Problems that can result from this are poor sleep, "doona eyes" (the term given to the darkening of skin around the eyes), facial and perioral dermatitis, Grover's disease (itchy red lumps on the upper chest), facial excrescences, atopic eczema, lank hair and scalp problems, acne, and dry itch (Molloy et al., 1993). The severity of many of these problems is reduced by using blankets instead of doonas, not using electric blankets or hot water bottles, wearing light bed clothing, not wearing bed socks, etc. Some early discomfort from an initially cool bed is better suffered than the later skin problems that can result.

Workplaces: The modern work environment can be another source of skin troubles. Air-conditioning systems in the office extract water vapor from the air before it is circulated, thereby creating a "hostile" environment for the skin

(Morris-Jones et al., 2002). "Screen dermatitis" is the name given to an increasingly common skin problem associated with electric equipment, and particularly video display terminals (Johansson et al., 2001). General symptoms can include dizziness, tiredness, and headache, and effects on the skin can be erythema, or skin "flushing." The effect may be simply due to dryness of the office environment, giving rise to irritation of the skin and consequent facial rubbing.

Unlike at home, it is often difficult to change the work environment. Some changes that may be helpful include having large bowls of water around the office, opening windows where possible instead of using air-conditioning, taking regular breaks outside the office, avoiding cigarette smokers, and regularly waterproofing the skin through the use of a moisturizer.

Traveling: Long-haul flights mean sitting for hours in planes where the relative humidity can get below 5% (the ideal for skin is around 40–65%). This, together with dehydration (or at least lack of hydration), can dry out the skin and mucous membranes and helps explain the washed-out, tired look when arriving at a destination. The local traveler is also not excluded. Cars, which used to have side windows to allow air for cooling while the main window was closed, now have air-conditioning for summer and heaters for winter, both of which dry out the interior of the car and cocoon it from the natural atmosphere. One way of rectifying this is a fine spray of water into the air every now and then from a spray bottle left in the car over long journeys, particularly those in winter when the air is drier (Lu et al., 2007).

In many large hotels, the windows can't be opened and the room becomes a veritable drying pit—even though it might be cool. The dryness requires turning off the air-conditioning (if you can find it), which then often means getting cold while sleeping. As with many hotels, the only sleeping cover is a doona or continental quilt, which increases overheating (see following) and can't be peeled off in stages. Once uncovered, the body then gets too cold and the cycle is then repeated, making the modern hotel room not only a blight on the environment (see Chapter 22) but a potential insult to skin. Propping open the bathroom door and putting a 2-cm base of water in the bathtub may go some way toward relieving this, and regular use of a skin moisturizer both while in flight and on the ground can help. Otherwise, it is prudent to advise traveling patients to look for hotels with windows that can open, and blankets that can be peeled off, to break into the cycle.

Modern Technology

Changes in living environments for humans over the past century have been caused and accompanied by big advances in technology. And while there's no doubt that this has improved the living conditions of the human race, it also has a downside, some of which can be seen in insults to the skin. Among the causes are:

Heating and air-conditioning: As discussed earlier, the dryness of heating and cooling through artificial means can lead to a dermatitis with features of roughness, flaking, scaling, fissuring, and apparent lack of moisture

(Sunwoo et al., 2006). Avoiding excessively dry places, minimizing air-conditioning and heating can help reduce these problems. The most common cause of physical irritant dermatitis in one major English study was low humidity due to air-conditioning, which caused dermatitis of the face and neck in office workers due to the drying out of skin (Morris-Jones et al., 2002).

Clothing: Synthetic fibers are now incorporated into much modern clothing and bedding. They are cheaper than natural fibers, wear better, are lighter, less bulky, and are often better insulators. Over 5–10% of these fibers, even when mixed with natural fibers, tend to cause a buildup of heat and this is often seen in different types of clothing.

Synthetic exercise clothing, for example, can reduce the natural functional effects of sweat i.e., cooling of the body. If there is a buildup of heat on the skin that is not dissipated, this can cause itchiness and irritation, which is then made worse by washing and scrubbing to get rid of the itch. Exercise clothing is best derived from natural fibers with the extremities (arms and legs) kept bare and a bare midriff. It should also be as loose as possible so as not to cause frictional damage (Ricci et al., 2006).

The same applies to children's wear. Parents often fail to understand that heat loss from the hands and feet of a child is important for the control of body temperature and that covering these with gloves, bootees, and bonnets in many parts of the country where this is not necessary can be potentially damaging to the child's skin. In a normally placid climate, it is possible that this can upset the normal thermoregulatory mechanisms. The head, and more specifically the face, becomes the main site for heat loss from the overclothed child, and this is made worse if the head is covered in an already overclothed child. Atopic eczema in infants frequently presents on the cheeks and chin between the ages of 6–12 months. Although not proven, it is feasible to suggest that overclothing/heating may play a part in this.

Modern footwear can also present problems for the skin (Springett, 2002). Contemporary shoes are often "hot boxes" from which heat has difficulty in escaping. The uppers are often made of synthetic material or, if made from leather, are sprayed with "antiscuff" applications to reduce the need for polishing. As a result, air is unable to circulate and the inside of the shoe is turned into a kind of "steam bath" providing an ideal environment for all forms of skin problems, including tinea pedis and pitted keratolysis. In obese patients the problem is exacerbated by the extra weight transferred to the feet, reducing the natural arch, which may otherwise allow an "air pocket," and causing an irritating itch under the feet (often a reason why big men wear sandals or "thongs" instead of shoes). This can be overcome by ventilating the shoes through a few small holes in the instep, approximately 1 cm above the welt. This allows air to circulate in the shoe and can reduce the internal temperature by up to 3°C.

Chemicals: Modern household chemicals also have great potential for skin damage. Detergent for washing dishes, for example, is usually overused, and this is complicated by the use of occlusive gloves made of synthetic materials that allow the hands to be inserted into much hotter water than usual. The resultant overheating and excessive sweating of the hands within the gloves can cause dermatitis of the hands that will not respond to the usual therapies. A more preferable option is to use minimum detergent in comfortably warm water, without gloves, but with the hands protected by a layer of moisturizer.

The Obsession With Cleanliness

If the modern environment and modern technology are problems for the skin, what we do to counter these often makes the problem worse. The immediate reaction for someone with dry skin, greasy hair, or an itchy scalp is to have regular showers, scrub down with soap, and shampoo the hair repeatedly, often not realizing that this is only drying the skin and scalp out more and leaving it open for the vicious cycle of "dermatitis." There are a number of things we do that cause this to happen (Wolf et al., 2004).

Showers and cleaning: Regular showering is obviously necessary and good for the skin, provided showers are brief and not too hot. Strong soaps, however, tend to take away the surface layer of "natural" grease that serves a barrier function on the skin. This is a common cause of perianal dermatitis. Washing with water only, on the other hand, dries out the epidermis and causes chapping, A simple and inexpensive moisturizing lotion can provide a solution to this (Draelos, 2005), used both in the shower or bath and applied to the skin after as an "'all-over" moisturizer (Simion et al., 2005). A relatively cheap mix of Sorbolene and glycerol is usually as good as the more expensive versions, although different moisturizers may be required for different reasons such as forming a barrier function or reducing dryness (Loden, 2003).

Shampoos: While the ads tell us that shampoos are a must for lovely hair, the famous French dermatologist Dr. Aron-Brunetiere has likened shampoo to industrial strength detergent, not unlike that used to "scrape" the grease off plates in dishwashing. It "scrapes" the natural grease out of hair and can cause damage to the hair, scalp, face, shoulders, and upper trunk and hands.

Hair can be easily managed by using conditioner alone. Conditioners contain surfactants (spreading agents), some of which have mild detergent capacities, enough to remove the normal grease in the hair but not enough to remove a thick tacky emulsion that can result from overheating at night, and which can be prevented by reducing this nighttime heating. Conditioners can be used as often as desired because the level of detergent activity is much less than in shampoo. During the transition period from shampoo to conditioner, some increase in greasiness may be noted, perhaps along with a little itching and flaking, but this usually settles in a matter of weeks. The condition of the hair becomes noticeably better by the end of 6–8 weeks—provided overheating at night is avoided.

Soap: Soap is made from oil and alkalines. It works by washing the grease, or sebum, off the skin, and in the process changing it from its usual slightly acid state to one of alkalinity. Unfortunately, no soap made has been able to differentiate between the skin's natural oils and exogenous dirt. Thus, washing with soap always involves the removal of the uppermost layer of oil from the skin, as well as changing the skin from predominantly acid to alkaline. Yeasts, fungi, and some forms of bacteria, especially gastrointestinal bacteria, grow better in alkaline situations, and hence this is a recipe for bacterial "insult" to the skin. The use of moisturizing soaps is of intermediate value only. They should also be used minimally and are less effective than the use of moisturizing cream and water (Wolf et al., 2004).

Scrubbing and drying: It is better to treat skin with tender loving care than to scrub it, and to waterproof it rather than to dry it out. If skin is adequately waterproofed, it can look after itself. Drying talcum powder tends to collect in skin creases and become abrasive (although some special powders such as diphemanil methylsulfate can be useful for sweat rashes); scrubbing the skin with a towel and the aggressive use of scourers like loofahs, belong in the field of masochism and have no real role in skin care.

Hair dryers: Although hair dryers are not as damaging as may be thought, their excessive use can contribute to hair damage. One useful way of reducing this is to firstly dry the hair with a synthetic chamois, like a sports towel or car washer. This will remove the bulk of the moisture quickly, particularly if the hair is long and thick. If a dryer is still necessary, it can be used relatively briefly at a mild temperature so it is less likely to set up later damage in the hair follicle.

Deodorants and antiperspirants: Where a new approach to caring for the skin, such as discussed here is undertaken, there is less need to use deodorant in temperate climates (although admittedly, these may be necessary in warmer parts of the country). Those who sell deodorant are usually those who sell soap. It is possible that the regular use of soap brings about minor alterations in the composition of the bacteria that are normally resident on the skin, in turn giving rise to changes in body odor. And a deodorant may be the obvious solution to counteract this effect. It may be wiser to not produce the effect in the first place, if this is the case.

Antiperspirants work by blocking the outer portion of the sweat duct. This forms a physical barrier from which sweat is unable to escape to the outer surface of the skin where it would normally be evaporated, and hence aid in heat loss. Although there are some idiosyncratic smells associated with individual sweat, the main offensive smells that come from it result from all those approaches to soaping, drying, and cleansing that we have discussed so far, and which make the body respond in an "unnatural way."

Cosmetics: In general, cosmetics have little overall effect on the skin. But the damage inflicted on the skin by some cosmetics or medicaments can be a problem for some women (Orton and Wilkinson, 2004). Various combinations of showering, shampooing, cleansing, and drying mentioned before are capable of aggravating, augmenting, and prolonging any instability already present in the skin. Adding a cosmetic that may consist of dozens of ingredients can raise the chances of one or a combination of ingredients to which the skin may be sensitive, and result in dermatitis in some form. Quite often this leads to medicines being prescribed for the skin, but unless the causes are considered in their own right, medicines are unlikely to have any long-term effect. Indeed, they may just add to the problem because of the addition of yet another potential skin irritant. And while few would support the total elimination of cosmetics in women, there is a definite case to be made for careful and perceptive use.

"Natural" cosmetics: Most so-called "natural" products are mixtures of many ingredients. Lanolin, from sheep's wool is claimed to have particular benefits for the skin because it is a "natural" product. Yet lanolins contain over 60 different components and thus have a high potential for skin reactions. Most plant extracts also consist of a lot of ingredients and the fact that these are "natural" is little different from a product consisting of exactly the same chemicals made up in a laboratory. A good example of this is the commonly used tea tree oil, with which a significant incidence of skin sensitivities is associated.

What Can Be Done in a Brief Consultation?

- ☐ Observe skin disturbances that may be related to lifestyle.
- ☐ Observe skin type and possible environmental effects (e.g., sun).
- ☐ Use the skin behavior check list (see appendix) if appropriate.
- ☐ Ask about use of shampoo, soap, astringents, etc.
- ☐ Check obese/overweight patients for diabetic dermopathy, acanthosis, etc.
- ☐ Discuss effects of constant dryness/overheating on skin.
- ☐ Investigate lifestyle habits (scratching, medicating) as a cause of specific skin insult.

Lifestyle

As well as the modern environment and technology, lifestyle behaviors leading to the health problems discussed in this book can have an effect on skin. The most obvious is obesity, which is responsible for changes in skin barrier function, sebaceous glands and sebum production, sweat glands, lymphatics, collagen structure and function, wound healing, microcirculation and macrocirculation, and subcutaneous fat (Yosipovitch et al., 2007). It is also a cause of a wide spectrum of dermatological diseases including acanthosis nigricans, intertrigo, acrochordons, keratosis pilaris, hyperandrogenism and hirsutism, striae distensae, adiposis dolorosa, and fat redistribution, lymphedema, chronic venous insufficiency, plantar hyperkeratosis, cellulitis, skin infections, hidradenitis suppurativa, psoriasis, insulin resistance syndrome, and tophaceous gout. Diabetic dermopathy is the most common cutaneous marker of diabetes and presents as single or multiple well-demarcated brown atrophic macules, predominantly on the shins. This has also been shown to be associated with diabetic retinopathy (Abdollahi et al., 2007) supporting the need for optical care in diabetic patients. Smoking can also affect skin and enhance wrinkling, giving the appearance of aging.

A Lifestyle Skin Care Package

For best effect, the skin care techniques discussed here should be seen as a package. If some suggestions only are taken up, and the rest ignored, the results are likely to be less than totally satisfactory. It should also be realized that this is unlikely to provide an overnight remedy. Patience and perseverance are necessary. An effective outcome can take at least 3–4 months (e.g., see Molloy and Egger, 2007). Under the principle of "do no harm," the following are not likely to have any adverse effects, even if not highly evidence-based in some cases.

Showering/Bathing

If the skin is irritable, no soap should be used and oatmeal baths should be substituted.

Moisturizing cream mixed with water may be used as soap. It will adequately clean ALL areas of the skin.

Keeping soap to a minimum will reduce the need for deodorants, which themselves can have a damaging effect on the skin.

Showers may be taken as often as desired, but they should be brief and tepid. Very hot water should be avoided. Any soap used should be rinsed off well before drying.

Avoid spa baths and other heat treatments such as saunas.

Avoid bubble baths or bath salts and do not use antiseptics such as Dettol in the bath water.

If desired, use an oatmeal bath (not oatmeal soap), by doing the following:

1. Put one cup of rolled oats in an old nylon stocking or muslin bag;
2. Fill a bath with tepid water so the whole body can be immersed;
3. Swish bag around in the water until it goes white;
4. Soak in the bath and squeeze bag against infected skin (10–15 mins);
5. After the bath, pat the skin dry gently.

Drying

After each shower, half dry the skin by patting, rather than rubbing.

Soak up most of the moisture from the hair with a chamois or absorbent towel.

It is not necessary to have the skin completely dry.

Moisturizing

Use a simple, inexpensive moisturizer (such as Sorbolene cream with glycerin) regularly after showering or washing.

Half dry the skin by patting before applying moisturizer.

Take approximately one teaspoon of moisturizing cream and four teaspoonsful of water. Mix these well in the hands until a nice simple lotion is formed and then rub this on all over the body.

Since the cream neither smells nor stains (except silk), and washes out of clothing by normal means, it is unnecessary to wait for the skin to dry before dressing.

Note:

Occasionally, when very dry skin is being moisturized, small red spots, some with minute pustules in the center may develop, especially on the thighs and forearms

Continued

A Lifestyle Skin Care Package—cont'd

due to blockage of hair follicles. Such spots should be ignored. They are not infectious. Squeezing or rubbing will make them worse.

Clothing

Aim to keep comfortably cool both during the day and at night. When the body is warmed up enough, consider taking some clothing off. (After some weeks of doing this the body adjusts, and it becomes easier to cope with temperature changes.)

Punch holes in the insteps of leather shoes to enable them to "breathe."

Sports Clothing

Avoid synthetic, full body clothing such as Lycra. Wear light cotton clothing that "breathes" to allow sweat to be evaporated off the body.

Do not overdress. There is no such thing as "sweating weight off" by overdressing during exercise.

Where possible keep the extremities, such as arms and legs, and the midriff bare to allow for maximum sweat evaporation.

Wear chunky acrylic socks during exercise to "wick away" moisture from overheating of the feet.

Hair Care

Avoid shampoos. Use one to two tablespoons of hair conditioner for washing hair.

Avoid the frequent use of hair dryers, especially if these are on hot.

Skin Care

Avoid the use of aftershave, toners, tonics, astringents, and cleansers.

Do not try to exfoliate the skin, particularly if exposure to sunlight is imminent.

Avoid chemicals and overuse of antiseptics and "medicated" skin care items.

Use moisturizing cream as "soap," hand cream, shaving cream, facial cleanser, moisturizer, and shampoo if desired (mix with a little water and run into the scalp at night, then rinse out the next day).

Reduce body fat (not necessarily weight).

Reduce the amount of chemicals (in cosmetic products) applied to the skin.

Don't be fooled by cosmetics that claim to be "natural."

Use basic lipsticks, lip creams for moisturizing and protection of the lips.

Learn techniques of stress management.

Indoors

Where possible, avoid situations of constant overheating such as air-conditioned rooms and cars.

Avoid rubber gloves while washing up.

Don't use excessive detergent in washing up.

Use a bowl of water to increase humidity in an air-conditioned office.

Take frequent breaks outdoors.

A Lifestyle Skin Care Package—cont'd

Outdoors

Use a maximum protection sunscreen at all times if exposed to the sun.
Move around if it is necessary to stay in the sun. Do not sunbake.
Use sun protective clothing if exposed to the sun over long periods.
Regularly wear a sun-protective hat.
Use moisturizer after sun and/salt water or chlorine exposure.
Wear sunglasses when in the outdoors, but choose a brand based on UV protection level (preferably 100%), rather than price.

Nighttime

Use blankets and sheets instead of a doona, or quilt.
Avoid electric blankets, hot water bottles, or heated water beds.
Use light, natural fiber blankets such as wool (but not angora or mohair).
Avoid overusing blankets or continental quilts that cause nighttime overheating.
Become used to "sleeping light."
Do not use bed socks.
If cold feet are a problem, heat a damp towel in a microwave and place under the feet.
Wear light night attire all year round.

Traveling

If possible in a hotel/motel room, turn of the air-conditioning and open the window. If this is not possible, fill the bath full of water and leave overnight.
If spending long periods in an air-conditioned or heated car, use a light spray mist to increase the humidity.
Carry a moisturizer in hand luggage for use on long-distance airplane flights.
Ask the hotel if it is possible to have blankets instead of a doona or continental quilt.

SUMMARY

Skin remains one of the body's largest organs. As such it reacts to "insult," both physical and psychological, which is reflected in the general term "dermatitis." While much of this may come from physiological sources, a good deal of modern skin damage is currently caused by the modern environment, particularly through excessive heating, drying, modern technology, and our obsession with cleanliness. A skin care package, incorporating a range of mild lifestyle changes can help overcome many modern environmentally induced skin care problems.

Practice Tips

Skin Care

	Medical Practitioner	Practice Nurse
Assess	• Check for signs of skin cancer and other nonlifestyle related skin damage • Check ABCDE[a] of moles and nevi • Consider lifestyle/environmental causes • Consider underlying disease (e.g., diabetes with acanthosis; obesity with heat rashes, etc.).	• Evaluate environmental causes of skin insult in detail • Check ABCDE[a] of moles and nevi • Conduct a lifestyle check list (see "measures" following) • Measure/assess possible underlying diseases identified
Assist	• Provide medication for skin treatment where required • Advise on possible influences of lifestyle factors	• Discuss modifying lifestyle factors that could be related to skin insult as a "package" (see following) • Provide ongoing
Arrange	• Referral to dermatologist where indicated • Identify other health professionals for a care plan (i.e., dietitian)	• Prepare care plan • Contact and coordinate other health professionals (dermatologist, dietitian)

[a]*ABCDE = Asymmetry, Border irregular, Color, Diameter.*

Key Points

Clinical Management

- Consider lifestyle factors that may affect other aspects of health as a potential cause of skin problems i.e., smoking, sleep, office/home environment, etc.
- Investigate possible modern causes of skin "insult"—physical or psychological.
- Explain the relationship between the modern environment and possible skin problems.
- Consider treatment for modern environmental skin problems as a "package," requiring attention to all factors.
- Consider obesity/overweight, smoking, and the sleep environment as a cause of skin problems.

REFERENCES

Abdollahi, A., Daneshpazhooh, M., Amirchagmaghi, E., Sheikhi, S., Eshrati, B., Bastanhagh, M.H., 2007. Dermopathy and retinopathy in diabetes: is there an association? Dermatology 214 (2), 133–136.

Adams, B.B., 2002. Dermatologic disorders of the athlete. Sports Med. 32 (5), 309–321.

Balognia, J.L., Schaffer, J.V., 2014. Dermatology Essentials. Elsevier, NY.

Draelos, Z.D., 2005. Concepts in skin care maintenance. Cutis 76 (6 Suppl), 19–25.

Gloster, H.M., Neal, K., 2006. Skin cancer in skin of colour. J. Am. Acad. Dermatol. 55 (5), 741–760.

Johansson, O., Gangi, S., Liang, Y., Yoshimura, K., King, C., Liu, P.Y., 2001. Cutaneous mast cells are altered in normal healthy volunteers sitting in front of ordinary TVs/PCs—results from open-field provocation experiments. J. Cutan. Pathol. 28 (10), 513–519.

Loden, M., 2003. Do moisturizers work? J. Cosmet. Dermatol. 2 (3–4), 141–149.

Lu, C.Y., Ma, Y.C., Lin, J.M., Li, C.Y., Lin, R.S., Sung, F.C., 2007. Oxidative stress associated with indoor air pollution and sick building syndrome-related symptoms among office workers in Taiwan. Inhal. Toxicol. 19 (1), 57–65.

Mason, R.S., Reichrath, J., 2013. Sunlight, vitamin D and skin cancer. Anticancer Agents Med. Chem. 13 (1), 83–97.

Molloy, H.F., Egger, G., 2007. Good Skin, second ed. Allen and Unwin, Sydney.

Molloy, H.F., LaMont-Gregory, E., Idzikowski, C., Ryan, T.J., 1993. Overheating in bed as an important factor in many common dermatoses. Int. J. Dermatol. 32 (9), 668–672.

Morris-Jones, R., Robertson, S.J., Ross, J.S., White, I.R., McFadden, J.P., Rycroft, R.J., 2002. Dermatitis caused by physical irritants. Br. J. Dermatol. Br. J. Dermatol. 147 (2), 270–275.

Okamoto-Mizuno, K., Mizuno, K., Michie, s, Maeda, a, Iizuka, S., 1999. Effects of humid heat exposure on human sleep stages and body temperature. Sleep 22 (6), 767–773.

Orton, D.I., Wilkinson, J.D., 2004. Cosmetic allergy: incidence, diagnosis, and management. Am. J. Clin. Dermatol. 5 (5), 327–337.

Ricci, G., Patrizi, A., Bellini, F., Medri, M., 2006. Use of textiles in atopic dermatitis: care of atopic dermatitis. Curr. Probl. Dermatol. 33, 127–143.

Scherer, P.E., 2006. Adipose tissue: from lipid storage compartment to endocrine organ. Diabetes 55 (6), 1537–1545.

Simion, F.A., Abrutyn, E.S., Draelos, Z.D., 2005. Ability of moisturizers to reduce dry skin and irritation and to prevent their return. J. Cosmet. Sci. 56 (6), 427–444.

Springett, K., 2002. Introduction to some common cutaneous foot conditions and their management. J. Tissue Viability 12 (3), 100-1–104-7.

Sunwoo, Y., Chou, C., Takeshita, J., Murakami, M., Tochihara, Y., 2006. Physiological and subjective responses to low relative humidity in young and elderly men. J. Physiol. Anthropol. 25 (3), 229–238.

Wolf, R., Orion, E., Parish, L.C.A., 2004. Scientific soap opera and winter itch. Skinmed 3 (1), 9–10.

Yokota, M., Shimizu, K., Kyotani, D., Yahagi, S., Hashimoto, S., Masaki, H., 2014. The possible involvement of skin dryness on alterations of the dermal matrix. Exp. Dermatol. 23 (Suppl. 1), 27–31.

Yosipovitch, G., DeVore, A., Dawn, A., 2007. Obesity and the skin: skin physiology and skin manifestations of obesity. J. Am. Coll. Dermatol. 56 (6), 901–916 2007.

Professional Resources

A Skin Behavior Checklist

Do you:

- Try to stay as warm as possible at all times?
- Work in an air-conditioned office?
- Drive an air-conditioned car?
- Sleep with a doona and/or electric blanket or heated water bed at night?
- Shampoo your hair daily or every two days?
- Frequently use soap, deodorant, astringents, aftershave, tonics, or toners?
- Regularly stay in hotels/motels without external ventilation?
- Travel regularly in airplanes?
- Work, play, or lie in the sun?

Affirmative answers to all or most of these questions could suggest the need for an environmental approach to skin care problems as outlined before in Practice Tips.

Chapter 29

Lifestyle-Related Aspects of Gastrointestinal Health

Ross Hansen

Man should strive to have his intestines relaxed all the days of his life.
Moses Maimonides (AD 1135–1204)

INTRODUCTION: EXPOSING THE GUT

The gastrointestinal tract is one of the largest body systems, which interacts extensively with the nervous and endocrine systems. Nutrient and fluid intake, physical activity, and psychosocial stress impact significantly on gut function, affecting the blood flow, neurohormonal balance, muscle tone, and motility of many gastrointestinal organs. Recent research has provided strong evidence for the significance of communication between the brain and the gut (the brain–gut axis) in health and disease, underpinning observations that implicate psychological factors in gut complaints. This all means there are many aspects of lifestyle with the potential to impact positively or negatively on digestive function. Consequently, anyone who Googles "lifestyle and digestion" will be inundated with Internet advice on gut health. This chapter provides an evidence-based, albeit brief, overview of lifestyle-related aspects of common upper gut (reflux, dyspepsia), small bowel (irritable bowel syndrome), and large bowel (constipation, diarrhea) dysfunction.

THE TOP END: REFLUX AND DYSPEPSIA

Reflux

Troublesome reflux ("heartburn") and regurgitation are a common cause for a consultation with a general practitioner (GP). These symptoms and (less commonly) difficulty in swallowing, atypical chest pain, and unexplained cough can be manifestations of gastroesophageal reflux disease (GERD). Although severe GERD can be associated with esophagitis, many people with GERD symptoms do not have visible changes in the esophagus. Recent advances in characterizing the nature and extent of reflux include impedance-pH monitoring over one or two days utilizing a catheter positioned in the lower esophagus (Bredenoord et al., 2007).

Lifestyle Medicine. http://dx.doi.org/10.1016/B978-0-12-810401-9.00029-2

Reflux from the stomach to the esophagus often results from transient relaxations of the lower esophageal sphincter, which normally prevents reflux. These relaxations can be triggered by distension of the stomach and by ingesting fatty food; this involves a nerve reflex from the stomach to the hindbrain and back to the lower esophagus (Fig. 29.1). Various lifestyle-related factors appear to contribute to excessive reflux. Obesity, smoking, excessive alcohol intake, eating too much, too rapidly, or too late at night, gassy or irritating food or drink, and even pregnancy are amongst the suspected culprits (Cremonini et al., 2009). In addition, food allergies should be considered in infants with reflux symptoms (Heine, 2009).

So, which of these factors should be considered in the management of GERD? One evidence-based review concludes that although there are limited studies to support it, reflux symptoms may improve with weight loss, avoiding late evening meals, sleeping with the head elevated, limiting alcohol intake, and stopping smoking (Kaltenbach et al., 2006). These can be done either before, or to complement, antireflux medication (usually proton pump inhibitors) or surgery.

Dyspepsia

Dyspepsia, involving upper abdominal pain or discomfort considered to originate from the gastroduodenal region, is another common upper gut complaint,

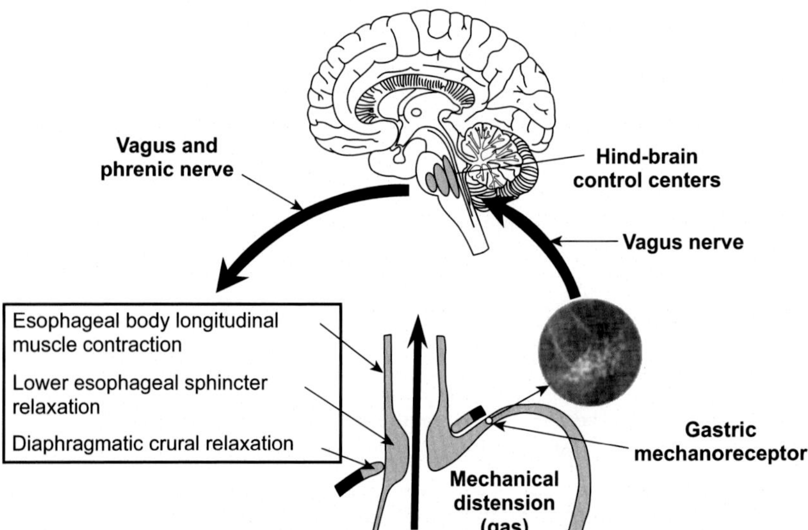

FIGURE 29.1 The lower esophageal sphincter and the crural region of the diaphragm act as a barrier to gastroesophageal reflux. Transient relaxation of both occurs when the stomach is distended. Fat ingestion can also trigger this nerve reflex. *Reproduced from Dent, J., 2009. Landmarks in the understanding and treatment of reflux disease. J. Gastroenterol. Hepatol. 24 (S3), S5–S14.*

with an estimated prevalence of 20–30% (Tack et al., 2006). Specific symptoms experienced include postprandial fullness, early satiation, and epigastric pain or burning. Physiological (delayed gastric emptying, altered gastric or duodenal sensitivity), pathological (*Helicobacter pylori* or salmonella infection) and psychological factors (anxiety, depression, stress) are implicated in this disorder (Tack et al., 2004; Talley and Choung, 2009). There is a huge overlap (approaching 50%) with reflux symptoms, and a majority with this disorder experience "functional dyspepsia," in which neither structural nor biochemical abnormalities can be readily demonstrated (Kellow, 2010). On the basis of findings from several large factor analysis studies, functional dyspepsia can be further defined as postprandial distress syndrome and epigastric pain syndrome (Tack et al., 2006).

Symptoms of functional dyspepsia can be aggravated by high-fat meals (Houghton et al., 1993) and coffee (Elta et al., 1990). There is anecdotal evidence that spicy foods, smoking, and alcohol can also contribute to symptoms (Talley et al., 1994). Epidemiological studies indicate that binge eating can exacerbate dyspepsia (Cremonini et al., 2009). Several pharmacological treatments for functional dyspepsia have been trialed, but these generally have limited efficacy beyond the typically high placebo response rates demonstrated (Tack et al., 2006). Providing reassurance and an explanation of symptoms should therefore be the starting point for management. Suggested lifestyle modifications include avoidance of "trigger" foods and stressors, eating several small low-fat meals per day, limiting caffeine intake, and ceasing smoking and alcohol consumption. Again, there is not a strong evidence base for all of these approaches (Tack et al., 2006; Talley and Choung, 2009). Preliminary studies with hypnotherapy show marked improvement in quality of life, symptom reduction, and medication usage; psychotherapy may also prove useful (Tack et al., 2006).

THE MIDDLE BIT: IRRITABLE BOWEL

Irritable bowel syndrome (IBS) is a functional bowel disorder featuring abdominal pain that is associated with defecation or a change in bowel habit. Prevalence is 10–20% in adults and adolescents. IBS has a considerable impact on work absenteeism and quality of life. The nature and severity of IBS symptoms frequently fluctuate and overlap with dyspeptic symptoms. Although two main subtypes of IBS are recognized (IBS with constipation and IBS with diarrhea), a wide spectrum of bowel habit exists (Longstreth et al., 2006).

IBS represents a classic model of interaction between the central nervous system and the enteric nervous system, between the "big" and "little" brains. Several underlying pathophysiological mechanisms have been implicated, including altered motility, visceral hypersensitivity, autonomic dysfunction, and psychosocial, genetic, and environmental factors. Gastric and small bowel dysmotility have been demonstrated in IBS sufferers following meals, fatty acid ingestion, and physical or psychological stress (Drossman et al., 2002), and altered intestinal motility may correlate with symptoms. Interestingly, IBS patients appear to display both visceral and cutaneous hyperalgesia, with lower pain thresholds to gut distension than symptom-free individuals (Verne et al., 2001). Autonomic disturbances have also been observed (Ng et al., 2007). Psychological factors that may contribute to IBS include stressful life events, depression and anxiety, negative coping strategies, and sleep disturbances (Longstreth et al., 2006). A history of physical or sexual abuse may lead to IBS, although the relationship is unclear (Blanchard et al., 2002). It is also important to note that up to 30% of people who experience bacterial gastroenteritis develop IBS. Carbohydrate-rich, fatty, or spicy foods, artificial sweeteners, coffee, and alcohol have also been implicated (Longstreth et al., 2006).

IBS is obviously a complex syndrome. As dependable pharmacological treatment options are quite limited (Gaylord et al., 2009), many IBS sufferers feel helpless, anxious, and frustrated. Their response to the disorder may in fact be more significant than the symptoms they experience. Patients can benefit greatly from a clear explanation of why they are likely to be experiencing their symptoms; this is reflected in a high placebo response in many trials. GPs most commonly care for the IBS patient and have good opportunities to identify the potential factors involved in each case. The GP can therefore play a vital role in reassuring the IBS sufferer, helping to cope and alleviating the stress that is often a feature of the disorder (Longstreth et al., 2006). In addition to this important role of education and reassurance, other lifestyle-oriented approaches with potential for stress and symptom relief in IBS include:

- dietary elimination trials (e.g., artificial sweeteners sorbitol and mannitol; fructose)
- increasing soluble fiber intake
- minimizing intake of carbohydrate-rich, fatty, or spicy foods, coffee, and alcohol
- probiotic therapy
- relaxation techniques, meditation, yoga
- hypnotherapy and psychotherapy
- cognitive-behavioral therapy
- biofeedback therapy

Longstreth et al. (2006) discuss the various approaches to IBS management in detail, and the specific applications of biofeedback to constipation and diarrhea are covered in the next section. A feasibility trial on mindfulness training showing some effects of this was carried out in 2009 (Gaylord et al., 2009), and Chinese herbal mixtures have also been shown to reduce symptoms (Longstreth et al., 2006). Zijdenbos et al. (2009), in their review of the efficacy of psychological treatments for IBS, found that few are superior to placebo in terms of improving quality of life or reducing symptoms, and the sustainability of any effects is questionable. Despite methodological shortcomings with the available data, there is reasonable consensus among expert clinicians that most IBS patients do respond to psychological support (Longstreth et al., 2006).

THE BOTTOM END: CONSTIPATION AND DIARRHEA

Constipation

"I want to but I can't" is a common scenario; constipation has long been the butt of many jokes. It seems we all know the accountant who "works it out with a pencil"! One in five middle-aged women suffers from constipation, which affects all age groups but increases with age. Many factors appear to exacerbate constipation, including physical inactivity, low fiber intake, usage of certain medications (e.g., opiates, psychotropics, calcium channel blockers, anticholinergics), poor general health, specific diseases (e.g., diabetes, hypothyroidism, pseudo obstruction), psychological conditions, and a history of sexual abuse (Longstreth et al., 2006). From a biopsychosocial perspective, personality, stress, infant toilet training technique, and delaying defecation can impact on aspects of constipation (Wald et al., 1989; Klauser et al., 1990).

Patients with constipation who do not have underlying disease and do not respond to changes in medication are usually given a trial of fiber supplementation. Those who do not respond to this require imaging or physiological tests to determine whether they have poor colonic transit or anorectal dysfunction.

Anorectal dysfunction is an important contributor to constipation. Patients with functional constipation or IBS with constipation often display altered anorectal physiology, including rectal hyposensitivity, inadequate rectal straining pressures, abnormal anal sphincter relaxation, and prolonged expulsion of a balloon (Suttor et al., 2010). Anorectal dysfunction may contribute to troublesome bloating and distension (Shim et al., 2010). Comprehensive biofeedback programs have been developed on the basis of these underlying physiological abnormalities; patients are taught to recognize appropriate toileting positioning, diaphragmatic breathing and coordination of rectal strain with anal sphincter relaxation with the aid of computer-based biofeedback. Several trials have demonstrated the efficacy of this type of program, including an early Australian study in which both functional constipation and IBS patients showed significant improvement in bowel symptoms and anorectal physiology (Malcolm et al., 2008).

Diarrhea

We all know jokes about this, too, but it's no fun if the condition is chronic. As with chronic constipation, many factors can cause or exacerbate diarrhea, and infection, biochemical abnormalities, lactose intolerance, Crohn's disease and cancer must be considered (Longstreth et al., 2006). In the absence of any of these pathologies, functional diarrhea, involving the passage of loose or watery stools without pain or discomfort, is debilitating and can severely limit quality of life. There is an obvious overlap between this condition and IBS with diarrhea, but the presence of abdominal pain suggests IBS.

In the treatment of functional diarrhea, avoidance of fructose, sorbitol, mannitol, and caffeine can be helpful, and some clinicians recommend a relatively low-fiber diet and limiting alcohol, especially beer, intake (Skoog and Barucha, 2004; Longstreth et al., 2006). An association with binge eating has recently been shown (Cremonini et al., 2009). When fecal incontinence is occurring, sphincter exercises and biofeedback therapy directed toward normalizing bowel habits and improving contraction of the external anal sphincter have shown promising results (Norton et al., 2003).

What Can Be Done in a Brief Consultation?

- ☐ Check for alarm features such as weight loss, severe vomiting, and blood in stools.
- ☐ Identify key symptoms of functional disorders and their relationship to potentially aggravating dietary behaviors and psychosocial factors.
- ☐ Provide education and reassurance (see the Website of the International Foundation for Functional Gastrointestinal Disorders www.iffgd.org.
- ☐ Plan simple lifestyle modifications to address key symptoms.
- ☐ Consider referral to a gastroenterologist, dietitian, or psychologist if symptoms persist despite these modifications.

EXERCISE: GOOD AND BAD NEWS

"Moderation in all things" can be related to exercise and the gut. Regular low- to moderate-intensity exercise is good for gut health, protecting against colon cancer, constipation, diverticular disease, and gallstones (De Oliveira and Burini, 2009). Strenuous exercise, however, can provoke acid reflux in GERD patients and 30–50% of endurance athletes are affected by gastrointestinal symptoms such as heartburn or diarrhea (Peters et al., 2001). Abdominal pain, nausea, vomiting, and gastrointestinal bleeding can also occur. These symptoms may result from altered motility, mechanical effects, neurohormonal modulation, or gut ischemia (De Oliveira and Burini, 2009).

SUMMARY

Many common gut complaints are functional in nature, lacking a clearly defined pathological basis to fully explain symptoms. Lifestyle-related factors are believed to exacerbate symptoms in most of these conditions, and the potential for lifestyle modifications to enhance their management is summarized in Table 29.1.

TABLE 29.1 Lifestyle Factors Believed to Reduce Common Gut Disorders

GI Disorder	Lifestyle Aspect(s) in Management	Level of Evidence
Gastroesophageal reflux disease	Weight loss; avoid late evening meals; elevate bed head; limit alcohol intake; smoking cessation	Physiological studies; few randomized control trials (RCTs)
Dyspepsia	Avoid "trigger" foods and stressors; eat small, frequent low-fat meals; limit caffeine intake; no alcohol; smoking cessation	Observational and epidemiological studies; few RCTs
Irritable bowel syndrome	Dietary elimination trials; increase soluble fiber; minimize carbohydrate-rich, fatty, or spicy foods, coffee and alcohol; probiotics; behavioral approaches including cognitive behavior therapy, hypnotherapy, psychotherapy and biofeedback	Physiological/ observational studies; emerging RCTs
Functional constipation	Review fiber and water intake; tailored biofeedback therapy for anorectal function	Several RCTs
Functional diarrhea	Avoidance of fructose, sorbitol, mannitol, and caffeine; low-fiber diet; limiting beer intake; biofeedback therapy for fecal incontinence	Physiological studies; few RCTs

REFERENCES

Blanchard, E.B., Keefer, L., Payne, A., Turner, S.M., Galovski, T.E., 2002. Early abuse, psychiatric diagnoses and irritable bowel syndrome. Behav. Res. Ther. 40, 289–298.

Bredenoord, A.J., Tutuian, R., Smout, A.P.J.M., Castell, D.O., 2007. Technology review: esophageal impedance monitoring. Am. J. Gastroenterol. 102, 187–194.

Cremonini, F., Camilleri, M., Clark, M.M., Beebe, T.J., Locke, G.R., Zinsmeister, A.R., Herrick, L.J., Talley, N.J., 2009. Associations among binge eating behavior patterns and gastrointestinal symptoms: a population-based study. Int. J. Obes. 33, 342–353.

Dent, J., 2009. Landmarks in the understanding and treatment of reflux disease. J. Gastroenterol. Hepatol. 24 (S3), S5–S14.

De Oliveira, E.P., Burini, R.C., 2009. The impact of physical exercise on the gastrointestinal tract. Curr. Opin. Clin. Nutr. Metab. Care 12 (5), 533–538.

Drossman, D.A., Camilleri, M., Mayer, E.A., Whitehead, W.E., 2002. AGA technical review on irritable bowel syndrome. Gastroenterology 123, 2108–2131.

Elta, G.H., Behler, E.M., Colturi, T.J., 1990. Comparison of coffee intake and coffee-induced symptoms in patients with duodenal ulcer, nonulcer dyspepsia and normal controls. Am. J. Gastroenterol. 85, 1339–1342.

Gaylord, S.A., Whitehead, W.E., Coble, R.S., et al., 2009. Mindfulness for irritable bowel syndrome: protocol development for a controlled clinical trial. BMC Complemant Altern. Med. 9, 24.

Heine, R.G., 2009. GORD in infants and children – when to investigate and when to treat? Med. Today 10 (6), 14–20.

Houghton, L.A., Mangnall, Y.F., Dwivedi, A., et al., 1993. Sensitivity to nutrients in patients with nonulcer dyspepsia. Eur. J. Gastroenterol. Hepatol. 5, 109–113.

Kaltenbach, T., Crockett, S., Gerson, L.B., 2006. Are lifestyle measures effective in patients with gastroesophageal reflux disease? An evidence-based approach. Arch. Int. Med. 166, 965–971.

Kellow, J.E., 2010. A practical, evidence-based approach to the diagnosis of the functional gastrointestinal disorders. Am. J. Gastroenterol. 105, 743–746.

Klauser, A.G., Voderholzer, W.A., Heinrich, C.A., et al., 1990. Behavioral modification of colonic function: can constipation be learned. Dig. Dis. Sci. 35, 1271–1275.

Longstreth, G.F., Thompson, W.G., Chey, W.D., Houghton, L.A., Mearin, F., Spiller, R.C., 2006. Functional bowel disorders. In: Drossman, D.A., et al. (Ed.), Rome III. The Functional Gastrointestinal Disorders, third ed. Degnon Associates, Inc, McLean, VA, pp. 487–555.

Malcolm, A., Hansen, R.D., Suttor, V.P., Prott, G.M., Simmons, L., Kellow, J.E., 2008. Anorectal biofeedback therapy improves bloating in irritable bowel syndrome and functional constipation. Gastroenterology 134 (4 Suppl. 1), A906.

Ng, C., Malcolm, A., Hansen, R., Kellow, J., 2007. Feeding and colonic distension provoke altered autonomic responses in irritable bowel syndrome. Scand. J. Gastroenterol. 42, 441–446.

Norton, C., Chelvanayagam, S., Wilson-Barnett, J., Redfern, S., Kamm, M.A., 2003. Randomized controlled trial of biofeedback for fecal incontinence. Gastroenterology 125, 1320–1329.

Peters, H.P.F., Devries, W.R., Vanberge-Henegouwen, G.P., et al., 2001. Potential benefits and hazards of physical activity and exercise on the gastrointestinal tract. Gut 48 (3), 435–439.

Shim, L., Prott, G., Hansen, R.D., Simmons, L.E., Kellow, J.E., Malcolm, A., 2010. Prolonged balloon expulsion is predictive of abdominal distension in bloating. Am. J. Gastroenterol. 105, 883–887.

Skoog, S.M., Barucha, A.E., 2004. Dietary fructose and gastrointestinal symptoms: a review. Am. J. Gastroenterol. 99, 2046–2050.

Suttor, V.P., Prott, G., Hansen, R.D., Kellow, J.E., Malcolm, A., 2010. Evidence for pelvic floor dyssynergia in patients with irritable bowel syndrome. Dis. Colon Rectum 53, 156–160.

Tack, J., Caenepeel, P., Corsetti, M., Janssens, J., 2004. Role of tension receptors in dyspeptic patients with hypersensitivity to gastric distention. Gastroenterology 127, 1058–1066.

Tack, J., Talley, N.J., Camilleri, M., Holtman, G., Hu, P., Malagelada, J.-R., Stanghellini, V., 2006. Functional gastroduodenal disorders. In: Drossman, D.A., et al. (Ed.), Rome III. The Functional Gastrointestinal Disorders, third ed. Degnon Associates, Inc, McLean, VA, pp. 419–486.

Talley, N.J., Choung, R.S., 2009. Wither dyspepsia? A historical perspective of functional dyspepsia, and concepts of pathogenesis and therapy in 2009. J. Gastroenterol. Hepatol. 24 (S3), S20–S28.

Talley, N.J., Weaver, A.L., Zinsmeister, A.R., 1994. Smoking, alcohol, and nonsteroidal anti-inflammatory drugs in outpatients with functional dyspepsia and among dyspepsia subgroups. Am. J. Gastroenterol. 89 (4), 524–528.

Verne, G.N., Robinson, M.E., Price, D.D., 2001. Hypersensitivity to visceral and cutaneous pain in the irritable bowel syndrome. Pain 93, 7–14.

Wald, A., Hinds, J.P., Caruana, B.J., 1989. Psychological and physiological characteristics of patients with severe ideopathic constipation. Gastroenterology 97, 932–937.

Zijdenbos, I.L., de Wit, N.J., van der Heijden, G.J., Rubin, G., Quartero, A.O., January 21, 2009. Psychological treatments for the management of irritable bowel syndrome. Cochrane Database Syst. Rev. (1), CD006442.

Section IV

The Future of Health

Chapter 30

The Next Chapter: The Future of Health Care and Lifestyle Interventions

Michael Sagner, Amy McNeil, Ross Arena

Intellectuals solve problems, geniuses prevent them.

Albert Einstein

INTRODUCTION

Most diseases currently being treated in the health care system are related to lifestyle factors and could be prevented with effective, individually tailored, lifestyle interventions. However, health care systems continue to predominantly function under a reactionary model that oftentimes delays initiation of care until risk factors manifest into symptoms or a confirmed diagnosis of chronic disease. To optimally treat the current chronic disease health crisis, a proactive model focused on the underlying mechanisms of physiologic dysfunction (e.g., genomic markers, the management of unhealthy lifestyle factors, etc.) will need to be created and universally adopted to promote health and prevent diseases. Ideally, the health care system of the future will strive to prevent chronic disease risk factors from ever manifesting (i.e., primordial prevention). Research has shown that the primordial prevention model, focused on lifelong health behaviors (e.g., being physically active, maintaining a healthy body weight, consuming a nutritious diet), can prevent or significantly delay the manifestation of chronic disease in a way that no drug or surgery can rival (Akesson et al., 2014; Larsson et al., 2014).

Ten years ago, the proposition that health care must evolve from a reactive to a proactive health care system that is predictive, preventive, personalized, and participatory was not a major priority. Today, while implementation is far from optimal, the core elements of that concept are widely accepted as vital steps on the path forward and are being promoted by leading institutions and governments around the world. With new technologies, from wearable lifestyle trackers to metagenomic sequencing and direct-to-consumer genetic testing,

Lifestyle Medicine. http://dx.doi.org/10.1016/B978-0-12-810401-9.00030-9

health care professionals and the individuals they care for will be able to monitor lifestyle and health more precisely and personalized than ever before. New technologies will further develop modes of communicating health information in manners that are easy to understand and instantly integrate throughout the day. These new developments will allow us to optimize the effectiveness of preventive medicine and, more specifically, lifestyle interventions, which will be integral to modern, proactive health care systems.

In addition to advancements in proactive lifestyle interventions, we will move away from generic advice to more personalized and effective interventions to prevent and treat chronic diseases. The location of the delivery of the majority of lifestyle interventions will ideally shift from hospitals and clinics to the places individuals live, work, and go to school, thereby, immersing individuals in a culture of health where appropriate lifestyle choices become innate. Recently, an international policy statement extended the list of stakeholders in a proactive lifestyle model to include not only traditional health care and government but also community, education, media industry, professional organizations, food industry, insurance, technology companies, employers, along with family to create a culture of health surrounding an individual (Arena et al., 2015). Health messaging and information will be readily available for individuals throughout their day and in multiple locations. For a proactive lifestyle model to emerge and be successful, entities must work together in creating this new culture of health.

HEALTH CARE SYSTEMS AND SOCIETY AT LARGE MUST FOCUS ON PROACTIVE PREVENTION

Predominantly, in the current health care system, diseases are classified, diagnosed, and treated primarily on the basis of a select set of signs (e.g., hypertension, dyslipidemia, etc.) and more typically symptoms (e.g., angina, shortness of breath, etc.) rather than on a deep understanding of the molecular and cellular origins of disease and health of an individual patient (National Research Council (US) Committee 2011). This long-established approach is central to the reactive health care system that currently exists. There is both a quality of care and an economic case for shifting away from this reactionary model.

THE QUALITY OF CARE CASE

Health care, in its entirety as a global entity, should continually strive for delivery of the highest quality of care that lends itself to achieving the highest quality of life outcomes for all individuals across the spectrum of health. Chronic diseases and associated risk factors are clearly linked to poor quality of life (Centers for Disease Control and Prevention, 2003). Thus, improving lifestyle behaviors and minimizing the risk/impact of chronic diseases and associated risk factors intuitively improves quality of life. One way to improve adherence to desirable lifestyle behaviors would be to establish two-way conversations about health status of individuals throughout their day and their life. Individuals will need to be able to easily retrieve information supporting their healthy choices, and this information will need to be presented in a manner that is understandable to the individual receiving and hopefully using it. Practitioners will then need to be trained to vary their communication style (both spoken and written) to specific audiences, and individuals will need to be educated on how to read and understand health information as early as childhood. This paradigm shift will create a new participatory language and communication style where health professionals and individuals receiving care will find common ground in achieving health communications leading to health goals.

Moreover, we are now in an era that evaluates excessive hospital readmissions for chronic disease as a central indication of poor quality of care and, in some instances, carries a financial penalty to health care systems (Centers for Medicare and Medicaid Services, 2016). Clearly utilizing established, evidence-based, lifestyle interventions such as cardiac rehabilitation substantially reduces the risk of readmissions (Oldridge, 2012). Therefore, lifestyle interventions used proactively to avoid hospitalization in the first place serve as a primary means to optimizing quality of care.

THE ECONOMIC CASE

In the coming decades, preventive medicine approaches will soar, justified not only from the perspective of providing high quality of care but also from an economic standpoint. The treatment of complex chronic diseases has become a major burden for health care systems worldwide. Regardless of the payment model for health care in a given country, the current reactionary model is unsustainable. Treating chronic diseases once they manifest is the most costly approach, and funds directed toward providing care in this model siphon away funds from other areas. In this way, the entire economy of a given country will suffer and, given the ever-increasing interconnectivity amongst countries, the negative implications for the global economy are significant. Most of these diseases are preventable through tailored lifestyle interventions that cost much less than the current reactionary model that we currently use. The simple truth is that keeping individuals healthy and thriving within their community, minimizing the need for high-cost medical care (e.g., diagnostic testing for chronic disease, surgery, hospitalization, etc.), has enormous cost-savings potential (Anand and Yusuf, 2011; Carlson et al., 2015; Castellano et al., 2014; Chaker et al., 2015; Dobbs et al., 2014; Haussler and Breyer, 2015; Jaspers et al., 2015; Khang, 2013).

EMBRACING PREDICTIVE MEDICINE

Research allowing for our refinement of identifying those at greatest risk for one or more chronic diseases is continually emerging. Through our knowledge of baseline characteristics (i.e., age, sex, race/ethnicity, family history, etc.), genomics, and lifestyle characteristics, as well as their complex interaction, we will be able to increasingly identify phenotypes at greatest risk for chronic disease and design interventions specifically targeted to ameliorate these risks (Mozaffarian et al., 2015; Lloyd-Jones et al., 2010; Roura and Arulkumaran, 2015; Folsom et al., 2015; Sankar et al, 2015; Havranek 2015; Abraham, 2015). Ideally, predictive modeling using baseline characteristics, genomics, biomarkers, etc., will be optimized and communicated in a primordial prevention model, before unhealthy lifestyle characteristics develop and further compound chronic disease risk.

Historically, successful interventions have been identified and promoted by comparing mean values between/amongst groups enrolled in randomized controlled trials (i.e., t-testing or analysis of variance). A p-value of <0.05 for these mean comparisons implies all individuals in an experimental group receiving an intervention derive benefit. This generalized approach is flawed given that mean comparisons do not assess individual responses within a cohort. There is a growing body of research describing the "nonresponder" to interventions that have demonstrated high value and are current standards of care (i.e., exercise training, healthy diets, weight loss, etc.) (Unick et al., 2016; Anjo et al., 2013; De Schutter et al., 2015; Herrmann et al., 2015). Developing models to accurately and preemptively identify nonresponder phenotypes and their motivations will further advance predictive medicine. Identification of those at risk for being "nonresponsive" to established preventive medicine may allow for earlier interventions that reduce actual and perceived barriers to the individual's ability and desire to incorporate health advice. Further, early identification of likelihood of response will open opportunities for differentiated communication that meet the same overarching goals, but allows health care providers to tailor messaging to the individual's goals, preferences, or health literacy level. Identification of nonresponders and a proactive titration of care increases the hope of optimizing outcomes for an even larger percentage of the population. Addressing the nonresponder phenomenon is a central component of the highly refined predictive approach to medicine that is needed moving forward.

PERSONALIZING LIFESTYLE INTERVENTIONS

Historically, health care professionals offer generic advice on a healthy lifestyle such as "eating balanced meals," "exercising more," and "sleeping enough." Evidence-based recommendations for physical activity and nutrition, defining ideal behaviors, are often used as the basis for this generic advice/counseling (Lloyd-Jones et al., 2010). While adherence to these guidelines certainly portends tremendous health benefits, attainment of these behaviors may seem impossible to the individual receiving advice, particularly for those who are currently leading a lifestyle that emulates poor health. Given the incongruence between poor lifestyle profile in a large percentage of the population and the current recommendations for leading an ideal lifestyle, generically relayed from the health professional to the individual receiving the intervention, it is not surprising that poor compliance with recommendations persists.

Often, individual psychosocial and physiological parameters are not taken into account. New research has highlighted highly significant interindividual differences in responses to lifestyle interventions such as physical activity and nutrition (Zeevie et al., 2015). Understanding specific and individual responses to lifestyle interventions will help to create more efficient preventive and therapeutic approaches. Research will need to be conducted to elucidate language variables that encourage transactional communication that values both the practitioner and patient intentions, goals, and motivations. Understanding not only what individuals should do but also what will motivate actions toward a healthy lifestyle built on shared decision making and authentic behavior modifications may positively improve individuals' responses to general guidelines. For example, narrative collection from individuals leading sedentary lives that shed light on their resistance to advice on physical activity and nutrition may offer authentic insight to the practitioner to persuade the individual to take the first step toward improved health.

Personalizing lifestyle interventions is a crucial next step in disease treatment and prevention that takes into account individual variability in genes, environment, and lifestyle for each person. The general concept of personalized medicine—prevention and treatment strategies that take individual variability into account—is actually not new; blood typing, for instance, has been used to guide blood transfusions for more than a century. But the prospect of applying this concept to lifestyle interventions has been dramatically improved by powerful methods for characterizing patients (such as proteomics, metabolomics, genomics, diverse cellular assays, and even mobile health technology) and computational tools for analyzing large sets of data (Hood et al., 2012).

For instance, using complex multidimensional biological patient data (lifestyle, blood biomarkers, microbiome, and others), scientists have been able to develop an algorithm that provides highly personalized and effective dietary recommendations to control blood glucose levels (Zeevie et al., 2015).

Traditional medical records will need to be combined with genomic, metabolomic, transcriptomic, and proteomic data, along with the participation of the patient who can provide personal lifestyle data about sleep, activity, and diet. Data from the health-lifestyle industry, particularly self-monitoring, is integrated with data generated by clinical institutions to improve care. Ever-evolving technology will offer unlimited possibilities for instantaneous communication between the clinicians and their patients, as well as provide platforms for individuals to self-monitor according to healthy lifestyle recommendations.

MEDICINE WILL BECOME MORE PARTICIPATORY

Unlike the health professionals (e.g., physician, nurse, pharmacist, dietician, etc.) of previous generations, today's health professional will not be able to keep up with the current demand for health care related to chronic diseases. For example, to treat the 21st century's problems with a 20th-century reactionary approach to health care would require an unachievable number of physicians, both from an infrastructure and economic perspective. These factors will drive a participatory approach where individuals receiving care will collaborate with health professionals to create common goals and care plans. This collaborative mindset has already started to take hold (Milani et al., 2015).

It is obvious that lifestyle interventions, however effective they can be, rely on the participation of the individual receiving care and there is only so much a health professional can do to ensure adherence to a preventive intervention. Individual behavior in homes, schools, and workplaces affects health, prevention, and disease management. The participation of the individual receiving care is crucial for lifestyle interventions to be effective, and a multilevel and multistakeholder approach is needed to improve participation in a healthy lifestyle. A culture of health, previously mentioned, will need to be created to support the individual's efforts toward consistent participation in a healthy lifestyle (Arena et al., 2016a).

Modern health care systems must maximize the effectiveness of lifestyle interventions by increasing participation, involving all stakeholders in the individual's environment, expanding out from hospitals and clinics into homes, workplaces, and schools (i.e., K–college education). A broad array of health professionals, trained as "healthy lifestyle practitioners," speaking a consistent language can be highly effective in supporting lifestyle interventions (Arena, et al., 2016b). Broadening the number of individuals, inside and outside of traditional health care, is needed to make the impact on lifestyle behaviors and change the trajectory of chronic disease and associated risk factors on a population level.

Using technology, the ability to directly observe the impact of lifestyle decisions on health and biology will allow individuals to more actively participate in prevention and treatment of chronic disease and associated risk factors (Franklin, 2015; Milani and Lavie, 2015). A new generation of young, networked, and activated consumers are increasingly using technology that has the potential to improve their health and prevent diseases. One in three Americans has gone online to investigate a medical condition and lifestyle advice (Fox et al., 2013). A growing number of "quantified self" enthusiasts are exploring methods, tools, and analytical procedures for a better understanding of the effects of their lifestyle on their health. There are now thousands of mobile health care applications, and the number is growing. Leading technology companies like Apple and Google are now driving direct participation of individuals in their personal health care by, for example, adding dedicated functionality to their smartphones to monitor lifestyle and health parameters (e.g., Apple Health).

Many new businesses have been developed to meet this demand, including direct-to-consumer genetics companies (e.g., 23 and Me) and information services (e.g., WebMD). In addition, there has been a deluge of new consumer products (e.g., Fitbit All in One wireless activity and sleep monitor, Withings

Smart Body Analyzer, Jawbone UP, Nike + FuelBand) that track personal health-related lifestyle information, ranging from physical activity metrics to sleep patterns. Individuals will move more and more to the center of their own health care, and with the help of new technology they will have a better ability to adjust their lifestyle to hopefully prevent and if necessary treat chronic diseases.

Progressive health insurance companies are already moving toward a more proactive and participatory approach. Health insurance providers (e.g., Vitality: https://www.powerofvitality.com/vitality/login) are offering innovative health care plans based on healthy lifestyle behaviors. Using the data from wearable lifestyle tracking devices and smartphones they aim to promote and incentivize healthy lifestyle choices to prevent diseases.

For research purposes, a more active participation of the individual and new quantities and forms of data from lifestyle self-monitoring and self-assessments can be aggregated and mined to generate new insight into health and disease. These insights will drive the development of new technologies, analytic tools, and forms of care that further enhance outcomes.

The possibilities to improve lifestyle interventions using personalized health technologies appear endless. Consumers can quantify their health remotely to motivate behavior changes, health professionals can deliver precision medicine to individuals receiving care, and academics can expose variations in health behaviors among diverse populations. The rapid innovation of these devices and use of associated data have also generated concerns about data privacy issues. In fact, the first guidelines and recommendations on how to use this health and lifestyle data have been published (Vitality Guidelines for Personalized Health Technology, 2016).

The direct and active participation of a given individual in disease prevention and a healthy lifestyle requires a basic knowledge about which lifestyle factors affect health. Individuals who are health literate have the capacity to interpret basic health information and make informed decisions about promoting and maintaining their health. Despite, or because of, constant and often conflicting reports in the media, the World Health Organization estimates that nearly half of all Europeans have inadequate or problematic health literacy, and in the United States as many as 9 out of 10 adults lack these skills needed to manage their health (World Health Organization, 2013). Poor health literacy and numeracy contribute to poor health. This is because the relationships between risk factors that underpin a healthy life (mainly lifestyle behaviors) and various health outcomes may not be understood or communicated in a useful manner. The US Centers for Disease Control and Prevention defines health literacy as the ability to receive and understand health information in a meaningful manner, and they recommend that both the individual receiving and delivering health information should be trained in health literacy. In this regard, health literacy becomes a two-way transaction of information that does not presuppose the patient will understand complex medical information that took the physician years to attain. The transaction of information must be personalized so that the individual has the chance to participate and prevent the initial onslaught to disease or self-modify the disease once they have been diagnosed. Future technology must help to improve health literacy and, consequently, the active participation of individuals receiving lifestyle interventions.

CONCLUSIONS

Based on current developments it can be seen that preventive medicine, and more specifically lifestyle interventions, will undergo several transformations and improvements in the coming years:

1. Science- and evidence-based health care is moving beyond disease care in the clinic to include the active preservation and enhancement of health by consumers in their homes, schools, and workplaces;
2. Lifestyle interventions will be accepted as a cornerstone of chronic disease treatment and prevention in modern medicine;
3. Lifestyle interventions will be used in synergy with biopharmaceutical interventions;
4. Reliance on averaging data from limited lifestyle interventions and test cohorts is being eclipsed by mathematically sophisticated "big data" analyses of billions of data points generated for each individual in the population;
5. Lifestyle-related chronic diseases are being predicted, diagnosed, and treated with far greater cost-effectiveness based on their molecular and cellular origins in each individual rather than categories of symptoms (e.g., the DNA sequencing of tumors to search for driving mutations that might effectively be treated by known drugs);
6. A new lifestyle-health industry is beginning to emerge that will become a major source of economic growth in the 21st century. Some predict that in 10–15 years, the lifestyle-health industry will become larger than the current health care industry (Bousquet et al., 2011).

The health care system of the future must evolve into such as system in order to better address the current state and future trajectory of chronic diseases around the world. Using a preventive approach that is predictive, personalized, and participatory, bringing together stakeholders from multiple sectors, with the individual receiving care at the center of this transaction, has the potential to make lifestyle interventions more impactful and efficient.

REFERENCES

Abraham, G., August 2015. Genomic risk prediction of complex human disease and its clinical application. Curr. Opin. Genet. Dev. 33, 10–16. http://dx.doi.org/10.1016/j.gde.2015.06.005 Epub 2015 July 24.

Akesson, A., et al., September 30, 2014. Low-risk diet and lifestyle habits in the primary prevention of myocardial infarction in men: a population-based prospective cohort study. Am. Coll. Cardiol. 64 (13), 1299–1306. http://dx.doi.org/10.1016/j.jacc.2014.06.1190.

Anand, S.S., Yusuf, S., 2011. Stemming the global tsunami of cardiovascular disease. Lancet 377 (9765), 529–532.

Anjo, D., Santos, M., Rodrigues, P., Sousa, M., Brochado, B., Viamonte, S., et al., 2013. Who are the non-responder patients to cardiac rehabilitation? Eur. Heart J. 34 (Suppl. 1).

Arena, R., Guazzi, M., et al., August 14, 2015. Healthy lifestyle interventions to combat noncommunicable disease–a novel nonhierarchical connectivity model for key stakeholders: a policy statement from the American Heart Association, European Society of Cardiology, European Association for Cardiovascular Prevention and Rehabilitation, and American College of Preventive Medicine. Eur. Heart J. 36 (31), 2097–2109. Epub 2015 July 1.

Arena, R., et al., 2016a. Transforming cardiac rehabilitation into broad-based healthy lifestyle programs to combat noncommunicable disease. Expert Rev. Cardiovasc. Ther. 14 (1), 23–36. http://dx.doi.org/10.1586/14779072.2016.1107475 Epub 2015 October 29.

Arena, R., Lavie, C.J., et al., January 2016b. Who will deliver comprehensive healthy lifestyle interventions to combat non-communicable disease? Introducing the healthy lifestyle practitioner discipline. Expert Rev. Cardiovasc. Ther. 14 (1), 15–22. http://dx.doi.org/10.1586/14779072.2016.1107477 Epub 2015 November 2.

Bousquet, et al., July 6, 2011. Systems medicine and integrated care to combat chronic noncommunicable diseases. Genome Med. 3 (7), 43.

Carlson, S.A., Fulton, J.E., Pratt, M., Yang, Z., Adams, E.K., 2015. Inadequate physical activity and health care expenditures in the United States. Prog. Cardiovasc. Dis. 57 (4), 315–323.

Castellano, J.M., Narula, J., Castillo, J., Fuster, V., 2014. Promoting cardiovascular health worldwide: strategies, challenges, and opportunities. Rev. Esp Cardiol. (Engl. Ed.) 67 (9), 724–730.

Centers for Disease Control and Prevention, 2003. Notes & Reports, vol. 16 No. 1. Winter.

Centers for Medicare and Medicaid Services, 2016. Readmissions Reduction Program (HRRP). https://www.cms.gov/medicare/medicare-fee-for-service-payment/acuteinpatientpps/readmissions-reduction-program.html.

Chaker, L., Falla, A., van der Lee, S.J., Muka, T., Imo, D., Jaspers, L., et al., 2015. The global impact of non-communicable diseases on macro-economic productivity: a systematic review. Eur. J. Epidemiol. 30 (5), 357–395.

De Schutter, A., Lavie, C., Jahangir, E., Menezes, A., Dinshaw, H., Shum, K., et al., 2015. Abstract 16117: nonresponders to cardiac rehabilitation and outcomes. Circulation 132 (Suppl. 3), A16117.

Dobbs, R., Sawers, C., Thompson, F., Manyika, J., Woetzel, J., Child, P., et al., 2014. Overcoming Obesity: An Initial Economic Analysis. McKinsey Global Institute.

Folsom, A.R., Shah, A.M., Lutsey, P.L., Roetker, N.S., Alonso, A., Avery, C.L., et al., 2015. American heart Association's Life's simple 7: avoiding heart failure and preserving cardiac structure and function. Am. J. Med.

Fox, et al., 2013. Health Online. Pew Research Center: http://www.pewinternet.org/2013/01/15/health-online-2013/.

Haussler, J., Breyer, F., 2015. Does diabetes prevention pay for itself? Evaluation of the M.O.B.I.L.I.S. program for obese persons. Eur. J. Health Econ.

Havranek, E.P., Mujahid, M.S., Barr, D.A., Blair, I.V., Cohen, M.S., Cruz-Flores, S., et al., 2015. Social determinants of risk and outcomes for cardiovascular disease: a scientific statement from the American Heart Association. Circulation.

Herrmann, S.D., Willis, E.A., Honas, J.J., Lee, J., Washburn, R.A., Donnelly, J.E., 2015. Energy intake, nonexercise physical activity, and weight loss in responders and nonresponders: The Midwest Exercise Trial 2. Obesity (Silver Spring) 23 (8), 1539–1549.

Hood, L., et al., August 2012. Revolutionizing medicine in the 21st century through systems approaches. Biotechnol. J. 7 (8), 992–1001. http://dx.doi.org/10.1002/biot.201100306. Epub 2012 July 20.

Jaspers, L., Colpani, V., Chaker, L., van der Lee, S.J., Muka, T., Imo, D., et al., 2015. The global impact of non-communicable diseases on households and impoverishment: a systematic review. Eur. J. Epidemiol. 30 (3), 163–188.

Khang, Y.H., 2013. Burden of noncommunicable diseases and national strategies to control them in Korea. J. Prev. Med. Public Health 46 (4), 155–164.

Larsson, S.C., et al., November 4, 2014. Healthy diet and lifestyle and risk of stroke in a prospective cohort of women. Neurology 83 (19), 1699–1704. http://dx.doi.org/10.1212/WNL.0000000000000954 Epub 2014 October 8.

Lloyd-Jones, D.M., Hong, Y., Labarthe, D., Mozaffarian, D., Appel, L.J., Van Horn, L., et al., 2010. Defining and setting national goals for cardiovascular health promotion and disease reduction: the American Heart Association's strategic Impact Goal through 2020 and beyond. Circulation 121 (4), 586–613.

Milani, R.V., Lavie, C.J., April 2015. Health care 2020: reengineering health care delivery to combat chronic disease. Am. J. Med. 128 (4), 337–343. http://dx.doi.org/10.1016/j.amjmed.2014.10.047. Epub 2014 November 22.

Mozaffarian, D., Benjamin, E.J., Go, A.S., Arnett, D.K., Blaha, M.J., Cushman, M., et al., 2015. Heart disease and stroke statistics—2015 update: a report from the American Heart Association. Circulation 131 (4), e29–e322.

National Research Council (US) Committee, 2011. Toward Precision Medicine: Building a Knowledge Network for Biomedical Research and a New Taxonomy of Disease. National Academies Press, MD, USA.

Oldridge, N., 2012. Exercise-based cardiac rehabilitation in patients with coronary heart disease: meta-analysis outcomes revisited. Future Cardiol. 8 (5), 729–751.

Roura, L.C., Arulkumaran, S.S., 2015. Facing the noncommunicable disease (NCD) global epidemic – the battle of prevention starts in utero – the FIGO challenge. Best Pract. Res. Clin. Obstet. Gynaecol. 29 (1), 5–14.

Sankar, A., Beattie, W.S., Wijeysundera, D.N., 2015. How can we identify the high-risk patient? Curr. Opin. Crit. Care 21 (4), 328–335.

Unick, J.L., Dorfman, L., Leahey, T.M., Wing, R.R., 2016. A preliminary investigation into whether early intervention can improve weight loss among those initially non-responsive to an internet-based behavioral program. J. Behav. Med. 39 (2), 254–261.

Vitality Guidelines for Personalized Health Technology. 2016. http://thevitalityinstitute.org/projects/personalized-health-technology/.

Vitality™. https://www.powerofvitality.com/vitality/login.

World Health Organization, 2013. Health Literacy, the Solid Facts. http://www.euro.who.int/__data/assets/pdf_file/0008/190655/e96854.pdf.

Zeevie, D., et al., November 19, 2015. Personalized nutrition by prediction of glycemic responses. Cell 163 (5), 1079–1094. http://dx.doi.org/10.1016/j.cell.2015.11.001.

Index

'*Note*: Page numbers followed by "f" indicate figures, "t" indicate tables, and "b" indicate boxes.'

C

D

E

F

K

L

M

N

O

P

R

S

Printed in the United States
By Bookmasters